COMPREHENSIVE BIOCHEMISTRY

COMPREHENSIVE BIOCHEMISTRY

SECTION I (VOLUMES 1-4)
PHYSICO-CHEMICAL AND ORGANIC ASPECTS OF BIOCHEMISTRY

SECTION II (VOLUMES 5-11)
CHEMISTRY OF BIOLOGICAL COMPOUNDS

SECTION III (VOLUMES 12-16)
BIOCHEMICAL REACTION MECHANISMS

SECTION IV (VOLUMES 17-21)
METABOLISM

SECTION V (VOLUMES 22-29)
CHEMICAL BIOLOGY

SECTION VI (VOLUMES 30-39)
A HISTORY OF BIOCHEMISTRY

COMPREHENSIVE BIOCHEMISTRY

Series Editors
ALBERT NEUBERGER
*Former Chairman of Governing Body, The Lister Institute
of Preventive Medicine, University of London, London (U.K.)*

LAURENS L.M. VAN DEENEN[†]
*Professor of Biochemistry, Biochemical Laboratory,
Utrecht (The Netherlands) (deceased September, 1994)*

VOLUME 39

SELECTED TOPICS IN THE HISTORY OF BIOCHEMISTRY

EXPLORING THE CELL MEMBRANE: CONCEPTUAL DEVELOPMENTS

Volume Editor
A. KLEINZELLER
*Department of Physiology, University of
Pennsylvania School of Medicine,
Philadelphia, PA 19104-6085, U.S.A.*

ELSEVIER
AMSTERDAM·LAUSANNE·NEW YORK·OXFORD·SHANNON·TOKYO
1995

Elsevier Science B.V.
P.O. Box 211
1000 AE Amsterdam
The Netherlands

ISBN 0-444-81253-9 (Volume)
ISBN 0-444-80151-0 (Series)

Library of Congress Cataloging-in-Publication Data

Exploring the cell membrane : conceptual developments / volume editor,
A. Kleinzeller.
    p.   cm. -- (Comprehensive biochemistry ; vol. 39. Section VI,
A history of biochemistry)
    Includes bibliographical references and index.
    ISBN 0-444-81253-9. -- ISBN 0-444-80151-0
    1. Cell membranes.   I. Kleinzeller, Arnošt.   II. Series:
Comprehensive biochemistry; 39.
QD415.F54 vol. 39
[QH601]
574.19'2 s--dc20
[574.87'5]   95-17268
                                                          CIP

This book is printed on acid-free paper

Printed in The Netherlands

PREFACE TO VOLUME 39

The suggestion for this collection of essays originated in part from a course: 'Methods of Scientific Inquiry in Biological Sciences' given to graduate students at the University of Pennsylvania School of Medicine. In sections of this course, the conceptual developments in the fields of membranes and transport (A. Kleinzeller) and cellular respiration (D.F. Wilson) were traced to illustrate general aspects of the development of ideas in a scientific field. Discussions with Dr. M.M. Civan on the topic also greatly contributed to the idea of embarking on this project. Once this difficult decision was reached, the concepts of approaching this task were to a major extent influenced by my recollections of a course on the development of knowledge in the biochemistry of muscular contraction given by Dorothy Moyle Needham in 1942 to Part II Biochemistry students at Cambridge (later published in the scholarly volume Machina Carnis, Cambridge University Press, 1971). My many discussions as a young post-doc with my preceptor, Joseph Needham, on the history of science also helped to fashion my ideas as to how to approach an evaluation of the development of a field as broad as that of cell membranes. An appreciation of progress in a scientific field is possible only in the context of an understanding of its development.

Given the breadth and depth of the field covered by each essay, one individual cannot hope to be comprehensive; hence, collaboration was sought and found for several topics. An evaluation of the development of any scientific field of necessity reflects the personal views of the author. Therefore, difficult choices had to be made as to an appraisal of the impact of a given discovery for progress in the field. Moreover, limitations of space dictated the still more demanding judgement which discovery or idea *not* to include. These limitations of a historical evaluation become more apparent the more recent the development. The authors endeavored to mitigate this pitfall by extensive consultations with colleagues, and are indebted to them for their help and comments.

The encouragement I found in the Department of Physiology at the University of Pennsylvania, and its chairman Dr. Paul De Weer, is gratefully acknowledged. And my actual search for pertinent information, including many interviews with colleagues, would have been close to impossible without grants-in aid from the University of Pennsylvania Research Foundation and the American Physiological Society (G. Edgar Folk, Jr. Senior Physiologist Award). The help of Prof. S. Silbernagl (University of Würzburg, Germany) and his colleagues in obtaining materials on E. Overton and some German scientists was invaluable. The devoted support of the staff of the Biomedical Library of the University of Pennsylvania, and in particular of its Interlibrary Service has to be recorded; they surely have sighed with relief when I signalled to see light at the end of the tunnel. The skills and patience of Mr. K. Ray in drafting of the various Figures were valuable. In spite of all this support, it took considerably longer than originally envisaged to bring this volume to some conclusion. Work on a historical evaluation of a scientific field is *never* finished. The Publisher's patience is to be commended.

Finally, this volume reflects concentrated work of many years (not counting the many decades of interests in the field and the unending files of pertinent reprints. Without the devoted understanding of my wife I would never have succeeded to spend the countless hours on this volume; hence, it is dedicated to Lotte.

Philadelphia, 1993.   *A.K.*

CONTRIBUTORS TO THIS VOLUME

D.E. GOLDMAN
63, Loop Road, Falmouth, MA 02540, U.S.A.

M.D. HOLLENBERG
*Department of Pharmacology and Therapeutics, University of Calgary
Faculty of Medicine, 3330 Hospital Drive, Calgary, Alberta,
Canada T2N4N1*

R.K.H. KINNE
*The Max-Planck Institute for Molecular Physiology, 44139 Dortmund,
Germany*

A. KLEINZELLER
*Department of Physiology, University of Pennsylvania Medical Center,
B-400 Richards Building, 3700 Hamilton Walk, Philadelphia
PA 19104-6085, U.S.A.*

DAVID F. WILSON
*Department of Biochemistry and Biophysics, University of Pennsylvania
Medical Center, 426 Anatomy-Chemistry Building, 3700 Hamilton Walk,
Philadelphia, PA 19104-6059, U.S.A.*

CONTENTS

VOLUME 39

A HISTORY OF BIOCHEMISTRY

Exploring the Cell Membrane:
Conceptual Developments

Preface to Volume 39   v
Contributors to this Volume   vii

Chapter 1. Exploration of the Cell Membrane
by A. KLEINZELLER   1

I.   Stages of conceptual developments in the domain of cell membranes   2
II. Evolution or revolution in the study of cell membranes?   11
III. The role of ancillary fields   14
Acknowledgements   23
References   23

Chapter 2. The Postulate of the Cell Membrane
by A. KLEINZELLER   27

Introduction   27
I.   The perception of a cell membrane as the osmotic and electrical
     barrier. Is there a cell membrane?   33
II. Properties of the postulated cell membrane   40
     II.1. Membrane structure   42
            II.1.1. Asymmetric structure of the cell membrane   45
            II.1.2. The cell membrane is a dynamic structure   47
            II.1.3. Membrane models   51
     II.2. Cell membrane permeability   51
            II.2.1. Model membranes   52

II.2.2. Permeability of cell membrane   56
  II.2.2.1. Permeability to water   56
  II.2.2.2. Permeability to non-electrolytes   58
  II.2.2.3. Permeability to electrolytes   60
II.2.3. The emerging new concept of the biological mem-
        brane   62
III. Membrane channels   64
III.1. Water pathways   64
III.2. Ionic channels   65
III.3. Channel-forming proteins   71
IV. Regulation of membrane function   71
V. The concept of the cell membrane towards the 100th anniversary
   of Overton's views   73
Acknowledgements   77
References   77

Chapter 3. The Concept of a Membrane Transport Carrier
by A. KLEINZELLER   91

Introduction   91
I. The postulate of membrane permeability pathways involving
   mechanisms other than simple diffusion   92
II. The phenomenology of carrier-mediated transport: kinetics and
    models   98
  II.1. Criteria of carrier-mediated transport processes   99
  II.2. Kinetics and models for the carrier-mediated transport of
        non-electrolytes   100
  II.3. Carrier-mediated coupling of fluxes   107
    II.3.1. The Na+-dependent transport of organic solutes . .   108
    II.3.2. Cation/cation exchange systems   110
    II.3.3. The anion exchange systems   111
    II.3.4. Cation/anion symport systems   114
    II.3.5. The carrier-mediated coupling of flux systems: terti-
             ary coupling   115
III. Towards the molecular mechanisms of carrier-mediated proc-
     esses   117
IV. Questions, questions   121
Acknowledgements   123
References   123

Chapter 4. The Concept of a Solute Pump
by A. KLEINZELLER 133

I.   The perception of the phenomenon 133
     I.1.   Electrolyte pumps 139
            I.1.1. Ionic pumps in plant cells 139
            I.1.2. The cation pump in erythrocytes and muscle 141
     I.2.   Active transport of non-dissociated solutes 146
II. The phenomenology of active transport 148
     II.1. Criteria for active transport 148
     II.2. Which ion is pumped? 152
     II.3. The nature of the coupling between solute fluxes and metab-
            olism 154
            II.3.1. Direct coupling 155
            II.3.2. Indirect coupling 156
     II.4. Critique of the active transport concept 161
III. Mechanism of active transport 162
     III.1. Towards the mechanism of the Na^+ pump 162
            III.1.1. Other ionic pumps 165
     III.2. Towards the mechanism of the secondary active solute
            transport 168
     III.3. Active transport systems and the membrane theory 172
IV. What next? 173
Acknowledgements 175
References 175

Chapter 5. Membrane Receptors
by M.D. HOLLENBERG and A. KLEINZELLER 187

I.   The receptor concept 187
     I.1.   Toxins, drugs and the role of receptive substances 187
     I.2   Early receptor models and drug action 189
     I.3   Membrane-localized receptors and the 'mobile' or 'floating'
            model of drug action 191
II. Receptor-mediated transmembrane signalling 194
     II.1. General mechanisms of transmembrane signalling and the
            mobile receptor paradigm 195
            II.1.1. Receptor mobility and transmembrane signalling:
                    the mobile receptor model of agonist action 196
            II.1.2. Receptor microclustering, receptor internalization
                    and agonist action 197

II.2. Ligand-gated ion channels 200
    II.2.1. The nicotinic cholinergic receptor 200
    II.2.2. Other ligand-gated channels 203
II.3. Ligand-modulated transmembrane enzymes 203
    II.3.1. Receptor tyrosine kinases 203
    II.3.2. Other receptors with intrinsic enzymatic activity . . 204
II.4. G-protein-coupled receptors 204
    II.4.1. G-protein-coupled receptors and the regulation of adenylate cyclase 207
    II.4.2. G-protein-coupled phospholipase activation 209
    II.4.3. The visual system 210
    II.4.4. G-protein-regulated ion channels 211
II.5. Other signal adapter-coupled receptors 211
II.6. Membrane receptors, methylation and signalling 212
III. Regulation of receptor function 213
IV. Summary and view to the future 215
Acknowledgements 218
References 218

Chapter 6. Energy Metabolism in Cellular Membranes
by D.F. WILSON 231

I. Introduction 231
II. Discovery and characterization of the respiratory chain 233
II.1. Early background 233
II.2. Tissue respiration in the period from 1900 to discovery of the respiratory chain 235
II.3. Identification of additional redox components of the respiratory chain 240
II.4. Redox components of the respiratory chain as structural elements of the mitochondrial membrane 245
III. Coupling of respiration to metabolic work 251
III.1. Early observations 251
III.2. Identification of mitochondria as the 'power plant' of cellular metabolism 255
III.3. Establishing the stoichiometric relationship of electron transport and ATP synthesis 256
III.4. Thermodynamics and mitochondrial oxidative phosphorylation 258
IV. Compartmentation of the matrix enzymes by the inner mitochondrial membrane 262

V. Mechanism in the coupling of respiration and phosphorylation . 264
   V.1. Chemical coupling mechanisms and mitochondrial oxida-
      tive phosphorylation 264
   V.2. Chemiosmotic mechanism and mitochondrial oxidative
      phosphorylation 266
   V.3. Chemiosmotic versus chemical coupling 269
References 272

Chapter 7. The Epithelial Membrane
by R.K.H. KINNE and A. KLEINZELLER 279

I. Introduction 279
II. The establishment of the existence of epithelial cell layers 280
III. The transcellular route in epithelial transport 281
   III.1. Driving forces 281
   III.2. Cellular asymmetry 284
IV. The direct demonstration of different transport properties of the
   apical and basal-lateral plasma membrane 291
   IV.1. The biochemical approach 291
   IV.2. The biophysical approach 294
V. The intercellular pathway 295
VI. What's next? 298
Acknowledgements 299
References 299

Chapter 8. The Role of Membranes in Excitability
by D.E. GOLDMAN 307

I. Excitability 307
   I.1. Development of the concept 308
   I.2. The role of the membrane 311
II. Excitation processes 314
   II.1. Excitation by physical stimuli 315
      II.1.1. Electrical excitation 315
      II.1.2. Electrical cell-cell communication 322
      II.1.3. Mechanical excitation 323
         II.1.3.1. Nerve 323
         II.1.3.2. Mechanical sensory mechanisms 325
      II.1.4. Visual excitation 326
      II.1.5. Other types of physical receptor mechanisms 327

II.2. The role of chemical substances in excitatory processes . . 327
    II.2.1. Chemoreceptors 329
    II.2.2. Chemical synaptic transmission 329
Acknowledgements 332
References 332

Name Index 341
Subject Index 355

A. Kleinzeller (Editor) A History of Biochemistry: Exploring the Cell Membrane.
(Comprehensive Biochemistry Vol. 39) © 1995 Elsevier Science B.V.

Chapter 1

Exploration of the Cell Membrane

A. KLEINZELLER

Department of Physiology, School of Medicine, University of Pennsylvania,
Philadelphia, PA, U.S.A.

The essays presented in this volume trace the development of concepts in various areas of the function and structure of cell (and subcellular) membranes. Some of the milestones in the 150 years of exploration of this field are sketched in the chart given in Figure 1. A detailed evaluation of the field is given in the respective chapters.

Mohl's and Nägeli's[1] concept of a cell membrane originated from studies of the properties of cells (cf. Chapter 2) which had been relatively recently described as the basic functional unit of living matter. Initially, the interest in phenomena at the cell surface was greatly stimulated by two well-defined physiological functions: (1) Cell (particularly nerve and muscle) excitability (Chapter 8); and (2) secretory and absorptive processes (Chapters 4 and 7). (3) Subsequently, when it was realized by Ehrlich that chemical agents (antigenic proteins and drugs) acted on the cell surface, and Clark then showed an involvement of cell membrane receptors (Chapters 2 and 5), interest in physiological and pharmacological regulation of cell function became an additional powerful stimulus. The

---

1 Here and henceforth in this Volume, where only the names of authors are mentioned, the full references, indicated by the date in ( ), are given in the indicated Chapters.

[1]

studies of du Bois-Reymond and subsequent investigators clearly demonstrated the localization of bioelectrical phenomena at the interface between cells and their environment (*cf.* Chapter 8). On the other hand, the physiology of absorption and secretion, having defined the role of cell activity in these processes (Chapters 4 and 7) gradually also led to a specific involvement of cell surfaces. This applied both to animal and plant physiology: whereas in animals the basic question was: how do cells selectively absorb (or secrete) solutions of substances (Heidenhain's studies were crucial), plant physiologists were puzzled by the mechanism by which plants absorbed nutrients from the soil and transported the sap to the top of high trees (*cf.* Sachs, 1865). Studies directed toward the understanding of drug action on cells (*cf.* Chapter 5) became crucial for the development of chemotherapy. A role of the cell surfaces in these phenomena was soon detected. These developments added to the basic interest in the role of cell membranes in fundamental processes of living matter.

I. Stages of conceptual developments in the domain of cell membranes

When analyzing the logic[2] of development of a particular scientific concept, i.e. the cell membrane, several stages of discoveries are apparent:

1. Identification (discovery) of a qualitatively new phenomenon (e.g. cell membrane);

2. Quantitative evaluations of the properties of the putative membrane (i.e. measuring the phenomenon);

3. The (biochemical and/or biophysical) mechanism of membrane phenomena (modelling; causal relationships).

---

2 Some sociological aspects affecting the development in this field are mentioned below (Sect. III).

Essentially similar conceptual stages were characterized by Racker (1965) and Dixon (1970) for the development of our knowledge on oxidative phosphorylation, and enzymes, respectively[3].

Once the mechanism of a biological phenomenon has been defined, two new options open:

4. (a) Analysis (biochemical and/or biophysical) of the constituent parts of a membrane phenomenon; (b) regulatory mechanisms characterizing specific membrane phenomena.

The first stage aims at answering questions about the existence of a membrane or membrane process, both functionally and with regard to its structural correlate. Once experimental evidence, accumulated particularly around 1850, suggested the concept of a cell membrane, the crucial questions in the exploration of membrane structure and function were: (*a*) Is there a cell membrane which, by separating the protoplasm from its external environment, could explain the phenomenology of osmotic and electrophysiological properties of cells[4]; (*b*) If so, what is the chemical and/or physical nature of such membrane? The qualitative aspects of the phenomenon were revealed at this stage of investigations, starting with the simple models of Traube and Pfeffer (Chapter 2). As indicated by the chart (Fig. 1), it took practically a century to answer convincingly the above questions: the description of the osmotic properties of cells, postulating a membrane, led eventually to the demonstration of a cell membrane by electrical and optical techniques and finally to its isolation, description of its bio-

---

3 The described logic of progress of knowledge in the membrane field is consistent with Langmuir's (1929) statement: 'The progress in modern science depends largely upon (1) giving to words meanings as precise as possible; (2) definition of concepts in terms of operations; (3) development of models (mechanical or mathematical) which have the properties analogous to those of the phenomena which we have observed'. For Langmuir, words are basic concepts which 'logically should be defined in terms of operations'.

4 At the time when this question was first raised only the cell membrane was considered but later studies led to a broadening by postulating also membranes producing a compartmentation of intracellular organelles.

C. Nägeli (1817–1891)

E. du Bois-Reymond (1818–1896)

A. Fick (1829–1901)

W.F.P. Pfeffer (1845–1920)

R.P.H. Heidenhain (1834–1897)

J.H. van 't Hoff (1852–1911)

P. Ehrlich (1854–1915)

W. Nernst (1864–1941)

E.W. Reid (1862–1948)

J. Bernstein (1839–1917)

E. Overton (1865–1933)

R. Höber (1873–1953)

5

F.G. Donnan (1870–1956)   I. Langmuir (1881–1957)   A.J. Clark (1885–1941)

M.H. Jacobs (1884–1970)   J.F. Danielli (1911–1984)   T. Teorell (1905–1992)

H.H. Ussing (b. 1911)   A.L. Hodgkin (b. 1914)   A.F. Huxley (b. 1917)

J. Monod (1910–1976)   J.C. Skou (b. 1918)   P.D. Mitchell (1921–1992)

R.K. Crane (b. 1919)   *W.F. Widdas (b. 1916)*

Fig. 2. Some eminent contributors to the cell membrane concept.
Permissions to reproduce the following photographs are gratefully acknowledged: E.W. Reid (The Royal Society, London); W. Nernst and J. Monod (The Nobel Foundation, Stockholm); F.G. Donnan and A.J. Clark (The Royal Society, London and Prof. P.L. Dutton, F.R.S., Philadelphia); I. Longmuir (The Natl. Academy of Sciences, USA and Prof. R.E. Forster, Philadelphia).

chemical nature and reconstitution of at least some of its properties from isolated components.

At the first stage, the characterization of the phenomenon may require further clarification, and possibly, classification. Thus, exploration of the concept of permeability[5] required further operational definition (see Chapters 2, 3 and 4) in terms of 'physical' (diffusion) and 'physiological' permeabilities (i.e. facilitated diffusion, active transport and group translocation).

The second stage is characterized by efforts to describe membrane phenomena in quantitative terms, aiming at a description of the involved processes in concrete physical and chemical events. Techniques were developed to quantitate the observations, and this enabled investigators to examine the extent to which membrane (*e.g.* osmotic, and electrical) phenomena were consistent with ideas stemming from the already known laws of diffusion of uncharged and charged solutes across idealized and artificial membranes. At the same time, information was sought concerning the structural elements of the membrane (*e.g.* chemical composition) which

---

5 Derived from the Latin *permeare, i.e.* to pass through.

would be consistent with its function. By analyzing predictions of existing hypothetical models and uncovering inconsistencies, the second stage of exploration led to concepts of the lipid, and later the mosaic membrane with semipermeable properties (including the model of charged membranes). The broad and vague concepts of membrane permeability (including 'pores' and 'leaks') had to be gradually replaced by more specific notions. This development (see Chapter 3) was triggered by a wealth of information showing that a variety of (mostly kinetic) phenomena related to solute passage across membranes was inconsistent with the assumption of simple (physical) diffusion of a solute across a membrane.

The exploration of the quantitative aspects of a defined phenomenon also permits pinpointing the biological preparation (species, tissue, cell type) displaying the given phenomenon most clearly (Krogh's principle; [Krogh, 1929]).

In the course of this development, new facets of membrane function were discovered, including membrane receptors (Chapter 5), energy (Chapter 6) and signal (Chapters 2 and 5) transduction, cell–cell communication (Chapter 2), etc.

While extending the understanding of membrane properties, new information gained on the quantitative properties of the given system added further support for the perception of the membrane (and associated functional aspects) as a qualitatively new phenomenon, thus reinforcing stage I.

The third stage of exploration witnessed the development of concepts of membrane channels, carriers, active transport, etc., as specific mechanisms of the observed phenomena.

The above chronological chart has basic limitations in several respects: First, major milestones in the development of the field could not readily be fitted into a simple chart, e.g. the discoveries of the asymmetry and dynamic nature of cell membranes, or (in part) membrane receptors. Second, the chronological description did not permit elaboration on the interplay between different experimental and conceptual approaches. It was the establishment of the concept of a semiper-

meable membrane which triggered the need for an analysis of
the way in which the membrane selectivity is brought about.
The realization that this membrane property would be difficult to explain in terms of a lipid membrane led to experiments
and ideas of a mosaic membrane, with patches capable of
bringing about a selective passage of solutes, culminating in
the concepts of solute carriers, channels, receptors and active
transport. And the development of the channel concept
evolved, apparently independently, from physico-chemical
studies of model membranes (Chapter 2), the exploration of
nerve excitability (Chapters 2, 8), and an analysis of the
spatial organization of membrane proteins (Chapter 2). These
new concepts specified the broad physiological notions of excitation, or absorption and secretion. And finally, luckily for science, scientific investigations do not follow simple linear logic.
Hokin's search for the driving force operative in the sodium
extrusion (the phosphatidic acid hypothesis, Chapter 4) led to
the discovery of the rapid turnover of phospholipids in the cell
membrane, thus disclosing the dynamic nature of the cell
membrane (Chapter 2); and it was the study of regulatory phenomena localized at the cell surface which shed a new light on
the structure and function of cell membrane. It is only in hind-
sight that the logic of the development of a field emerges from
the mosaic of separate pieces of information.

Once a better understanding of the mechanism of a biological phenomenon has been reached, two strategies are open to
investigators: the analytical approach favors a study of the
constituent parts of the phenomenon in greater detail (and by
differing techniques) at a more basic level: thus, the mechanism of the original phenomenon becomes the phenomenology
of its constituent parts.

In transport phenomena, at this stage the investigators aim
at isolating the protein(s) involved in the transport process,
and at elucidating its structural (conformational) changes related to its function. In the field of membranes and transport,
this was reflected in the transition from kinetic studies to-

wards an exploration of molecular mechanisms. On the other hand, the physiological (more synthetic) approach aims at an illumination of the mechanism(s) by which the original phenomenon is regulated (modulated) to meet physiological demands. The explosive development of the understanding of regulatory mechanisms based on membrane receptors and their role in signal transduction (Chapter 5) is evidence for the success of this strategy. The fourth stage thus actually represents the first stage of the development of a new set of concepts: Once it was realized that electrolytes pass across membranes via channels (Chapter 2), this qualitatively new phenomenon raised new questions which could be answered only by quantitative studies defining channel properties, in order to reach the level at which the channel phenomenon is defined in terms of interactions of an ionic species with a specific structure within the protein molecule(s) constituting the channel. The exploration of the function of membrane receptors in turn led to a new level of insight into the organization and dynamics of membrane proteins (and their carbohydrate moieties) in membrane channels, enzymes and other structures (Chapter 5).

The essays presented in this volume are organized to reflect the above stages of developing knowledge and concepts. In a field as broad as membrane phenomena, the individual stages took many decades, and involved many allied fields. Milestones in development were reflected in reviews, symposia and textbooks: Overton (1899) made a clear case for the concept of a lipid membrane; Höber's texts[6] (1926, 1945) summarized the wealth of evidence (qualitative and quantitative) of membrane phenomena; the Davson and Danielli (1943) mono-

---

6 The influence of his texts is illustrated by the personal experience of this author. When applying in 1934 for the privilege of working in his laboratory, Vl. Morávek (the first Czech professor of biochemistry who made a name for himself by studies on the cation distribution in various tissues) pressed Höber's (1926) (Sixth Edition) textbook in my hand with the comment: 'Study this. When you think you understand the contents, you can start working with me'.

graph pinpointed the data for the structure of the biological membrane as a bimolecular lipid leaflet coated by proteins; the 1954 Bangor Symposium of the Society of Experimental Biology (Brown and Danielli, 1954) clearly established active transport conceptually and experimentally, while the Prague 1960 Symposium (Kleinzeller and Kotyk, 1961) focused on ideas of mechanisms involved in this process. In the course of analyzing active transport, the term had to be redefined to encompass both direct and indirect coupling of the active solute transport to metabolism. The understanding of the broad phenomenology of solute transport as defined in the early sixties, allowed a new focus in the field of membrane processes. Major advances in techniques were instrumental for this progress, *e.g.* the availability of radio-labelled materials (solutes, physiological agents and inhibitors), sophisticated electrophysiological methods (*e.g.* the patch-clamp), as well as the powerful tools of molecular genetics. The use of these tools opened the door for the understanding of the mechanisms of transport processes and their regulation (including the isolation of participating proteins). By 1978, 5 volumes summarized existing knowledge (Giebisch et al., 1978) on membrane phenomena, reflecting the exponential growth of interest in the field. Moreover, information on, and analysis of progress was facilitated by new venues in scientific literature, including review publications (e.g. *Current Topics in Membranes and Transport; Biomembrane Reviews*, etc.). The study of membrane processes began its full integration with many other fields of biology, including the exploration of the process of learning as a modulation of membranes at nerve synapses (*cf.* Kandel et al., 1991).

At the present stage of development the amount of new information generated each day is overwhelming. Hence, of necessity any analysis of the field is far from comprehensive.

The study of the development of concepts in the membrane fields permits an enquiry into several pertinent questions.

II. Evolution or revolution in the study of cell membranes?

Is it possible to discern a 'revolutionary' change of views in the sense of Kuhn's definition (1962), *i.e.* a sudden change of accepted paradigms? Perceptions of the existence of a cell membrane both functionally and as a structural entity (Nägeli, 1855 → Overton, 1899) may actually represent the start of a revolution in biology, culminating in the realization of the multifaceted role of membranes:

> 'It is important to recognize that the plasma membrane is not a dead structure or a purely passive partition, but represents in reality a portion of the living protoplasm, characteristically modified in its structure and physical properties by the surface conditions. Hence it is the seat of metabolic and other activities which influence its physical properties. Evidently it is that part of the cell which comes into most direct relations with the surroundings... it thus exerts a far-reaching control over the metabolic and other processes occurring in the cell interior' (Lillie, 1922).

The emergence of this concept certainly meets the criteria for scientific revolutions put forward by Cohen (1985), *i.e.* (a) the judgment of scientists at the time[7]; (b) later documentary history of the subject of the revolution; (c) the judgment of competent historians of science; and (d) importance to the living scientific tradition. Subsequently, whereas it is possible that in the perspective of some 150 years of developing knowledge on membrane phenomena major changes of views become blurred (Kuhn's invisible revolutions), it is difficult to discern any 'revolution' in knowledge similar to that in physics or in genetics. The emerging notion is one of an incremental development, extending existing knowledge by addition of new fac-

---

7 This includes, of course, strenuous polemics directed at discrediting the new concept.

ets, and these changes appear to be related to the introduction of new techniques.

Two instances: In Chapter 2 it will be shown that Overton's concept of a (very thin) lipid membrane was extended to an idea of a composite membrane consisting of lipid and protein areas, the latter functionally responsible for the sieve-like properties for the passage of lipid-insoluble solutes. The sensitive electrophysiological techniques proved to be crucial for this development. The next step, the actual demonstration of the membrane as a structure, had to await the development of techniques such as electron microscopy (although there were many convincing, but more indirect pieces of evidence in favor of such a concept). And finally, the crucial step from the concept of a simple, very thin membrane to contemporary ideas about the complex nature of a cell membrane had to await the development of knowledge on the nature of receptor molecules (the role of glycoproteins at the outer membrane face), the use of radioisotopes to describe the dynamic nature of cell membranes, and electrophysiological (patch-clamp) techniques to demonstrate the existence of membrane channels.

Chapters 2, 3 and 4 demonstrate that the initial concept of solutes passing through the membrane only by diffusion had to be extended when the use of radioisotopes (and also the introduction of the flame photometer, as well as electrochemical and optical techniques) permitted an unequivocal demonstration of the existence of additional pathways, such as protein-mediated (facilitated) diffusion and active transport. In the course of this development, the term 'permeability' had to be redefined to encompass, in addition to diffusion, also specific conductive pathways and exchange processes. The concept of a 'simple' facilitated diffusion had to be extended further by the discovery of several carrier-mediated systems for some substrates (e.g. the now known five (or possibly seven) transport systems for glucose); in addition, cotransport (symport) and countertransport (antiport) systems were described.

As to channels, the simple notion of one channel for one ionic species had to be extended, leading to the present view of multiple channels with differing pharmacological (receptor) characteristics. In this development, the role of discoveries of new preparations (cell ghosts, membrane vesicles, lipid membranes permitting a reconstitution of transport systems) and the techniques and concepts of molecular biology cannot be overstated. The described development does not fit the concepts of progress as the result of a series of scientific revolutions. The present models of cell membrane phenomena are far more complex than the earlier ones, while in the Kuhnian view new theories (with new paradigms) are 'more suitable' or 'simpler[8]'. A more restrictive use of the term 'scientific revolution' appears appropriate. In any case, an analysis of developing concepts in the membrane field also documents that often strenuously contested opposing views on a given topic are eventually resolved by a consensus on a synthetic view proving both sides partly 'right'. Thus, the energetically defended views on membrane 'pores' vs specialized structures (cf. Chapters 2, 3, 4) are now muted in the light of evidence that channels, carriers and pumps are in effect an arrangement of the component peptide(s) leaving a cleft for the passage of the solute, but allowing mutual interactions in a variety of ways.

In one respect there might have been a rather silent 'revolution'. The classical dichotomy in biology, structure versus function, did not fit the observations in the membrane field. While the fathers of the membrane concept, particularly Nägeli and Overton, considered an integration of both structure and function of cell membranes, in their thinking the cell membrane was an inert structure. The present knowledge recognizes the membrane as a dynamic system where both structure and function have to be considered when describing a

---

8 Whitehead's (cf. Price, 1954) admonition: 'Seek simplicity and then distrust it' should be kept in mind.

process. Ritchie (1935) pointed out that 'in the end, the distinction (between structure and function) is quantitative and rests upon the time scale we are using'. The student of membrane phenomena now accepts the notion that transient changes in the structure of the involved molecules may actually reflect the investigated function.

III. The role of ancillary fields

Did existing concepts in other fields affect the conceptual development of our knowledge of membrane phenomena?

In the initial phase, it was the recognition by biologists of the existence of membrane phenomena which opened this field to physical chemists. The towering contributions of van 't Hoff, Ostwald, Nernst, Planck, and Gibbs and Donnan laid the foundations for understanding the passage of solutes across physical membranes, and associated transfers of charges for the movement of ions. Biologists readily accepted these physico-chemical concepts in the interpretations of their observations.

The conceptual framework of colloid chemistry appealed to many biologists who sought possible mechanism for the broad area of ionic effects on physiological processes. Correlations between the phenomenology of colloid chemistry, e.g. the Hofmeister (1888) series of cation and anion effects on the swelling of proteins, sol-gel transitions, etc., and emerging knowledge of permeability phenomena were certainly striking. The colloid-chemical approach postulated the cell surface as a simple boundary separating protoplasm[9] from its environment; the properties of the protoplasm were the prime determinants of the differences in the electrolyte composition of

---

9 At its extreme, the unit of living matter was viewed as a giant, labile colloid molecule, called by Verworn (1899) 'biogen'; this concept was strongly criticized by Hopkins (1912).

cells and their environment, signalled around 1850 (Liebig; Schmidt; *cf.* Chapters 2 and 4), and metabolic reactions took place after adsorption on the colloid surface. On the other hand, the emerging membrane theory visualized the cell membrane as a structurally and functionally defined entity differing from the enveloped protoplasm, and responsible for the observed phenomena. Overton (1899) warned against 'erroneous interpretations of processes of absorption and secretion' in terms 'of the osmotic behavior of colloid substances in solution'. He was over-optimistic when believing that in the second half of the 19th century 'a number of facts became known (particularly concerning the erroneous belief that colloids have a special capacity to absorb water) that showed the falseness of the above mentioned assumptions'. Davson (1989) may be right in suggesting that the paradigms of colloid chemistry actually retarded the understanding of membrane phenomena. In his popular text on colloid chemistry for biologists, Handovsky (1922) stated that the membrane concept, postulated by physiologists, is far from compelling. He mentioned three possible membrane theories: (1) The protoplasm is separated from its environment only by a physicochemical boundary, and surface forces are responsible for the uptake of solutes. (2) Membranes are dense aggregates of protoplasm; its permeability would then reflect that of the protoplasm. (3) A semipermeable membrane differing in its properties from that of the protoplasm. The physicochemical concept of a membrane responsible for bioelectric phenomena, first emphasized by Ostwald (*cf.* Chapter 2) was not deemed to be acceptable in the light of evidence that electrical potential differences may arise at the interphase of two immiscible fluids, and 'phase boundary forces' were invoked as an explanation. In this context it may not be surprising to note that Ehrlich (1900) did not mention the cell membrane when discussing the existence of receptors for antigens at the cell surface, although he later quoted Overton's views about the role of cell lipids for the uptake of solutes.

By concentrating on studies of how ions affected biological processes, questions concerning the ionic distribution between cells and their environment, and mechanisms involved in this phenomenon (including effects of ions thereon, as well as how do ions pass across biological membranes) were often neglected.

The apparent antagonistic action of various cations on physiological phenomena attracted considerable attention, particularly between 1880–1930 (cf. Höber, 1926). Ringer (1881) was the first to show that the toxic effect of a pure, isotonic solution of NaCl on the contraction of the heart ventricle was abolished by small (physiological) concentrations of K^+ and Ca^{2+}; finding a similar effect of these cations on the swelling of the polysaccharide laminarin (1886), he interpreted his results in terms of ionic effects on the structure of protoplasmic components. Overton's observation in 1902 that Na^+ (or Li^+) was essential for muscle excitability, while K^+ (and its congeners) were ineffective, was interpreted by him as membrane effects (cf. Chapter 2). Lillie (1903) noticed that divalent and trivalent cations abolished the toxic effect of NaCl solutions on the movement of mussel cilia, and offered two possible explanations of this antagonistic effect: (a) on the state of aggregation of protoplasmic colloids; or (b) changes in the surface tension at various semipermeable surfaces of the cilia (this in the light of experiments of Overton, Hamburger, Gryns and Koeppe, cf. Chapters 2 and 4). Subsequent authors emphasized such results in analogy with Hofmeister's (1888) series $Na^+ > Li^+ > Cs^+ > Rb^+ > K^+$, and $Ca^{2+} > Sr^{2+} > Mg^{2+}$ cation effects on colloid systems. The above appraisal should not detract from the significant role of the colloid-chemical approach to the membrane field. The contribution of Donnan (Chapter 2) and the general concept of proteins and other cellular components as polyelectrolyte systems have to be mentioned here. The paradigm that catalytic (metabolic) processes occur at interfaces of colloids after adsorption on their surfaces began to fade only in the 1920s, with a major input by Loeb (1922) (cf. also Fruton, 1976). Physiologists could not avoid questions of ion movements across membranes in connection with the properties of red blood cells (cf. Hamburger, 1890; Koeppe, 1897, *cf.* Chapter 2) and of nerve and muscle excitation (Bernstein, 1902; Overton, 1902; *cf.* Chapters 2, 8) and based their research squarely on concepts of membrane phenomena started by biologists and developed by physical chemists. This is witnessed by many classical physiology texts[10] which based their exposition of physiological phenomena on their cellular basis, such as those of Schäfer[11] (1898), Bayliss (1912), Starling (1912) or Tigerstedt (1919); Richet's (1897) dictionary of Physiology specifically mentioned the existence of a cell membrane in some animal cells, and most plant cells. A notable exception was Verworn (1901) who avoided discussion of the work of Nägeli or Pfeffer, terms such as diffusion or permeability, or the unequal distribution of sodium and potassium between cells and their environment; all this was related to his views on protoplasm (*cf.* note[7]). Also, Foster's (1891) text, avoiding discussion of diffusion phenomena, specifically rejected the concept of a membrane for red blood corpuscles, maintaining that hemoglobin was bound to the cell stroma.

The development of our understanding of the mechanism of membrane phenomena may also have been retarded by an ap-

---

10 Actually, Ludwig, in his Textbook of Human Physiology (1858) discussed in some detail diffusion phenomena and their role in physiological processes (Vol. 1, pp. 59–84) and concluded: '...the solid animal parts serve to partition diverse body fluids. This... allows the exit from a mixed solution of one or another dissolved substance with exclusion of others. In view of the great importance of hydrodiffusion, regretfully those fluids and membranes which are most important for life have not yet been studied from the point of view of their diffusion properties.'
11 Schäfer not only embraced the emerging views of Overton, but postulated: 'Whether permeability be a function of physical or chemical nature, it is obvious that in the case of a living membrane ... the term 'physiological condition' must affect the property, so that one and the same membrane ... may ... be more or less permeable by the same substance' (p. 276).

parent distaste of many biochemists for this field, particularly in the period 1945–1955 (cf.[12]). Some of the founding fathers of modern (dynamic) biochemistry were not uncomfortable with notions of organized chemical reactions. Hofmeister (1901), in a visionary statement, considered a compartmentation of protoplasm, allowing a spatial separation of enzymic reactions while permitting diffusion of their substrates and products between the compartments. In this context, he also emphasized the tendency of colloid systems to form membranes at interfaces. F.G. Hopkins (1912) accepted cell architecture and its dynamic equilibrium as the basis of life phenomena:

'Life ... is the property of the cell as a whole, because it depends upon the organization of processes, upon the equilibrium displayed by the totality of the co-existing phases'.

However, Peters (1929), was 30 years ahead of his time[13] when emphasizing (perhaps reflecting his more biological background[14]) the role of cell architecture with its surface structure as an integrating element of cell activity, and he did not find understanding amongst most biochemists[15], although he did elicit support amongst physiologists and morphologists. A survey of classical texts of biochemistry from 1902 to 1958 is revealing as to the relative lack of interest in questions relat-

---

12 In this respect, the training of students of biochemistry may have been crucial, with an emphasis on an analysis of life processes in terms of organic reactions at a time when little was known about the chemistry of structural elements beyond the more general statements of physiology (K.V. Thiman and S. Weinhouse, personal communications). Attending different meetings, and reading different journals also contributed to the apparent compartmentation of sciences focusing on cell function.
13 Peters was not aware of Hofmeister's views.
14 Krebs' (1967) comment on differences in the biologist's and the chemist's approach to biochemical problems may be relevant here: 'If there is a difference in the outlook of the biologists on the one hand, and that of the chemist and physicist on the other, it is the urge of the biologist to look upon every property of living material as part of a complex system and to enquire into the fundamental significance of this property. Time and again this has proved to be a most fruitful working hypothesis – and it is no more than a working hypothesis'.
15 Even Donnan (1930) did not favor the tentative ideas of Peters.

ing the entry of nutrients into cells as a component of cell metabolism[16]: Bunge's (1902) and Mathews' (1915) textbooks essentially accept the colloid-chemical approach (but deal with the physical-chemical aspects such as osmotic phenomena). The advent of dynamic biochemistry, with emphasis on the intracellular metabolism of nutrients, led mostly to disregard the entry mechanisms (the authors mostly restricting themselves to a discussion of osmotic phenomena and the Gibbs-Donnan equilibrium, and postulating a free permeability of cell membranes for small solutes (e.g. urea, glucose etc., *cf.* Cantarow and Shepartz (1954), with occasional reference to texts of physiology (*cf.* Bodansky, 1938). The well-known textbook of Fruton and Simmons (1958) ignored questions of nutrient entry into cells. Biochemists obviously disregarded Hill's (1924) challenge:

'... we have arrived here at one of the most pertinent enquiries which biochemistry can make: that of the velocity at which, and the path by which, chemical change (*i.e.* in the cells) goes on'.

Volumes I–VI of the *Annual Reviews of Biochemistry* had chapters on permeability. After 1937, the emphasis shifted particularly to reports on metabolic processes, and new knowledge on membrane phenomena was not reported. Evidence for a lack of interest of biochemists in problems of permeability stems also from comments by many prominent biochemists. In Germany, Otto Warburg[17] blamed particularly R. Höber for all this messy thinking (*'An all dem is der Höber schuld'*, H.

---

16 On the other hand, texts focusing on physico-chemical aspects of biochemistry, i.e. by McClendon (1925), Bull (1964), or the six editions of Höber's classic (*cf.* 1926), dealt in considerable detail with then known aspects of cell membranes and their permeability.
17 Warburg in his studies 1919–1920 on the effects of narcotics on the HCN inhibition of cell respiration (*cf.* Warburg, 1948) was well aware of Overton's postulate a lipid cell membrane but chose to ignore this by speaking of cell lipids. However, prior to that, Warburg (1912) did recognize the organizing role of cell structure in cell respiration: 'possibly, membranes mediate the coordination of chemical reactions' (*cf.* *also* Chapter 6).

Holzer, personal communication). And Meyerhof (1924), in the light of Warburg's experiments on the role of narcotics on cell respiration, firmly rejected Overton and Meyer's views on the role of lipid membranes in the theory of narcosis. In Great Britain, the views of Marjorie Stephenson:

> 'Don't talk to me about permeability. That is the last resort of the biochemist who can't find a better explanation' (cf. Gale, 1971)

prevented her prominent student, E.F. Gale, from recognizing the accumulation of amino acids in bacteria as a major discovery. And although the scientific atmosphere at the Department of Biochemistry at Cambridge at that period encouraged extensive exchange of ideas, there was a failure of communication between J.F. Danielli and the Cambridge metabolic enzymologists (Mitchell, 1981)[18]. Hans A. Krebs, in spite of several notable contributions of his laboratory to transport phenomena, was never at ease with the role of cell membranes in metabolism (H.L. Kornberg, personal communication):

> 'Hans used to ring me repeatedly with the question: What is the *real* evidence that the entry of a substrate may be the rate-limiting step for its cellular metabolism'?.

In the USA, many colleagues (e.g. H.N. Christensen, A.K. Solomon) recall difficulties in having their research published in biochemical journals[19]. There may have been a subconscious rationale for this experience: particularly in the period 1945–1955, biochemistry basked in the success of unravelling metabolic pathways, and thus biochemical research in the still undeveloped field of membrane phenomena did not promise easy

---

18 An aversion of transport-oriented investigators to the biochemical approach should also be noted. P.D. Mitchell (personal communication) points out that Jim Danielli disapproved of his efforts in the 1950s to obtain a fusion of classical enzymic catalytic ideas with those of transport catalysis by means of his specific vectorial ligand conduction concept.
19 This doubtless accounted for the great influence of the *Journal of General Physiology* at that time.

recognition. Moreover, the logic of studies of intermediate metabolism was that of organic chemistry (Parnas, 1937), while the rationale for investigations of transport phenomena is based on the physical chemistry of heterogeneous systems, and, more specifically, the organization of chemical and physical processes at interfaces of heterogeneous systems. The available information on transport phenomena around 1950 was essentially kinetic in nature, and may have been thought to be more appropriate to physiology than to biochemistry; at that time, the study of membrane phenomena did not offer many concrete clues relating to structure or concrete enzyme-like processes, tempting a biochemical exploration, and the necessary techniques were not available. The indicated hesitation of biochemists to tackle problems understood to be 'physiological' recalls a comment of the great biochemist F.G. Hopkins (1936) that towards the end of the 19th century there

'was a tendency (in the biological thought) which in itself discouraged attempts to probe the secrets of living cells by chemical methods'.

Interest in a biochemical approach to membrane phenomena was generated when investigators realized that the phenomenology of many transport processes showed elements similar to those found in enzyme systems (Chapter 3). The involvement of biochemists was provoked by several separate advances: First, Claude (1948) showed that it was possible to separate subcellular components by differential centrifugation. This discovery opened the door to a biochemical exploration of the structure (and composition) of membranous subcellular components, first mitochondria, and later also the plasma membrane. Secondly, at the functional level, interest in membrane transport phenomena was stimulated by studies on the effects of insulin, particularly when Levine (1950) provided solid experimental evidence indicating that this hormone affected the entry of sugars into cells; this view was at variance with ideas held particularly by Gerty Cori[20] (cf. Colowick et al., 1947). Later, Carl Cori (cf. Crane et al., 1957; C.R. Park, personal communication) fully accepted the concept that studies of tissue permeability (and the effect of insulin thereon) could be tackled by biochemical techniques. The emerging interest of biochemists in transport phenomena, may be related to the studies of Cori, but particularly of Monod (cf. Cohen and Monod, 1957) and Hodgkin (1958), coupled with their personal brilliance and expository skills. Skou's (1957) discovery that the ATPase from crab nerve shows a dependence on Na^+ and K^+, reflecting the postulated properties of the sodium pump, provided an additional impetus for the biochemical approach to transport phenomena. The rather slow recognition of membrane phenomena by biochemists also reflects the leadership role of involved personalities. It was the analysis of processes associated with membranes using biochemical concepts of enzyme action, protein chemistry and molecular biology which facilitated the present state of understanding of the field.

An analysis of the development of views on the cell membrane reinforces the view that in this complex topic a multifaceted comprehensive scientific characterization was required. In this pluralistic approach, information gathered from many fields, and a broad spectrum of experimental techniques, produced the present level of our understanding of the cell membrane.

An analysis of the development of concepts in this field leads this student to the conclusion that milestones were characterized when investigators first raised questions, and then directed their research by considering *both structural and functional aspects* of the phenomena. This point should be borne out by Chapters 2–8.

------

20 Gerty Cori realized that hormone effects on metabolic processes could be demonstrated only in whole cells, but studiously avoided considering the cell membrane in this context (M. Cohn, personal communication).

Acknowledgements

The author is indebted to Drs. M.M. Civan, R.K.H. Kinne and P. Mueller for their critical comments on the draft of this chapter. The advice of Drs. Br. Chance, H.N. Christensen, M. Cohn, R.K. Crane, Z. Domotor, P.D. Mitchell, C.R. Park, A.K. Solomon, K.V. Thimann, S. Weinhouse and T.H. Wilson is greatly appreciated. The expert help of Mrs. W. Patriquin with Fig. 1. is gratefully acknowledged.

REFERENCES

Bayliss, W.M. (1912) *Principles of General Physiology*, 2nd. Ed., pp. 114–145. London, Longmans, Green and Co.
Bodansky, M. (1938). Introduction to Physiological Chemistry. New York, Wiley.
Brown R. and Danielli, J.F. (eds.) (1954) *Active Transport and Secretion.* Symp. Soc. Exp. Biol. 6. New York, Academic Press.
Bull, H.B. (1964) *An Introduction to Physical Biochemistry*. Philadelphia, F.A. Davis Comp.
Bunge, G. (1902) *Textbook of Physiological and Pathological Chemistry*. 2nd English ed. (Starling, F. and Starling, E.H., transl.) London, Kegan Paul, Trench, Trübner and Co.
Cantarow, A. and Shepartz, B. (1954) *Biochemistry*. Philadelphia, Saunders.
Claude, A. (1948) Studies on cells: Morphology, chemical constitution and distribution of biochemical functions. Harvey Lect. Ser. 43, 121–164.
Cohen, I.B. (1985) Revolution in Science. Cambridge, Mass., Harvard Univ. Press.
Cohen, G.N. and Monod, J. (1957) Bacterial permeases. Bact. Rev. 21, 169–194.
Colowick, S.P., Cori, G.F. and Slein, M.W. (1947) The effects of adrenal cortex and anterior pituitary and insulin on the hexokinase reaction. J. Biol. Chem. 168, 583–596.
Crane, R.K., Field, R.A. and Cori, C.F. (1957) Studies of tissue permeability. I. The penetration of sugars into Ehrlich ascites tumor cells. J. Biol. Chem. 224, 649–663.
Davson, H. (1989) Biological membranes as selective barriers to diffusion of molecules. In *Membrane Transport, People and Ideas* (Tosteson, D.C., ed.) pp. 15–49. Bethesda (MD), Am. Physiol. Soc.

Davson, H. and Danielli, J.F. (1943) *The permeability of natural membranes.* Cambridge, Cambridge Univ. Press.

Dixon, M. (1970) In *The Chemistry of Life* (Needham, J., ed.) p. 15–37. Cambridge, Cambridge Univ. Press.

Donnan, F.G. (1930) Comment made at Trans. Faraday Soc. 26, 815–816.

Ehrlich, P. (1900) On immunity with special reference to cell life. Proc. Roy. Soc. 66, 424–448.

Foster, M. (1891) *A Text-Book of Physiology,* (Fourth Amer. Ed.). Philadelphia, Lea Brothers.

Fruton, J.S. (1976) The emergence of biochemistry. Science 192, 327–334.

Fruton, J.S. and Simmonds, S. (1958) *General Biochemistry,* 2nd Ed. New York, Wiley.

Giebisch, G., Tosteson, D.C. and Ussing, H.H. eds. (1978) *Membrane Transport in Biology.* Heidelberg, Springer Verlag.

Handovsky, H. (1922) *Leitfaden der Kolloidchemie für Biologen und Mediziner.* Dresden, T. Steinkopf.

Hill, A.V. (1926) The physical environment of the living cell. In *Lectures on Certain Aspects of Biochemistry* (Dale, H.H., Drummond, J.C., Henderson, L.J. and Hill, A.V.), pp. 253–280. London, Univ. of London Press.

Höber, R. (1926) *Physikalische Chemie der Zelle und Gewebe,* 6. Aufl., Leipzig, W. Engelmann.

Höber, R., Hitchcock, D.I., Bateman, J.B., Goddard, D.R. and Fenn, W.O. (1945) *Physical Chemistry of Cells and Tissues.* Philadelphia, Blakiston Comp.

Hodgkin, A.L. (1958) Ionic movements and electrical activity in giant nerve fibres. Proc. Roy. Soc. Ser. B 148, 1–37.

Hofmeister, F. (1888) Zur Lehre von der Wirkung von Salzen. Arch. exp. Pathol. Pharmakol. 24, 247–260.

Hofmeister, F. (1901) Die chemische Organisation der Zelle. Naturwiss. Rundschau 16, 581–583; 600–602; 612–614.

Hopkins, F.G. (1912) The dynamic side of biochemistry. Presidential address (Physiology), Brit. Assn. Adv. Science, 652–668.

Hopkins, F.G. (1936) The influence of chemical thought on biology. Science 84, 255–260.

Kandel, E.R., Schwartz, J.H. and Jessel, T.M. (1991) *Principles of Neural Science,* 3rd Ed., pp. 1009–1031. New York, Elsevier Publ.

Kleinzeller, A. and Kotyk, A. (eds) (1961) *Membrane Transport and Metabolism.* New York, Academic Press.

Krebs, H.A. (1967) The biologist's and the chemist's approach to biochemical problems. In *Reflections on Biologic Research* (Gabbiani, G., ed.). St. Louis, Warren H. Green.

Krogh, A. (1929) Progress in Physiology. Am. J. Physiol. 90, 243–251.

Kuhn, T. (1962) *The Structure of Scientific Revolutions.* Chicago, Chicago Univ. Press.

Langmuir, I. (1929) Modern concepts in physics and their relation to chemistry. J. Am. Chem. Soc. 51, 2847–2868.

Levine, R., Goldstein, M.S., Huddlestun, B. and Klein, S.P. (1950) Action of insulin on the 'permeability' of cells to free hexoses, as studied by its effect on the distribution of galactose. Amer. J. Physiol. 163, 70–76.

Lillie, R.S. (1903) The relation of ions to cilliary movement. Am. J. Physiol. 10, 419–443.

Lillie, R.S. (1922) Protoplasmic Action and Nervous action, p. 108. Chicago, IL, Univ. of Chicago Press.

Loeb, J. (1922) Proteins and the Theory of Colloidal Behavior. New York, McGraw-Hill.

Ludwig, C. (1858) Lehrbuch der Physiologie des Menschen. 2nd Ed., Vol. 1, pp. 58–84. Leipzig, C.F. Winter Verl.

Mathews, A.P. (1915) Physiological Chemistry. New York, W. Wood & Co.

McClendon, J.F. (1925) Physical Chemistry in Biology and Medicine. Philadelphia, Saunders.

Meyerhof, O. (1924) Chemical Dynamics of Life Phenomena, pp. 13–14. Philadelphia, Lippincott Co.

Mitchell, P.D. (1981) Bioenergetic aspects of unity in Biochemistry: evolution of the concept of ligand conduction in chemical, osmotic and chemiosmotic reaction mechanisms. In Of Oxygen, Fuels and Living Matter, Part I. (Semenza, G. ed.), pp. 1–56, New York, Wiley & Sons.

Nägeli, C. and Cramer, C. (1855) Pflanzenphysiologische Untersuchungen. 1. Heft. Zürich, F. Schulthess.

Overton, E. (1899). Ueber die allgemeinen osmotischen Eigenschaften der Zellen, ihre vermutlichen Ursachen und ihre Bedeutung für die Physiologie. Vierteljahrschr. d. naturforsch. Ges. Zürich 44: 88–114.

Parnas, J.K. (1937) Der Mechanismus der Glykogenolyse im Muskel. Erg. d. Enzymf. 6, 57–110.

Peters, R.A. (1929) The Harben Lectures, 1929. Coordinative Biochemistry of the cell and tissues. Lecture I. 'Cell surfaces'. J. State Med. 38, 683–709.

Price, L. (1954) Dialogues of Alfred North Whitehead. Boston, Little, Brown & Comp.

Racker, E. (1965) Mechanisms in Bioenergetics. New York, Academic Press.

Richet, C. (1897) Dictionnaire de Physiologie, Vol. II, pp. 506 ff. Paris, F. Alcan.

Ringer, S. (1881) Concerning the influence exerted by each of the constituents of the blood on the contraction of the ventricle. J. Physiol. (Lond.) 3, 380–393.

Ringer, S. (1886) A further contribution regarding the effect of minute quantities of inorganic salts on organic structures. J. Physiol. (Lond.) 7, 118–127.

Ritchie, A.M. (1935) In The Natural History of Mind, p. 69. London, Longmans, Green and Co.

Sachs, J. (1865) Handbuch der Experimental-Physiologie der Pflanzen. Leipzig, W. Engelman.

Schäfer, E.A. (1898) Textbook of Physiology, pp. 264–280. Edinburgh, Young J. Pentland.

Skou, J.C. (1957) The influence of some cations on an adenosine tri-phosphatase from peripheral nerves. Biochim. Biophys. Acta 25, 394–402.

Starling, E.H. (1912) *Principles of Human Physiology*, pp. 145–153. Philadelphia, Lea & Febiger.

Tigerstedt, E. (1919) *Lehrbuch der Physiologie des Menschen*, Vol. I. pp. 40–48, Leipzig, s. Hirzel.

Verworn, M. (1899) *General Physiology. An Outline of the Science of Life.* English Edition (Lee, F.S., ed.), p. 481. London, Macmillan & Co.

Warburg, O. (1912) Über Beziehungen zwischen Zellstruktur und biochemischen Reaktionen. I. Pflüger's Arch. Ges. Physiol. 145, 277–282.

Warburg, O. (1948) *Schwermetalle als Wirkungsgruppen von Fermenten*, pp. 16–19. Berlin, W. Saenger.

A. Kleinzeller (Editor) A History of Biochemistry: Exploring the Cell Membrane.
(Comprehensive Biochemistry Vol. 39) © 1995 Elsevier Science B.V.

Chapter 2

The Postulate of the Cell Membrane

A. KLEINZELLER

Department of Physiology, School of Medicine, University of Pennsylvania,
Philadelphia PA 19104-6085, U.S.A.

Introduction

It is generally accepted that the concept of a cell membrane gradually developed in the middle of the 19th century on the basis of microscopic and physiological studies on plant cells (see e.g. the scholarly reviews of the subject by Pfeffer (1877), Overton (1895), Höber (1926), Jacobs (1962), Smith (1962) and Davson (1989). The *functional* concept of a cell membrane arose from careful observations of the botanists (von Mohl, 1846), but particularly of Nägeli (1855): In most plant cells they distinguished between the cell wall and the very thin layer of protoplasm (a slimy, 'nitrogenous' material dissolved, but not freely miscible with the surrounding aqueous medium) adhering to the cell wall and enclosing the cell sap (now called vacuoles). On placing plant cells into concentrated solutions of non-poisonous substances, e.g. sucrose, Nägeli and Cramer found that the protoplasmic layer could be separated from the cell wall, forming a sac or globe and this process was reversible on replacing the sucrose solution with water. (The phenomenon of separating the protoplasm from the cell wall was also described by Pringsheim (1854) and was denoted later by De Vries (1884) as plasmolysis.) If dyes were present in the vacuole, the sucrose-induced shrinking of the protoplasm permit-

ted a loss of water, but not of the dye, from the cell sap, while
the dye was permeable into and from the protoplasm of dead
cells. Nägeli interpreted these observations as the capability
of the protoplasmic layer to permit exosmosis[1] and endosmosis
of *water*, but not of the various solutes in the cell sap. Thus, the
protoplasm displays a property later defined by van 't Hoff
(1887) as a semipermeable system; this characteristic was also
reflected by the growing realization that the protoplasm of
plants and animal cells is capable of accumulating salts
against sizable osmotic gradients[2]. At that time, osmotic phenomena had already been described by Nolet in 1748 and particularly by Dutrochet (1827), using membranes such as the
pig bladder.

Nägeli was inclined to ascribe the osmotic properties to the
whole protoplasm (p. 5 of his paper); later studies of Traube
(1867) and Pfeffer (1877) localized them in a thin, clear protoplasmic layer (hyaloplasm[3]) separating the environment from
the granular portion of the protoplasm. On cutting the cell
wall, Nägeli noticed that the protoplasm often streamed out
into the surrounding fluid, occasionally breaking down into a
multitude of small droplets. This observation was taken as
evidence for the de novo formation of boundary layers, required to meet the need for increased surface/volume ratio. He
then generalized by stating that

> 'whenever ... protoplasm comes in contact with other aqueous media, a
> membrane-like layer is formed on the whole boundary by a process compa-
> rable to protein coagulation'.

---

1 Derived from ὠσμός (propulsion). The terms endosmosis, exosmosis and diosmosis
were coined by Dutrochet (1827), indicating osmotically-induced fluid flows into a
compartment, from the compartment, and bidirectional flows, respectively. He also
found that the flux of the solvent (water) could proceed in a direction opposite to that
of the solute (NaCl).
2 This point will be discussed in detail in Chapter 4.
3 Pfeffer specifically considered the possibility that the properties at the outer and
inner membrane surfaces may differ; the membrane may contain valve-like systems
permitting flow in one direction only (p. 160).

The term designating this boundary layer gradually changed, from cell skin (Zellhaut; the pellicle[4] of von Mohl, 1855, p. 701) to plasma membrane. Since cell growth had to entail an increase in membrane surface, membrane growth might occur by incorporation (intussusception) of membrane particles from the protoplasm (Pfeffer). Thus, operationally (*cf.* Langmuir, Chapter 1) the concept of a cell membrane encompassed *functional as well as structural* elements; such a membrane would physically and chemically differ from the cell protoplasm.

The above experimental basis for the cell membrane concept was actually preceded by earlier work, later largely ignored: Hewson (1773) reported the results of his studies on erythrocytes of a variety of species. In his experiment I, he demonstrated that red blood particles are disks with some rigidity, pointing out that liquid particles would assume a globular shape. Many types of red blood particles contained a central denser body (later defined by Schultz (1836) as the nucleus). Experiment II showed that on diluting the suspending fluid (serum) with water the particles gradually changed

'from a flat to a spherical shape'. ... 'The solid middle particle can be distinctly seen to fall from side to side in the hollow vesicle like a pea in a bladder' (p. 311).

Exp. III: Addition of neutral salts shrivelled fresh red blood particles. Moreover,

'When the vesicles have been made spherical, by being mixed with water, a small quantity of pretty strong solution of a neutral salt be added, they are immediately shrivelled, a few of them recover their former flat shape...'.

Hewson thus clearly qualitatively described what was later termed osmotic swelling and shrinking of red blood cells which

---

4 ... 'this seems to indicate that the outer layer of the primordial sac (corresponding to hyaloplasm) is solidified into a pellicle'. v. Mohl was very careful in discriminating between the inert cellulose wall and the pellicle.

he considered to be vesicles; by implication, a surface structure, containing the elastic fluid content of the vesicles, contributed to the 'solid' properties of the vesicles. Hewson's experiments were fully confirmed by Schultz (1836) and to some extent by Müller (1838) (who did not mention Hewson's work). Schultz actually described experiments (p. 17–20) to show that the vesicle envelope envelope (Bläschenhülle) was a water-insoluble, white structure with some rigidity and elasticity which could be visualized using iodine. Both Hewson and Schultz misinterpreted their studies on human erythrocytes by believing that these cells also were nucleated vesicles. Schwann in 1839 (*cf.* 1847), apparently unaware of Hewson's and Müller's studies, could not reproduce Schultz' statement about the rolling of the nucleus in swollen cells; he did, however, fully endorse the concept of blood corpuscles as vesicles. An extension of observations on the development of animal cells then led to the concept of a structural membrane as an *integral* component of each cell. Schwann's (1847, p. 199) visionary statement that:

> 'in order to explain the distinction between the cell content and the external 'cytoblastema' (meaning the immediate environment of cells), we must ascribe to the cell-membrane not only the power in general of chemically altering the substances which it is in contact with, or has imbibed, but also of separating them that certain substances appear on its inner, and others on its outer surface'

was then borne out by 100 years of subsequent investigations.

It now would appear that in the period 1773–1838 the basic knowledge about the osmotic phenomena observed by Hewson, Schultz, Schwann and Müller was insufficient to relate the strong indication of a structural membrane to the functional aspects of an osmotic response of cells, i.e. as a barrier. And Müller's misgivings concerning diversities in reports about the size and shape of red blood particles of different species may have contributed to the neglect of information concerning the cells as vesicles.

The above state of the art in the period 1773–1870 established osmotic phenomena as a crucial property associated with the cell membrane, with diffusion as the actual process; the laws governing diffusion were already known as the result of Fick's (1855) fundamental work. As defined by Pfeffer (p. 127 of his monograph):

'The protoplasm has to take up any substance which diosmoses through the plasma membrane, and ... each substance, provided it is not retained in the protoplasm by binding, also has to pass into the cell sap if the enveloping membrane permits diosmosis. Such uptake (or loss) must persist as long as a concentration gradient within and outside the cell produces a unidirectional diosmotic movement'.

Pfeffer also emphasized additional properties of the cell membrane: its distensibility (and elasticity) (p. 143) would account for the behavior of cells when changing external osmolarity; moreover, since the death of the cell made the membrane permeable for a variety of solutes, membrane integrity was related to the living cell (p. 169). The phenomenology of semipermeable membranes separating the intracellular compartment from its environment was foreshadowed by Traube in 1867 (cf. Loeb, 1903):

'The cell membrane makes a diminutive chemical factory of the contents of the cell by shutting it off from its surroundings and enables each to lead a specifically different life from the neighboring cells.'

Electrochemical phenomena were gradually associated with the membrane function as the result of several observations: Liebig (1847), having described the difference in inorganic components of tissue fluids (high K, low Na) and blood (high Na, low K), expressed the idea (p. 329) that such difference might produce electrical potential differences; since the concept of a cell membrane had not yet been developed, Liebig suggested that the permeability (Durchlässigkeitsvermögen) of the wall of blood vessels was responsible for the unequal ionic distribution (and electrical potential). He then instigated experiments demonstrating electrical potentials arising from simple experiments with tissue preparations. These observations were not immediately taken up by other investigators. (Graham, 1861) showed that the separation of permeant (crystalloid) solutes from the impermeant colloids could be achieved by skins (*e.g.* pig bladder) or even thin paper impregnated with gelatin. The use of precipitation membranes of Traube (formed at the interface of *e.g.* two colloids of opposite charge, or copper sulfate and potassium ferrocyanide solutions (1867), particularly when produced on or in a porous support (Pfeffer, 1877) then demonstrated their semipermeable properties[5], recalling such characteristic of biological systems. The semipermeable properties of living cells used as an osmometer were quantitatively explored particularly by De Vries (1884), and also by Hamburger. The accumulated data of Pfeffer, de Vries and Hamburger (*cf.* van 't Hoff, 1887) led to the development by van 't Hoff of the physico-chemical concept of osmotic pressure.

The next step was the study of the physico-chemical properties of idealized (infinitely thin) membranes separating two liquid phases. This led to the idea of an electrical double layer by Helmholtz (1879), and Ostwald (1890) then demonstrated electrical potentials across semipermeable membranes. The theory of these phenomena (diffusion of ions across junctions of two fluids and boundary layers, including conceptually inert membranes) was established particularly by Nernst (1889) and Planck (1890). Two basic views for the explanation of electrical potentials across membranes were put forward: While Ostwald favored the idea that membrane 'pores' selectively affected the mobility of cations and anions, Nernst (*cf.* in particular, 1908) preferred the view of selective solubility of ionic

---

5 Traube's copper-ferrocyanide membrane was shown to be impermeable to the sulface anion.

species in a non-aqueous interphase[6]. Both Ostwald (1890) and Nernst (1899) related their concepts to the existence of electrophysiological[7] phenomena first described by Matteucci, but particularly by Du Bois-Reymond (1848) (*cf.* Chapter 8). The physico-chemical abstractions, relating osmotic and electrical phenomena to junctions and boundary layers, were based on the idea of very thin membranes. Osmotic and electrical properties became the crucial phenomenology in the study of the cell membrane concept.

I. The perception of a cell membrane as the osmotic and electrical barrier. Is there a cell membrane?

The postulate of a permeability barrier at the cell surface was enhanced by numerous studies. Quincke (1888), on the basis of experiments on the physical properties at the interface of lipid droplets and aqueous solutions, arrived at the conclusion that the protoplasmic cylinder was enveloped by a thin fluid membrane (less than 0.1 μm thick) consisting of a fatty oil or soap[8], using the term 'protein-soap'. A summary of the extensive studies preceding Overton's membrane theory is given by Höber (1926).

The osmotic phenomenology was first systematically used by Overton to explore the basic properties (and nature) of the putative cell membranes. In the first of a series of three classical papers Overton (1895, 1896, 1899) demonstrated that the passage of solutes across plant cell membranes depended not

---

6 The question of simple pores for the diffusion of water and salt through inanimate membranes, suggested by Brücke (1843), was already criticized by Fick (1855): 'The endosmotic process may take place not through pores, but possibly through a real molecular interstition'.

7 The concept of small electrical currents passing through the nerve 'envelope' (Hülle) as the basis of excitation was actually put forward by Hermann in 1872 (*cf.* 1899).

8 Newton (1704) actually calculated the thickness of the 'black holes' in soap bubbles to be of the order of 95 Å.

only on their molecular weight, but also on chemical structure: solutes with great lipid solubility entered cells very rapidly, compared with solutes only slightly soluble in lipids. This characteristic was shared also by animal cells (experiments on tadpoles). He thus introduced the concept of selective permeability. Such observations led to the postulate (1899) that

'the general osmotic properties of the cell are due to the impregnation of the boundary layers of the protoplast by a substance ... corresponding to a fatty oil'.

According to this concept, and as opposed to the views of Nägeli, Traube and Pfeffer, the nature of the cell membrane differed qualitatively from that of the protoplasm. The lipid membrane was assumed to be absolutely permeable to water, and practically impermeable to salts. Physical forces (i.e. diffusion) would be responsible as the operative force for the transfer of substances across the membrane. Overton thus used Nernst's (1891) concept of the role of the partition coefficient in the passage of solutes through membranes. He also noticed that narcotics were characterized by their lipid solubility found by Meyer (1899) and permeability of cell membranes; this observation than served as the basis of the lipid theory of narcosis (Overton, 1901).

The idea of a lipid membrane as the osmotic barrier of cells had to be reconciled with the observation that the cell surface was wettable by, and permeable to, water[9]. Overton proposed that a membrane containing e.g. cholesterol esters (1899) or phospholipids (1900) would meet such requirement. The biologists were thus presented with two fundamentally differing concepts of a cell membrane: the (continuous) lipid membrane, as opposed to Pfeffer's semipermeable membrane. These two concepts were also reflected by the criteria which each hy-

---

9 The postulate of an absolute permeability of biological membranes for water was shown by Loeb (cf. 1903) not to apply to cells of some marine fish.

pothesis had to meet. For the semipermeable membrane, movement through 'pores' would be inversely related to the molecular volume of the diffusing solute (the square root of the molecular weight M according to Thovert (cf. Jacobs, 1935), the cubic root of M (at least for solutes of higher molecular volume) according to Stein (1967) and would not display solute specificity. On the other hand, lipid solubility would be the determinant for the passage of a solute across the lipid membrane. Since diffusion was assumed to be the actual mechanism by which solutes passed through the membrane, determinants of diffusion (e.g. temperature dependence, with a Q_{10} of 1.3[10] had to be obeyed for both membrane concepts.

Overton (1899) made abundantly clear that the above concept of a lipid membrane applied to the passive uptake of solutes and would not explain the observed permeability of cell membranes to salts or nutrients such as sugars:

'It is now no longer doubtful that the uptake by tissue cells of ... proteins, but probably also of sugars, is not regulated through purely osmotic processes; the cells must also play an active role in this uptake'.

Opponents of Overton's views focused on the fact that membranes do allow the entry of lipid-insoluble salts and nutrients, and exit of metabolic products. Vigorous proponents, e.g. Höber[11] (1907) emphasized the difference between physical (by simple diffusion) and physiological permeability, the latter denoting the hitherto unexplored mechanism for the passage of lipid-insoluble solutes across cell membranes. Voices denying the membrane concept by suggesting that the whole proto-

---

10 As pointed out by Danielli (Davson and Danielli, 1943), the value of Q_{10} for diffusion across biological membranes may actually be up to 1.4, owing to the concomitant changes of the viscosity of the system. The magnitude of the temperature coefficient per se does not permit to draw a distinction between the hypotheses of continuous lipid membranes, as opposed to discontinuous, sieve-like membranes.
11 Höber became impressed by Overton's views when both were working at the Dept. of Physiology of the Zürich University.

plasm was responsible for the osmotic phenomenology (*e.g.* Bütschli, 1894) were numerous (Chapter 1).

Several experimental approaches emphasized the role of the cell membrane as a permeability barrier:

(1) *Microdissection*. Nägeli (1855) and more clearly, Pfeffer (1877) demonstrated that crushing cells produced protoplasmic droplets still showing osmotic phenomenology, and provided indications that such observations were not consistent with simple interfaces between two immiscible fluids. Hertwig (1893) was the first to provide indications of the existence of a cell membrane in frog oocytes by a microsurgical technique. Subsequently, the studies of Chambers (*cf.* Chambers, 1926), started in 1915, showed that the cell surface film could be torn off, leading to a disintegration of the protoplasm.

'The protoplasmic surface film serves ... to retain within its confines the protoplasmic ingredients'.

Moreover, impermeant dyes, when injected into the protoplasm, rapidly diffused there, demonstrating the cell surface film as a permeability barrier. The torn protoplasmic surface could reform in the presence of Ca^{2+}; hence,

'calcium is needed for the formation and maintenance of the protoplasmic membrane'.

(2) *Electrophysiological phenomena*. The selective permeability of cell (membranes) to electrolytes, known to exist in plant cells since the Saussure's studies in 1805 (*cf.* Sachs, 1865), was demonstrated also in animal cells (Hamburger, 1890; Koeppe, 1897). Bernstein (1902) related measured electrical potentials across putative muscle and nerve membranes to the diffusion of K^+ across the boundary layer; the observed electrical potential was considered to be a Nernst potential of K^+. The generation of electrical potentials across membranes was demonstrated by Donnan (1911), showing that an unequal distribution of electrolytes, as well as a transmembrane electrical

potential may result from the presence of non-diffusible ions within cells[12]. The applicability of the Gibbs-Donnan concept to a biological system (red blood cells) was then provided by van Slyke (*cf.* 1926).

Höber (1910) (with an input of Nernst) introduced the measurement of cell conductivity as a tool in order to arrive at an answer concerning the existence of a cell membrane. The conductivity of intact red blood cells corresponded to a 0.002 N KCl solution, whereas cells made permeable by the presence of saponin showed a conductivity close to 0.1 N KCl. These results clearly indicated the presence of a non-conductive cell membrane which prevents electrolyte exit. Fricke (1925) then extended this experimental approach and calculated that the insulating thickness of the putative red cell membrane was 33 Å thick, assuming a monomolecular lipid film with a dielectric constant of 3. Once it proved possible to measure all the required electrical parameters of a membrane, i.e. potential, conductance and current, the cell membrane could be described in terms of an equivalent electrical circuit (Fricke, Blinks, Cole; *cf.* Cole, 1972): in this language, the potential corresponds to the driving force, the conductance to permeability, the current to the net ion flux across the membrane, and the capacitance reflects the insulating characteristics of the membrane.

In order to reconcile the passage of charged particles (including dyes) across plant cell membranes, Nathanson (1904) proposed the concept of a discontinuous membrane consisting of lipid and non-lipid areas. Such a view would also explain earlier observations (Hedin, 1897) that some lipid-insoluble non-electrolytes of low molecular weight, *e.g.* glycerol or erythritol, did enter cells. Höber (1907) also postulated a com-

---

12 The thermodynamic principles involved in the effect of a non-diffusible ionic species on the electrolyte distribution and generation of a pressure (and electrical potential) gradients was first deduced by Gibbs (1876), and were further elaborated by Donnan and Guggenheim (1932).

plex membrane, based on experiments analyzing the resting potential of muscle. Paying dues to the then prevalent paradigms of biological phenomena, he interpreted the correlation between electrolyte effects on colloid proteins (the Hofmeister series) and measured resting potentials as effects on the colloid-chemical properties of the portion of the membrane involved in the permeability to electrolytes. The inhibitory effects of narcotics on the resting membrane potential were then interpreted as effects of the lipid-soluble agents on the colloid properties of the cell membrane (at that stage of knowledge, Höber postulated membrane phospholipids as the target), thus proposing the permeability theory of narcosis[13]. An extension of this approach led later to the sieve theory of membranes (see, *e.g.* Collander and Bärlund, 1933). Harvey (1911) also established that the membrane of marine eggs, impermeable to alkalis, became rapidly permeable by fertilization; thus, membrane permeability could be affected by physiological phenomena.

(3) *Structural studies.* The membrane concept received further support when Gorter and Grendel (1925) related the calculated surface of red blood cells of various species to the area occupied by films of cell lipids and concluded that the putative cell membrane had a thickness corresponding to two lipid molecules[14]. Based on the work of Langmuir (1917), Gorter and Grendel supposed that

'.... every chromocyte is surrounded by a layer of lipoids, of which the polar groups are directed to the inside and the outside, in much the same way as the ... molecules of a soap bubble are according to Perrin'.

---

13 This general theory is quite acceptable to this day (cf. Krnjevic, 1991), once the rather vague concepts of permeability to electrolytes are translated in more concrete terms of ionic channels.
14 Subsequent reexamination of these data revealed several errors which in effect offset each other. More recent data indicate that the lipid bilayer represents about 75% of the cell surface area, with intrinsic membrane proteins representing 25% (*cf.* Zwaal et al., 1976).

Grendel (1929) analyzed the composition of erythrocyte lipids and found that in terms of percentage of cell surface cholesterol represented 36%, while the bulk of 'membrane' lipids were the phospholipids cephalin (50%) and sphingomyelin (13%). The calculated thickness of such lipid membrane would be of the order of 30–40 Å. Such data were consistent with Overton's views. These studies then directly led to the Danielli-Davson membrane model (1935). The concept of a lipid membrane was not universally accepted; Brooks (1937) still visualized a membrane model based (at least in part) on protein.

(4) *Cell–drug interaction.* Ehrlich (1900) located the interaction between antigenic proteins and cell receptors (sidechains of protoplasmic proteins) at the cell surface. Subsequently, he (1902) and also Langley (1906) indicated that chemotherapeutic agents produced their effects by first interacting with cell components at the cell surface. Further studies in the field demonstrated a complex phenomenology of drug effects on cells (*cf.* Chapter 5). An experimental milestone in relating cell surface receptors to the membrane was the observation of Cook (1926): Using the antagonistic effects of acetylcholine and methylene blue on heart muscle and nerve, he showed that a) the pharmacological effect of the dye could be reversed by washing out and hence was due to some *reversible* interaction between the dye and the superficial receptor; and b) the bulk of the dye inside the cells failed to produce an effect; hence the dye did not act from the inner face of the surface film. Clark (1937) produced a unifying hypothesis by

'the assumption that drugs react with specific receptors on the cell surface (membrane) and by occupying these receptors modify the functions of the cells in much the same manner as an enzyme poison can modify the functions of an enzyme by reacting with the active group of the enzyme'.

The membrane receptor concept included the postulate of a (membrane) mechanism coupling the receptor-solute complex to the cellular response.

The hypothesis of the cell membrane as the osmotic barrier separating protoplasm from the immediate environment withstood rigorous tests by a variety of approaches. The stage was set for a more detailed analysis of the nature of such membrane: quantitative measurements of permeability were sought to provide information as to the morphology and physiology of cell membranes.

II. Properties of the postulated cell membrane

The increasing sophistication in the study of membrane phenomena in 5 decades after Overton's views were published raised new questions as to the nature of the cell membrane, particularly when considering the passage of water-soluble solutes. Two basic concepts were considered: (1) Either the membrane consists of a homogeneous layer of a water-insoluble substance, or (2) it has a sieve-like structure similar to a filter. In the first case, the membrane permeability is due to the solubility of the permeating substance in the membrane material, and mechanisms had to be sought to explain how the lipid solubility e.g. of an ion could be increased; alternatively, diffusion takes place through holes which directly join the aqueous phases on both sides of the membrane (cf. Höber, 1936; Wilbrandt, 1938; Davson and Danielli, 1943). The analogy of these two concepts and the opposing views of Nernst and Ostwald on the generation of electrical potentials across physico-chemical phases and semipermeable membranes (see above) is patent. Such questions could be answered only by a more precise definition of the terms used and then obtaining quantitative measurements to provide information as to the extent to which cell membrane phenomena were consistent with data obtained using simple physico-chemical models. A third concept put forward to explain the phenomenology of permeability (Troshin, cf. 1976, and Ling, cf. 1984) assumed that both intracellular water and electrolytes are not 'free' (i.e.

their activities are decreased by interaction with cell protein)[15]. An extensive discussion of the opposing views took
place at the 1960 Prague Symposium (*cf.* Kleinzeller and
Kotyk, 1961). For a scholarly discussion of these points, the
review of Dick (1959) should be consulted. No unequivocal experimental evidence in favor of the 'association-induction hypothesis' of Ling (1984) has been presented and the suggestion
that intracellular electrolytes are immobilized and hence elec-
trochemically not active could not be substantiated (*cf.* Civan
and Shporer, 1989). Extensive studies did reinforce the membrane concept[16]; in particular, the reconstitution of membrane
function in liposomes using isolated transport proteins, and
the expression of single genes for transport processes in other
cells, *e.g.* the frog oocytes, can not be reconciled with Ling's
hypothesis.

Measurements of cell permeability by Overton and his con-
temporaries were essentially qualitative. Moreover, the employed experimental arrangements did not allow discrimination between the passage of the solute and solvent across the
membrane. Jacobs (*cf.* 1935), applying Fick's law of diffusion
to the more complex conditions of transport kinetics in cells,
derived the equations necessary for a quantitative evaluation
of solute permeability; his analysis also took into account that
both solute and solvent flows have to be considered. This then
supplied the basis for a comparison of the extent to which simple diffusion was consistent with the observed phenomenology. Furthermore, Teorell (1935) began to emphasize the need
of distinguishing between *permeability* (ability to penetrate
across the membrane, later refined further by the introduction
of the concept of the reflection coefficient by Staverman (1952)

---

15 In essence, the advanced arguments represented an extension of the colloid-chemical approach, considering protoplasm as a polyelectrolyte colloid system (*cf.* Chapter
I).
16 Although *some* cell water may not be available as solvent because of hydration of
cell proteins, and *some* electrolytes appear to bind to proteins.

and the *driving forces* for the solute flux: these include (a) the
thermodynamic gradient, often described as osmotic forces;
this acts on neutral molecules as well as on charged particles
(ions and colloids); (b) electrical potential gradients, acting on
charged particles; (c) hydrostatic gradients, involved in ultrafiltration, etc.; and (d) surface force gradients – differences
in interfacial tension. These developments greatly influenced
the next stages in understanding the concept of the cell membrane.

1. MEMBRANE STRUCTURE

Enquiries into the properties defining the structure of cell
membranes were deemed to be essentially concluded by 1940
(*cf.* the Cold Spring Harbor Symposium in 1940). Measurements of the tension at the surface of sea-urchin eggs (Cole,
1932), fish eggs and egg oil (*cf.* Danielli and Harvey, 1934) led
to the suggestion that the relatively low tension was due to a
layer of adsorbed protein at the oil-aqueous interphase; this
idea was then elaborated further by Davson and Danielli
(1943) in their now classical membrane model[17]. The presence
of globular protein adsorbed on the liquid lipid layer would
impart to such membrane physical stability while maintaining its flexibility, and presumably affect also its permeability
characteristics. Later models visualized a protein layer coating both sides of the lipid bilayer. As opposed to a simple lipid
bilayer, the protein component of the membrane would
contribute to the significant elasticity of the system (Frey-
Wyssling, 1953). The concept of the plasma membrane as a
lipid-protein film also arose from optical studies, such as the

---

17 For a detailed account of the development of ideas on the lipid membrane, the
monograph of Danielli and Davson (1943) and Davson's retrospective appraisal,
(1989) should be consulted. The concept was actually developed by Danielli (Davson,
personal communication), and without knowledge of the previous work of Gorter and
Grendel (*cf.* Danielli, 1962).

diffraction and histochemical analysis (Schmidt, 1935) of the myelin sheath of the nerve (lipids organized perpendicular to the sheath surface), the X-ray diffraction studies of Schmitt and Bear (1939) and birefringence measurements on red cell ghosts by Mitchison (1953), and from measurements of the electrical capacity of cell membranes (Cole and Cole, 1936): a capacity of approx. 1 μF/cm^2 would correspond to a lipid membrane thickness of 50–100 Å. On the other hand, optical studies suggested a composite structure of considerably greater overall thickness.

The chemical analysis of membrane components represented another approach to the study of membrane structure. The use of new techniques (cf. Parpart and Dziemian, 1940; van Deenen, 1965) extended the view of Grendel by showing that phospholipids and cholesterol represented the predominant part of membrane lipids. Again, the amount of these lipids in membrane preparations indicated at best a paucimolecular layer at the cell surface. However, the availability of purified membrane preparations also led to the recognition that proteins are a major component (cf. Zwaal et al., 1976). All the above studies indicated a membrane thickness of the order of 100 Å (or less).

Progress in the understanding of cell membrane structure provided only indirect evidence for the cell membrane concept. Two milestones in the new development were: (1) Straub (1953) (cf. Chapter 4) demonstrated that membrane preparations (erythrocyte 'ghosts') were capable of displaying some functional properties of cell membranes. (2) As predicted by Schmitt in 1939, electron microscopy supplied crucial direct evidence for the existence of a structured surface layer of cells. The resolution achieved by early electron microscopy was insufficient to detect the cell membrane, in spite of staining with OsO_4 (Porter, personal communication). In the period 1956–1957, several investigators (cf. Sjöstrand, 1956; Robertson, 1960) saw at the surface of a variety of cells a structure consisting of two parallel dense lines (each about 25 Å thick) separated by a light space about 75 Å wide. This structure was later named by Robertson (1960) the unit membrane. The dense layer was identified as the polar group of the bimolecular lipid leaflet (Stoeckenius, 1962). The freeze-etching technique of electron microscopy then allowed to see proteins spanning the lipid bilayer (Branton, 1966). Although problems persisted how to relate structures seen by electron microscopy to the molecular components of membranes[18], the existence of a cell membrane as a distinct part of the cell appeared to be established. The structural lipoprotein framework then had to be 'filled in' with a specific arrangements of its lipid components as indicated by Danielli (Davson and Danielli, 1943, p. 71) and elaborated later (cf. Finean, 1962)[19]. The concept of a continuous unit membrane had to be modified when evidence for some discontinuity was obtained by a variety of methods (cf. Finean, 1972). Such observations thus led to the idea of subunit concept, where the bilayer was interrupted to the extent of up to 30% by non-lipid material (proteins). These observations on the cell membrane structure approximately 100 Å thick had to be correlated with the developing knowledge of details on membrane permeability.

The described basic concepts of the (mammalian) cell membrane were subsequently found to apply to cell membranes in general as well as to subcellular membranous structures.

The views (and questions) held on the structure of the cell membrane in the early 60ties prompted further development. Several fundamental properties of biological membranes had to be addressed in structural terms, e.g. (a) the capability of large molecules (e.g. proteins) and particles passing through cell membranes (in the course of the phagocytic and pinocy-

---

18 At the Frascati meeting in 1965, prominent electron microscopists argued for a whole day, accusing each other of producing artifacts.
19 The membrane models of 1960 visualized an essentially static membrane. Physico-chemical studies on lipid membranes demonstrated (cf. Luzzati, 1968) that the lamellar phase of the lipid bilayer may be labile, permitting changes in its properties, including permeability (Seddon, 1990, see below).

totic processes), or protein entry into cells (*e.g.* antigens) or secretion; and (b) Nägeli's observation in 1855 that the breaking-up of cells produces a rapid 'de novo' formation of plasma membranes.

1.1. Asymmetric structure of the cell membrane

The models put forward by Dewey and Barr (1970) and Singer and Nicolson (1972) (the fluid-mosaic membrane) visualized membrane proteins coating both lipid faces (ecto- and endo-proteins), with some molecules spanning through the lipid bilayer (integral proteins); in addition to providing mechanical stability of the fluid structure, the integral proteins might be involved in the passage of solutes and solvents, as suggested by the Stein-Danielli model (1956). A better understanding was provided by several developments in the 70s:
(1) An asymmetry of the lipid bilayer of eukaryotes was signalled by Bretscher (1972) and further explored in van Deenen's laboratory (Verkleij et al., 1973): While the outer monolayer contains primarily sphingomyelin and phosphatidylcholine, the inner monolayer is constituted mainly by aminophospholipids; this structural asymmetry appears also to be reflected in an asymmetry of physical properties, the inner lipid leaflet being more fluid. In contrast, the outer leaflet of most bacterial cytoplasmic membranes contains mostly anionic phospholipids (*cf.* Voelker, 1985).
(2) Studies of the organization of proteins spanning the lipid membrane by three-dimensional electron diffraction techniques, using the purple membrane of *Halobacterium* as a model, led to the recognition (Henderson, 1977) that the hydrophobic portions of the peptide chain are imbedded in the lipid membrane in distinct rods of α-helical chains arranged to form a channel. Subsequently, an essentially similar arrangement was found for other proteins spanning the membrane (Unwin and Ennis, 1984). Such channels are not strictly per-

pendicular to the plane of the membrane, adding to some asymmetry.

(3) A distinct asymmetry of the proteins coating the lipid bilayer, was indicated. On the outer face of the lipid bilayer, glycoproteins (Klenk and Uhlenbruck, 1960a) are the cell receptors for antigens (Klenk and Uhlenbruck, 1960b; Winzler, 1969) and hormones (cf. Kahn, 1976); glycoproteins are also responsible for intercellular adhesion (Ocklind et al., 1983). By now, a family of transmembrane glycoproteins, cell adhesion proteins or integrins, is known (cf. Edelman, 1983; Quaranta and Jones, 1991) spanning the membrane. These proteins interact with a variety of extracellular ligands, including other cells as well as or matrices, while their intracellular domain is anchored to the cytoskeleton; recognition of the carbohydrate domain of the ligands has now been documented (cf. Brandley et al., 1990). The spanning of the membrane bilayer by some glycoproteins may be related to the discovery by Sutherland (cf. 1962) that the enzyme adenosyl cyclase was located in the cell membrane and was the target of various external signals (e.g. hormones) for cell function (Chapter 5). This then led to the concept that the interaction of agents with receptors at the cell surface starts a cascade of events in the membrane (second and third messengers) by which signals are transduced to produce intracellular events. At the protoplasmic side of the bimolecular lipid leaflet, an electron microscopically visible protein meshwork retaining the erythrocyte shape (Yu et al., 1973) substantiated earlier views (cf. Mitchison, 1953) that a rather thick 'structural' protein layer (now defined as the cytoskeleton, cf. Bennett, 1989) is anchored in the inner face of the lipid membrane. The cytoskeleton is responsible for the mechanical properties of the red blood cell: while the lipid bilayer of erythrocytes cannot withstand major pressure gradients (cf. Evans and Hochmuth, 1978), in many other cells the cytoskeleton attached to membrane appears to play a role in defining the tensile strength of the structure as well as its permeability (Kleinzeller and

Mills, 1989). Several enzymes are also associated with band 3 of the membrane polypeptides (Yu and Steck, 1975). Cytoskeletal elements at the inner face of the cell membrane have now been shown to be involved in the incorporation of cytoplasmic particles into the membrane (see below).

It is now increasingly realized (*cf.* McLaughlin, 1989) that the charge of the membrane phospholipid molecules, as well as that contributed by the sialic component of glycoproteins, may be reflected in the electrostatic potentials of the membrane, affecting *e.g.* the conductance of channels in membranes and ion permeability.

The asymmetry of membrane phospholipids was found not to be absolute, whereas that of proteins appears to be crucial. This new dimension in the understanding of the cell membrane structure, first discovered in erythrocytes, appears to be a general phenomenon (Rothman and Lenard, 1977; Rottem, 1982).

(4) The structural asymmetry of the cell membrane, particularly that due to its polar components, imparts to the membrane a charge polarity, and this in turn is reflected by an electrical asymmetry. In addition, at the steady state, the asymmetry of electrolyte distribution between the cell and its environment, produced by the operation of ionic pumps and leaks (Chapter 4), adds to the electrical asymmetry of the cell membrane[20].

1.2. The cell membrane is a dynamic structure

Cells respond dynamically to changes of the environment by

---

20 The electrical potential gradient across many cell membranes is of the order of some 100 mV. Given the average thickness of the lipid bilayer of some 75 Å, this translates to a gradient of some 10^5 V/cm. It is justified to wonder about the possible effect of a electrical potential gradient of this magnitude on the physical properties of the membrane and transported solutes.

modifying the chemistry and arrangement of their membrane complex lipids in order to maintain the average transition temperature of the bilayer well below the ambient temperature, thus assuring the fluid nature of the structure (Cronan and Gelman, 1975). Two major points arise from this new information. First, the structural (molecular) asymmetry of the membrane reflects its functional asymmetry. Second, for the lipid asymmetry to exist, it has to be relatively stable. While the lateral diffusion of phospholipids (Hubell and McConnell, 1968) and proteins (Blasie and Worthington[21], 1969; Frye and Edidin[22], 1970) in cell membranes is rapid, the movement of a phospholipid molecule from one side of the bilayer to the other is slow, and may take place by a protein-catalyzed exchange (flip-flop) mechanism (Kornberg and McConnell, 1971); in addition, an ATP-dependent aminophospholipid translocase may be involved (cf. Devaux, 1990). Thus, the asymmetry of the membrane structure has to be explained in terms of its assembly. These aspects are crucial for the understanding of the very fast de novo membrane formation (cf. Bishop and Bell, 1988).

Cell membranes are characterized by a fast metabolism. This was first demonstrated by Redman and Hokin (1959) for the turnover of phospholipids[23], and subsequently a rapid metabolism of other membrane components was described (cf. Cook, 1976). The role of protein kinases and phospholipases in signal transduction by membranes is now well established (cf. Berridge, 1987). By now, evidence for the dynamic nature of cell membranes can be extended also to the cytoskeletal proteins attached to the inner face of the plasma membrane (cf. Kleinzeller and Mills, 1989). Events occurring during cell

---

21 Using low-angle X-ray diffraction methods.
22 This point was established in fusion experiments using cells from two different species: within minutes the antigens of both cells were intermingled.
23 An exchange of phospholipids between cellular (e.g. mitochondrial and microsomal) membranes has been first established by Wirtz and Zilversmit (1968) and has been amply documented (cf. Voelker, 1991).

growth when the membrane area has to increase, provide another illustration of membrane metabolism. Although proteins and phospholipids may be incorporated in the cell membrane independently of each other (see above), functionally active membranes require their insertion in concert (Pasternak, 1975).

Danielli (Danielli and Davson, 1943) postulated that the penetration of the continuous lipid membrane by large molecules and particles, such as proteins or viruses could not take place by diffusion, and hence, a specially differentiated part of the cell membrane would have to be responsible for such process. Two mechanisms appear to be involved: (a) The experiments of Palade (1959) on protein secretion by the pancreatic cell provided information on this aspect of cell function by focusing on membrane fusion in the course of exo- and endocytotic processes, and during cell–cell fusion. Palade postulated a continuous recirculation of the cell membrane in order to account for the steady state of protein secretion by the gland. This view is now generally accepted (Cohn and Steinman, 1975; Steinman et al., 1983). (b) New evidence now shows that intact proteins can transverse from the endoplasmic reticular membrane (Blobel and Dobberstein, 1975), the nuclear envelope (see below, p. 63) and the mitochondrial membrane (cf. Wienhues and Neupert, 1992): An unfolding of the protein molecule appears to be the first step in a process involving a protein 'pore' (Simon et al., 1992). The mechanism responsible for the translocation of proteins may be identical with that involved in the insertion of integral proteins into the bilayer (Singer and Yaffe, 1990)[24].

For membranes to fuse, they have first to come into contact (apposition), interact (by a temporary destabilization of the

---

24 Judging from the ease with which some membrane proteins (e.g. the excitability-inducing material (see p. 66) interacts with membrane lipids, the insertion of relatively small integral proteins may not require a special mechanism and may be based exclusively on the properties of the hydrophobic region of the proteins.

50   A. KLEINZELLER

bilayer, possibly involving phase transition), and then fuse.
Fusion of lipid membranes has been known since soap bubbles
were first seen, but the actual mechanism of this process is as
elusive as that of the fusion of membranes (cf. Pethica, 1984).
An involvement of membrane phospholipids was postulated,
and membrane proteins also appear to be involved (see below).

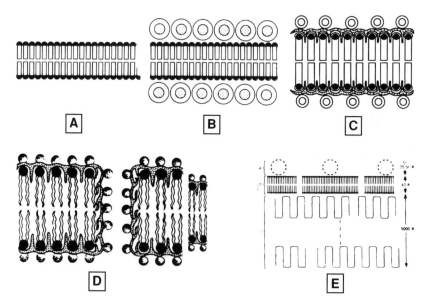

*Fig. 1. Evolution of membrane models 1925–1956: a) Gorter and Grendel
(1925) – the lipid bilayer; b) Danielli and Davson (1935): globular proteins
with their hydrating shells on the lipid bilayer ('these protein films may in
any case be capable of showing selective permeability towards molecules of
different size'); c) Danielli (cf. Davson and Danielli, 1943): a layer of unrolled
protein and globular proteins on the lipid bilayer; the possible protein monolayer was suggested by Langmuir's (cf., 1939) experiments. d) Stein and
Danielli (1956): Proteins from 'pores' in the lipid bilayer. e) Mitchison (1952):
the thickness of the cytosolic side of the membrane. Permission to reproduce
Figs. 1 B and C (Cambridge University Press) and 1 E (J. Exp. Biol.) are
gratefully acknowledged.*

1.3. Membrane models

Figure 1 shows the development of ideas on the cell membrane structure from the simple lipid bilayer to Danielli's model and then to ideas including the concept of a pore and a rather thick shell of proteins at the inner face of the membrane. At first view, this development signals a return to simple ideas held when the concept of the cell membrane was developed some 140 years ago (see above). Advancing knowledge has replaced the thin membrane of some 100 Å thickness by a structure which may come close to what Nägeli saw and Pfeffer called 'hyaloplasm'.

2. CELL MEMBRANE PERMEABILITY

An answer to the basic questions on the nature of the cell membrane (see above), *i.e.* a continuous (lipid) layer, or a sieve-like structure, required quantitative data on the membrane permeability and its characteristics. Only in this way could the student of membrane phenomena decide whether the passage of a solute across the membrane proceeded by physical forces alone (diffusion) or required some additional mechanism[25]. Progress thus depended on a refinement of the term permeability, and this could be achieved only by new experimental data.

---

25 Such a possibility had to be invoked whenever the observed properties of cell permeability could not be explained in terms of physical mechanisms. Hence, Höber since 1899 (*cf.* 1926) used the term 'physiological', as opposed to 'physical' permeability, implying that the mechanism of the physiological permeability still had to be elucidated. Ussing (1954) points out that since the understanding of the basic distinction between diffusion and active transport developed only slowly (see Chapter 4), investigators raising such distinction, including Krogh, were considered by some to reintroduce vitalistic concepts.

2.1. Model membranes

The experimental strategy of biologists for studying the diffusion of solutes across 'ideal' artificial membranes was to throw
light on the physical mechanisms involved in solute transport,
thus exploring their possible involvement in biological systems[26]. As pointed out by Danielli in 1941 (cf. Davson and
Danielli, 1943), a continuous lipid layer of some 100 Å stabilized by proteins at both oil-water interfaces presents resistances to free diffusion of solutes (a) at the first interface:
water → oil; (b) in the membrane interior; and (c) at the second interface: oil → water. Diffusion of lipid solutes through
such membrane may be more complex than predicted by Fick's
law (Davson and Danielli, 1943, p. 342). On the other hand,
diffusion through an aqueous 'pore' may be limited by the pore
size, as recognized by Fick (1855). The choice of the nature of
the membrane proved to be crucial for the specific question
asked.

Collodion and gelatin membranes were used primarily as
models for diffusion through 'pores'. In the hands of Collander,
starting 1924 (cf. Collander and Bärlund, 1933) and Michaelis
(cf. 1928), these membranes showed electrical potentials dependent on concentration gradients of inorganic salts, as predicted by Nernst (1889). A dependence of the permeability of
such membranes on the molecular volume of solutes was also
found, consistent with the laws of diffusion. Observed inconsistencies when using some solutes, particularly the selectivity of biological membranes, could be explained by the assumption that as opposed to the concept of an inert membrane,

---

[26] The impetus for this experimental approach was given by studies of Loeb and
Beutner (1912) demonstrating that the electrical potential difference of the apple skin
against electrolyte solutions depends on the nature and concentration of the dissolved
electrolyte. Beutner (1912) then developed a useful model membrane, a water-immiscible organic liquid containing water-immiscible acids, and demonstrated electrical
potentials when this membrane was bathed with KCl solutions of differing concentrations.

the charge of the membrane exerted an effect on the diffusion of solutes through very narrow 'pores', acting as a molecular sieve. At first, Michaelis and Fujita (1925) suggested that an interaction of K^+ with acidic membrane components (fatty acids) was responsible for the selective permeability of biological membranes. The observation that even 'simple' collodion membranes displayed electrical potential asymmetry led to the idea of compound sieve membranes (Wilbrandt, 1935) with an emphasis of the role of membrane structure for selective permeability. The concept of an electrically charged (or ionic) membrane matrix, introduced by Teorell (1935) and by Meyer and Sievers (1936), extended the conceptual framework laid down by Nernst and Planck for the diffusion of ions across boundaries and membranes: An electrochemical, vs. the simple concentration potential across a membrane, had to be considered. The selective permeability for charged solutes was indicated previously for Traube's precipitation membranes (see above). An extension of these efforts led to an exploration of the properties of membranes with ion-exchange characteristics (Sollner et al., 1954). The existence of a membrane matrix with fixed anionic charges also provided some explanation for the selection of the nearest cationic neighbor, thus offering a concept for the physical basis of ionic selectivity of artificial and living membranes (Eisenman, 1961). An additional important result of using artificial membranes was the recognition that considerable ionic concentration gradients could be generated by extending the principles of the Gibbs-Donnan system (Teorell, cf. 1954); such mechanism was thought to account for some of the observed accumulation of ions in biological systems (see Chapter 4).

The exploration of ionic fluxes through artificial membranes, and related electrical parameters, improved our understanding of the physical properties of cell membranes. One milestone arose from studies of the relationship between membrane potential and the composition of the medium, leading to the assumption of a constant electrical field in the mem-

brane (Goldman's field equation; [Goldman, 1943]); the
Goldman equation, particularly in the form derived by
Hodgkin and Katz (see Chapter 8) provided an important conceptual tool for the analysis of the membrane electrical potential as a function of ionic permeabilities across cell membranes. Another major step was provided by further examination driving forces for ionic fluxes (*cf.* Schlögl, 1964). The
Nernst-Planck equation considered the diffusion of a given
ionic species across an *inert* membrane with the concentration
gradient as driving force. Teorell (1935) added a term describing the transfer of charged particles through membranes characterized by an ionic matrix[27]. Other terms describe the con-
vection of ions in membranes with very fine pores[28] (Schmid,
1950) and effects of pressure gradients. Schlögl (*cf.* 1964) then
succeeded in integrating the resulting differential equation.
In this development of the membrane concept the formalism of
irreversible thermodynamics was employed. Kirkwood (1954),
but particularly Katchalsky (Kedem and Katchalsky, 1958)
were instrumental in introducing this approach to students of
membrane biology. The phenomenological equations of irreversible thermodynamics proved particularly useful when analyzing the coupling between fluxes of individual solutes, as
well as the coupling between fluxes of solutes and solvent.
Finally, an analysis of ionic fluxes across membranes demonstrated that when diffusion forces alone are operative, the flux
ratio of a given ionic species equals the ratio of the ion activities on both sides of the membrane, with a term relating to the
electrical potential across the membrane:

---

27 Teorell (cf. 1953) later extended his analysis by including also the transport processes (ionic fluxes and conductivity).
28 minimal in systems (including cell membranes) with very fine 'pores' (cf. Schlögl,
1964); this term becomes rather considerable when considering major solvent fluxes
e.g. through intercellular spaces in epithelia (Chapter 7).

$$J_i/J_o = c_i/c_o \text{ (for non-electrolytes)} \qquad (1)$$

and

$$J_i/J_o = f_i \cdot c_i/f_o \cdot c_o \cdot e^{zFE/RT} \text{ (for electrolytes),} \qquad (2)$$

where J denotes the flux, c concentration, f activity coefficient, z valency, E transmembrane potential and F the Faraday constant; the subscripts i and o refer to the inner and outer compartments, respectively. This equation, now best known as the Ussing equation (1949), was derived in a more general form by Behn (1897) and in 1949 by Teorell (*cf.* 1953); it permitted to establish a new criterion for the passive permeability of membranes to solutes (Ussing, 1949): if the flux ratio obeys the equation, the mechanism of the studied transport phenomenon is passive, i.e. the driving forces are physical in nature (see also Chapter 4); Hodgkin and Keynes (1955) pointed out some limitations to such interpretation. Unfortunately, inherent problems of measuring the activity of the studied ionic species at the inner (protoplasmic side) of the membrane limited the usefulness of Ussing's equation especially to epithelial membranes.

Langmuir (1917) introduced mono- and multimolecular films for the exploration of membrane phenomena. In the hands of Rideal (1937) and Schulman (1937), the permeability of such films to solutes became an important tool for the understanding of membrane models.

The discovery of how to prepare artificial lipid bilayer membranes (Mueller et al., 1962) and lipid vesicles (Cohen and Bangham, 1972) as experimental tools led to the demonstration that the partition coefficients of organic solutes in the lipid phase were the major determinant of their permeability (Cohen and Bangham, 1972; Orbach and Finkelstein, 1980),

i.e. consistent with Overton's hypothesis[29]. The thickness of
these membranes is of the same order of biological membranes, and they display some permeability to water molecules[30], while that for electrolytes is low (*cf.* Haydon, 1979),
unless transport proteins are incorporated (see below).

2.2. Permeability of cell membranes

Discussions on the nature of cell membranes provoked strenuous efforts to explore whether the permeability of various solutes meets the requirement of the lipid or 'pore' hypotheses,
using the criteria mentioned above (p. 52). Exploration of solute kinetics became the main experimental tool.

2.2.1. Permeability to water

Overton's osmotic experiments actually reflected measurements of cell permeability to water since very little solute
passed the membrane in the observation period. Kinetics of
volume changes thus could be taken as a measure of the membrane permeability to water.

The concept of a continuous lipid membrane posed several
questions concerning the pathway by which water passes
through. (1) As compared with membranes of many cells, e.g.
the erythrocyte, the permeability of model lipid membranes
for H_2O is relatively low (*cf.* Haydon, 1979; Finkelstein, 1987),

---

29 The discovery that artificial lipid bilayers can display properties associated with
cellular membranes (*e.g.* action potentials) represented crucial evidence in favor of the
essentially lipid nature of the membrane. The difficulties encountered by Mueller et
al. (1962) in publishing their observations (P. Mueller, personal communication) may
have reflected the reluctance of many investigators at the time to accept the notion of
a lipid membrane.
30 The permeability of lipid layers to water appears to depend on their composition:
it may be practically zero, in spite of water being a component of the lipid-water
system (Lawrence, 1961). This observation was related by Ussing to the impermeability of trout eggs to D_2O.

although the lipid composition does play a role. (2) Another set of questions arose when the great physiological variability of water permeability was found. Thus, Lillie (1916) showed that fertilization of echinoderm eggs produced an immediate three-fold increase in the permeability to water (measured by the osmotic technique). The following specific questions had to be addressed (Lucké and McCutcheon, 1932): (a) Is the exchange of water between the cell and its environment subject to the laws of osmosis and of diffusion? (b) Are the changes in the volume of the cell, which can be brought about by changing the external concentration of the medium, governed the laws of osmotic equilibrium? and (c) Can the rate at which such changes take place be predicted from the laws of diffusion? These questions thus outline a quantitative study of the quali-tative observations of Nägeli, Pringsheim, de Vries, Ham-burger and Overton (see above). Membrane pores, wide and narrow, were suggested as an explanation of the arising ques-tions.

The introduction of isotopically labeled water demonstrated further discrepancies between prediction and observation: thus, Hevesy et al. (1935) drew attention to the fact that the osmotic permeability of the frog skin was 3–5 times greater than the diffusion permeability as determined with heavy water, and this phenomenon was then found in many cells. Koefoed-Johnsen and Ussing (1953) interpreted such discrep-ancies as indications for the concept that water may cross the cell membrane independently by two pathways: (a) as solute (by diffusion through the phospholipid bilayer, the small water molecules finding a way between the lipid molecules), and (b) as solvent by diffusion and bulk flow through mem-brane 'pores'; the variability of the permeability coefficient for water, noticed by Lucké and McCutcheon, may thus reflect the role which these pathways for water flow play in various cells and under varying physiological conditions[31]. A single-file movement of water through very narrow pores was suggested as an explanation of the above observations (Rosenberg and Finkelstein, 1978). The pore as a functional concept also had to take account of findings showing that its apparent radius could change: Thus, Koefoed-Johnsen and Ussing signalled that the flux of water through cell membranes was greatly increased by the neurohypophyseal hormone.

The actual measurements of the membrane permeability to water were based on a variety of assumptions: (a) In 1904, Nernst pointed out the possibility that unstirred layers close to both faces of the membrane could affect permeability parameters. The experiments of Teorell in 1937 (cf. 1953) then emphasized this point. While intensive mixing of the cell suspension might greatly reduce the unstirred layer at the outer membrane face, the development of our knowledge on the membrane structure at the protoplasmic side (see above) signals caution in the intrepretation of experimental values[32]. (b) It is generally assumed that all (or most) cell water is freely accessible to solutes; this view is challenged by some investigators (Ling, 1984). The scholarly review of Finkelstein (1987) signalled the persistent uneasiness concerning water movement across membranes.

2.2.2. Permeability to non-electrolytes

The studies of Hedin (1897) (see above) already drew attention

---

31 Following the conceptual lead of Pappenheimer (1948), measurements of the parameters for water diffusion and osmotic flow were used to calculate the equivalent radius r of a membrane pore as an idealized water-filled cylindrical structure crossing the membrane. In human erythrocytes, r was assessed to be below 5 Å (Paganelli and Solomon, 1957), wide enough to permit passage of solutes such as glycerol or monovalent inorganic ions.

32 Such an unstirred layer would also represent a major difficulty in assessing the activity of cellular ionic species at the membrane, thus preventing the use of the flux ratio equation for the evaluation of the driving forces involved in the permeability ionic species.

to the fact that some lipid-insoluble solutes did enter erythrocytes. Efforts of several laboratories in the period 1924–1935 focused on quantitative measurements of the permeability of cell membranes for a variety of solutes with some contradictory results: In the sulfur alga *Beggiatoa* (Ruhland and Hoffman, 1925) the dominant factor for permeability appeared to be the solute's molecular volume, pointing in favor of their ultrafiltration theory; some exceptions, indicating a role of lipid solubility were discovered later. On the other hand, the data of Collander and Bärlund (1933) showed that in a variety of plant cells the permeability constants of many solutes were linearly related to the oil/water[33] partition coefficient of the solute; this result was consistent with Overton's hypothesis. However, several solutes passed the membrane considerably faster than corresponded to their oil/water partition coefficient, notably water, and some organic non-electrolytes of small molecular volume (*e.g.* lower aliphatic alcohols, or the lipid-insoluble urea); evidently, here lipid solubility was not the dominant factor. Jacobs (1931) drew attention to the observation that in erythrocytes, major differences can be seen for the permeability of a given solute (for glycerol up to two orders of magnitude) in cells from different species. The fact that to a varying degree, both lipid solubility and molecular volume appeared to play a role in the permeability then led Collander (1937) to suggest a fusion of the lipid and pore theory into the 'lipoid-sieve' concept of the cell membrane: 'The plasma membrane seems to act both as a selective solvent and as a molecular sieve'. However, even this concept could not explain patent discrepancies between observation and hypotheses of membrane structure, such as the major differences in the entry of glucose and mannitol into human erythrocytes

---

33 Danielli (Davson and Danielli, 1943) suggested an improvement of the plot; however, this did not affect Collander's basic conclusions).

(both these lipid-insoluble saccharides have practically identical molecular volumes) (see Chapter 3).

For the permeability of biological membranes to proteins and other large solutes, see p. 49 of this Chapter.

2.2.3. Permeability to electrolytes

The analysis of the cell membrane permeability to electrolytes proved to be complex because of apparently contradictory observations. (1) It was known since the 1850s that animal (erythrocytes) and plant cells showed an unequal distribution of ions. In plant cells, electrolytes could accumulate against considerable concentration gradients, and the accumulated salts could not be leached out by placing the cells into distilled water (see detailed discussion of these points in Chapter 4). Such observations led to ideas that the cell membrane might be impermeable to (at least some) ions. It was Teorell who pointed out in 1935 that studies should deal with permeability, instead of impermeability, of cell membranes. Hence, an analysis of cell permeability to electrolytes had to take account of these observations and eventually led to a clear distinction between diffusion and active transport of electrolytes. (2) In the light of information obtained using artificial membranes, the interpretation of membrane permeability to electrolytes had to consider the transfer across the membrane of both the individual ionic species and their electrical charges (see above). As opposed to notions of some ions being accumulated in (or excluded from) the protoplasm against their concentration gradients, the transport against, or in accordance with, their respective electrochemical gradient had to be established; this put an additional burden on the investigator. (3) The possibility of some binding of ions to protoplasmic proteins could not be dismissed without specific control experiments (*cf.* Civan and Shporer, 1989). (4) Concepts derived from colloid chemistry, *e.g.* the role of the lyotropic (Hofmeister) series of ions, may have had a delaying effect on developing ideas when trying to analyze the specificity of ion transfers across cell membranes (Davson, 1989; *cf.* Chapter 1).

The membranes of many cells were shown to be permeable to some ions. While red blood cells were found to be permeable to anions (Koeppe, 1897), the permeability to cations, indicated by Hamburger and Bubanovic (1910), was not accepted for another 30 years. The permeability of muscle fibers to electrolytes was still more controversial (see Chapter 4). The membrane of muscle fibers was assumed by Bernstein (1902) to be permeable to K^+, but the free passage of Cl^- was demonstrated only 40 years later; Overton's (1902) data indicated a possible exchange of cell K^+ for Na^+ during muscle contraction. The unequal distribution of electrolytes between cells and their environment has been attributed mostly to the very low permeability of their membranes to cations. Permeability to electrolytes of the epithelial cell membrane was also indicated by observations on the absorptive or secretory processes (see Chapter 4). Since the passage of electrolytes across cell membranes could not be readily reconciled with the concept of a lipid barrier, the above observations had to be interpreted in terms of a selective permeability through an aqueous boundary; a simple inert 'pore' concept would not be consistent with the observations.

Several developments helped to shed better light on the ionic permeability of cell membranes, particularly the introduction of radioisotopes, and electrophysiological studies. Brooks was the first to report in 1937 (*cf.* 1939) that the entry of cations into algal cells was very fast and proceeded against the concentration gradient. The analysis of this (and similar) phenomena eventually led to the concept of an ionic pump (Chapter 4). Electrophysiological experiments, heralded 1902 by Bernstein and Overton, then opened the door for studies which, by meeting the criteria defined by the characterization of 'passive' fluxes of ions across artificial membranes, could shed light on such transport phenomena in cells.

2.3. The emerging new concept of the biological membrane

Quantitative studies of the structure and function of artificial and cell membranes provided compelling evidence in favor of the membrane concept as the structural barrier which imparts to cells its osmotic and electrophysiological properties. It was recognized that there was no irreconcilable difference between the concepts of a lipid membrane and the pore theories: a more comprehensive concept took hold where an asymmetric fluid lipid bilayer anchored specific membrane proteins, endowing the structure with phenomenologically defined pores and the observed selective permeability (Höber, 1936; Danielli, 1962; Steck, 1974). The fluid mosaic model of Singer and Nicolson (1972) (*cf.* also Dewey and Barr, 1970) took into account available evidence of the membrane lipid and protein structure, its organization and lateral mobility of its components. Bretscher (1972) emphasized the asymmetry of the lipid bilayer, while Nicolson (1976) added the role of glycoproteins and glycolipids at the outer surface of the membrane, and the attachment of cytoskeletal proteins at the inner face; he also emphasized

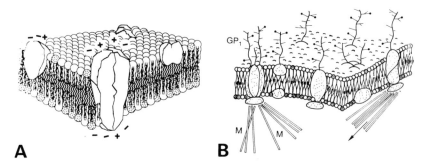

A   **B**

Fig. 2. Evolution of membrane models 1970–1976: a) Dewey and Barr (1970); Singer and Nicolson (1972): proteins embedded in the membrane, and spanning the membrane (some with 'pores'); the intrinsic asymmetry of the lipid bilayer (Bretscher, 1972) and of charges is indicated; b) Nicolson's (1976) model of a cell membrane (slightly modified): GP: Glycoprotein complexes; MT: microtubules; MF: microfilaments. The asymmetry of the lipid bilayer is indicated.

that the postulated membrane structure should allow signalling from receptor groups at the membrane surface to the cytoskeleton and the cytoplasm. These developments are shown in Figure 2. While differing in detail, this general concept of the cell membrane applies also to membranes of subcellular organelles (cf. de Duve et al., 1959). Thus, the nuclear envelope of eukaryotes is composed of two lipid bilayer membranes with pore proteins providing the connection between the cyto- and nucleoplasm (cf. Newport and Forbes, 1987); an ATP-dependent portion of the protein component is responsible for the control of the 'pore' diameter, allowing larger molecules (nucleic acids, proteins) to pass (Goldfarb and Michaud, 1991). Both the outer and inner membranes of mitochondria display structural elements of unit membranes (cf. Chapman and Leslie, 1970).

The emphasis now shifted to a clearer definition of the possible role of the individual structural components which represent only a very restricted portion of the membrane area. Two basic hypotheses were put forward to explain membrane selectivity: Osterhout et al. (1934) postulated the existence of a carrier or shuttle mechanism in the membrane which would bring about an interaction of ions with a membrane component for the transport process. On the other hand, pores lined by proteins capable of hydrogen bonding (Danielli and Davson, 1935) could offer a mechanism for the selective permeability. In addition, structural components responsible for receptors postulated by physiologists and pharmacologists also had to be identified (see Chapter 5). The success of this approach, documented particularly in Chapters 3 and 4, tended to reduce the appreciation of the role of membrane lipids: at its extreme, the lipid membrane is seen but as a matrix anchoring transport proteins. Such one-sided view neglects clear evidence for a crucial interaction between the lipids and transport proteins of a cell membrane: Starting with the demonstration that the lipid composition of liposomes (particularly their cholesterol content) affects their permeability to various solutes (cf. De

Gier et al., 1968), the cholesterol content of lipid bilayers is a determinant (by affecting membrane fluidity?) of the activity of the reconstituted glucose transporter (Conolly et al., 1985) and the (Na^+-K^+)-ATPase (Yoda and Yoda, 1987).

III. Membrane channels

The analysis of membrane selectivity got a new impetus by several discoveries, which pushed the field from concepts of relatively large, inert pores towards narrow structures, now usually designated as channels, which permitted molecular interaction between the solute and the channel lining. The lessons derived from studies of the physico-chemical properties of artificial membranes (p. 51 ff.) now made a major impact on the understanding of the function of biological membranes.

1. WATER PATHWAYS

Recent discoveries shed a new light on the pathways of water movement across cell membranes: In addition to the rather low H_2O diffusion across the lipid bilayer, several additional pathways for water transport exist:

1. A membrane-spanning protein with all the properties of a channel (relatively low activation energy, high osmotic permeability coefficient, and selective inhibition by mercurials) was isolated and cloned from erythrocytes (Preston et al., 1992) and renal epithelial cells (Verkman, 1989). This channel is selective, not permitting the permeability of *e.g.* urea. A shuttle mechanism, permitting the incorporation of internalized channel proteins in the membrane, as well as their retrieval, may be operative at least in some cells.

2. The regulation of water flow through the frog urinary bladder (see 2.2.1 above) led to the suggestion that the neurohypophyseal hormone changed the effective pore diameter.

A new understanding of the phenomenon was first introduced by the observation (Chevalier et al., 1974) of membrane-associated particles at the cytoplasmic side of the apical membrane of the epithelial cells; the neurohypophyseal hormone oxytocin increased the clustering of the particles; this phenomenon was related to the increase in permeability elicited by the hormone. Subsequently, Wade et al. (1981) demonstrated a hormone-stimulated shuttle mechanism by which these particles (later denoted 'aggrephores') are incorporated in the membrane; at the end of the hormonal challenge the aggrephores are internalized again in a cellular structure adjacent to the membrane. A role of the cytoskeleton in the antidiuretic hormone-induced water flux, first signalled by Hardy and DiBona (1982), has now been specified further by showing an involvement of F-actin in the shuttling movement of the particles (Ding et al., 1991). The aggrephore represents a water channel which in a sieve-like action appears to exclude particles as small as urea.

Thus, at the present level of our knowledge, water may pass across the cell membrane by a variety of independent pathways.

2. IONIC CHANNELS

The selectivity of a narrow pore might be due to differences in the diameter of the hydrated ionic species – the pore acting as a sieve (Conway, 1947). Alternatively, special properties of pores permitting discrimination between ionic species had to be invoked. Attention was drawn rather early to the fact that the permeability of cell (cat erythrocyte) membranes for Na^+ was sufficiently different from that for K^+ (see Chapter 4) to postulate different pathways: Davson and Reiner (1942) suggested that the membrane permeability for these ions was brought about by an enzyme-like system. Studies of the ionic basis of electrophysiological phenomena (see Chapter 8) provided a new insight. The now classical observations of

Hodgkin and Huxley (1952) provided unequivocal evidence that in the squid nerve the transient passive (down-hill) passage of Na^+ and K^+ in the course of the action potential proceeded independently for each ionic species; the movement of Na^+ displayed a distinct dependence on the transmembrane electrical potential (the gating phenomenon). The results were consistent with the view that 'a number of charges form a bridge or chain which allows sodium ions to flow through the membrane'. Moreover, the transient nature of the respective ionic fluxes in the course of the action potential (cf. Hodgkin et al., 1949) indicated that pathways of the down-hill Na^+ and K^+ fluxes can be modulated by physiological activity. The concept of regulated ion-selective channels in an excitable membrane was thus established (cf. Chapter 8). Hodgkin and Huxley also described a minor, apparently non-selective channel for electrolytes (the leak conductance). Saturability of the transport pathway for K^+ through the membrane channel was then demonstrated (Hodgkin and Keynes, 1955). The observed selectivities, as well as saturability and transmembrane voltage dependence were not consistent with ideas of an inert pore.

During the period 1962–1967, several discoveries opened the field to a better understanding of ion transport across cell membranes at a molecular level. 1. Lipid membranes are impermeable for electrolytes (cf. Haydon, 1979). Mueller et al. (1962) made the seminal discovery that a protein (excitability-inducing material) lowered the electrical resistance of the bimolecular lipid membrane by several orders of magnitude, 'possibly by penetration and/or formation of channels for electrolytes' in the membrane. These channels also made it possible to establish transmembrane potentials according to the Nernst equation, and contained a dual gating mechanism. 'The electrokinetic phenomena were indistinguishable from cellular action potentials'. Subsequent studies demonstrated that small amounts of the excitability-inducing material produced an increased conductance of lipid bilayers for ions in discreet steps (Bean et al., 1969) and the formed channels showed transitions between 'open' and 'closed' states (Ehren-stein et al., 1970). 2. In the period 1964–1967, investigators in Cambridge, England (Chappel and Crofts), Philadelphia (Pressman et al.; Mueller and Rudin), Durham (Tosteson et al.), and Leningrad (Lev and Buzhinski) arrived at the conclusion that some macrocyclic antibiotics (*e.g.* valinomycin, gramicidin) previously shown to affect mitochondrial energy metabolism, did so by increasing the passive permeability of the membranes for inorganic cations[34]. The idea that such compounds insert in the lipid membranes and provide a cation-selective channel took hold[35], particularly when evidence was obtained showing that the same phenomenon was seen when using artificial lipid membranes (*cf.* Pressman, 1985 for a historical account). The formed channels showed remarkable ionic selectivity dependent on the antibiotic (now designated ionophore) used which could be related to an interaction of the cation with dissociating bonding sites within the channel, as suggested for glass membranes by Eisenman (1961). Such views implied that the hydrating shell around ions be first stripped off (at least in part) so that the ions could then interact with bonding sites.

These new data established the concept that protein-lined channels impart to membranes of a variety of cells their selective ionic permeability. The kinetic properties of such passive[36] channels meet many of the criteria put forward for carrier-mediated transport processes (Läuger, 1980) earlier

---

34 Moore and Pressman (1964) were actually the first to show that the effect of the antibiotic valinomycin on mitochondrial metabolism was associated with net movements of H^+ and K^+.

35 The rapid development of our knowledge in this field also reflects a great degree of international cooperation in science: Shemyakin and his group in Moscow, as well as research laboratories of pharmaceutical companies, made freely available synthesized ionophores and their analogs; in the reverse direction, B. Pressman gave valinomycin to A.A. Lev of Leningrad when the two met in the authors laboratory in Prague, 1966.

36 The term passive denotes the fact that the transport of a give ion across such channels proceeds only in accordance with its electrochemical gradient.

68   A. KLEINZELLER

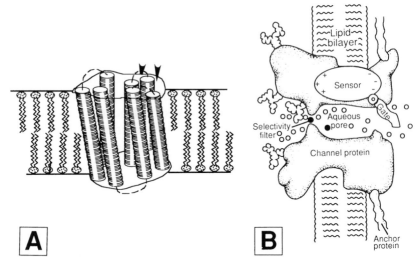

Fig. 3. Model of a channel: a) Arrangement of the peptide chain in the lipid membrane (Henderson, 1977; Unwin and Ennis, 1984); b) Hille's, (1984) model.

postulated for non-electrolytes (Chapter 3). One property distinguishes channels from carriers: while the permeability of ionic channels is characterized by a Q_{10} of about 1.4 (corresponding to diffusion values of 2–3) are found in carrier mediated processes. Läuger visualized ionic movement through a channel as a series of thermally activated processes in which the ion, after having shed some of its hydrating shell, then moves to a binding sites (or sites; two such sites have been considered) of the energy barrier (model shown in Fig. 3); the binding site(s) may exist in several conformational states. The existing knowledge opened the door to an exploration of ionic channels by two experimental approaches: a) Studies of the gating currents provided information for the modelling of channels for Na^+ and K^+ (*cf.* Armstrong, 1980); b) the discovery of specific inhibitors of individual ionic pathways offered the promise of isolating the protein component(s) of the channels and relating their structure to function. Such an approach led to the isolation of the Na^+ channel protein and reconstitution of its function in lipid vesicles (cf. Catterall, 1986).

A new chapter in the study of ionic channels in cells started with the discovery by Neher and Sakmann (1976) that single channel currents in cell membranes can be recorded by a miniaturization of the voltage-clamp technique developed in Cole's laboratory (Marmont, 1949; Cole, 1949) for measuring ion fluxes across the giant axon membrane; fluctuations in this current were assigned to changes in the open and closed states of the channel, as previously indicated by studies with lipid bilayers. A further refinement in the patch-clamping cell membranes (Hamill et al., 1983) opened the door to the characterization of ionic channels in cell membranes in terms of their specificity, kinetics, activation and inactivation: by now, specific channels for Na^+, K^+, Ca^{2+} and Cl^- have been described[37], and several channels for each ionic species, differing in their characteristics are known. The at first-sight bewildering number of channels for a single ionic species (particularly K^+, while only a few Na^+ channels have been described) may reflect the recognition that the regulation of many cellular functions by physiological and pharmacological agents is brought about by the opening or closing of ionic channels (cf. Hille, 1984). The considerable degree of homology of proteins forming K^+ channels reflects the encoding of these structures by one gene family (Stühmer et al., 1989), and suggests that the variety of such channels in biological systems represent but variations on one theme. There are increasing indications that there exist also cationic channels of small selectivity, possibly corresponding to the 'leak conductance' of Hodgkin and Huxley. Hille's (1984) scholarly monograph provides detailed

---

37 Recent progress indicates that the channel selectivity for a given ionic species is rather limited. Thus, the Ca^{2+} channel allows also the passage of Na^+ under some experimental conditions (Kostyuk et al., 1983). While the Na^+ channel appears to be highly selective, it can acquire the characteristics of a Ca^{2+} channel by a single mutation affecting the some negatively charged areas of the channel peptide (Heinemann et al., 1992).

information on the role of pores and channels in defining the selectivity of cell membranes as understood a decade ago.

The molecular aspects of the actual mechanism by which ions pass through the narrow channels have to be modelled on the basis of existing information. The view that channels can assume many conformational substages between the open and closed states (Läuger, 1985) appears to be borne out by more recent studies. The channel conductance appears also to be affected by physical factors, such as effects of the dipolar nature of the lipid bilayer on the electrical properties of ionic channels (Parsegian, 1969; Jordan, 1987). The prevalent view considers ion channels as water-filled pores across biological membranes, lined by a rosette-like arrangement of protein-subunits spanning the hydrophobic lipid bilayer; the permeability of such channel is then modulated (possibly by phosphorylation or dephosphorylation of component proteins) either by the transmembrane potential (gating) or various physiological agonists (see Chapter 5). Moreover, the similarities in the amino acid sequences of a variety of ionic channels may offer a unifying concept for the evolution of channels (*cf.* Hille, 1984). A detailed elucidation of the molecular aspects of channel operation may have to await a three-dimensional structural model of the channel protein imbedded in a lipid matrix.

The recognition that protein-lined channels impart the selectivity of biological membranes represents a new chapter in our understanding of membrane structure and function. Substantial evidence is now available for the view that amino acid residues of the hydrophilic regions of channel proteins (at the outer and inner mouths of the channel) are determinants of the ionic specificity, as well as receptor sites for channel blockers (*cf.* Numa, 1989; Mackinnon and Miller, 1989; Armstrong, 1992). The voltage gating channel properties appear to be determined by charged amino acid residues in the channel region imbedded in the lipid membrane (*cf.* Catterall, 1992). Channels are also involved in the actual fusion between membranes (i.e. during exocytosis; *cf.* Breckenridge and Almers,

1987) and communication between cells (gap junctions, first described by Lowenstein, 1967; cf. Lowenstein, 1984; Makowski et al., 1977). The emphasis is now shifting towards further details of the anatomy and function of channels in molecular terms, dissecting events at the entry of a solute to the mouth of the channel (including the stripping of the water shell), translocation of the solute within the channel, and the exit of the solute (associated with some hydration) into the aqueous phase. The exploration of the molecular properties of transport proteins has also yielded new information as to their specificity: The unexpected considerable homology found by Riordan et al. (1989) between the proteins of the Cl^- (cystic fibrosis) channel and that of the multidrug resistant P-glycoprotein (Chapter 4), as well as other membrane glycoproteins in bacteria, yeast and plasmodia, led to the postulate of a superfamily of transport proteins involved in widely different transport systems.

3. CHANNEL-FORMING PROTEINS

Studies of some bacterial toxins, (e.g. colicin E1, cf. Cramer et al., 1986) led to the recognition that their effects on target cells is due to water-soluble 'killer' proteins which, on insertion in the membrane of target cells, form non-selective channels ('pores') leading to depolarization of the cell membrane and cell death. The capability of producing this type of pore-forming proteins is not limited to bacteria; it is found e.g. in white blood cells (Young et al., 1986) and is associated with functions of the immune system (Tschopp et al., 1982).

IV. Regulation of membrane function

Nature uses a variety of means for meeting physiological demands by modulating cell function, and cell membranes represent a crucial site for signal transduction from the environment, as well as a direct target for modulation.

Shuttle mechanisms for transport proteins between their intracellular pool and the membrane allow a transient regulation of the transport of some solutes (see above and Chapter 3). Stretch-activated channels (Guharay and Sachs, 1984) involve the cytoskeleton for transduction of mechanical stimuli to control solute flow through ionic channels, and this mechanism appears to be responsible for an array of cell phenomena (cf. Sachs, 1992). The control of ionic channels by the transmembrane electrical potential (gating, see above) is analyzed in greater detail in Chapter 8. The first electrophysiological indication for the presence of light receptors in retinal cells (Ottoson and Svaetichin, 1953) was followed by evidence that light affects the function of ionic channels (Baylor and Hodgkin, 1973). As to chemical agents, the channel (and, generally, receptor) pharmacology (Chapter 5) represents a new chapter in the field.

The relationship between the function of channels and cellular metabolism is borne out by the observation that cell ATP is one of the regulatory agents for the operation of one of the K^+ channels (Noma, 1983) and a Ca^{2+} channel (Greenberg et al., 1988). A protein-kinase C-catalyzed phosphorylation of transport proteins (channels, but also pumps and carriers) appears to be an important modulating mechanism (cf. Nishizuka, 1986). The rather complex chloride channel protein requires ATP hydrolysis for its maintenance in the open conformation (Anderson et al., 1991). In addition, external (and probably also intracellular) ATP does induce the formation of non-selective pores in the membranes of various cells (cf. El-Moatassim et al., 1992).

V. The concept of the cell membrane towards the 100th anniversary of Overton's views

When Danielli in 1962 surveyed the basic concepts of a cell membrane, he considered a structure (see Fig. 1B, 1C) consisting of a bimolecular lipid leaflet coated on both sides by a protein layer; this structure was spanned by (protein) structures, imparting to the membrane special permeation processes which, because of their properties, could not take place by simple diffusion, i.e. facilitated diffusion (see Chapter 3) and active transport (Chapter 4).

Comparing the mechanisms responsible these special permeation processes to the catalytic properties of an enzyme E, he proposed a unifying concept of these special membrane processes by the following scheme: the subscripts o and i denote the outer and inner compartments, separated by a membrane in which the catalyst E is located):

enzyme[38] : $S_o \rightleftharpoons ES \rightleftharpoons S_o$;
carrier-mediated (facilitated)
diffusion : $S_o \rightleftharpoons ES \rightleftharpoons S_i$;

$$\text{active transport} : S_o \rightleftharpoons ES \underset{\searrow d^-}{\overset{\nearrow d^+}{\rightleftharpoons}} S_i$$

where d^+ is an energy donor;

receptor : $S_o \rightleftharpoons ES$[39].

---

38 This is consistent with the views of Mitchell (1962) suggesting that if the catalyst is suitably oriented in the membrane, this will result in the presence of S_o on one side of the membrane, and S_i (the product of enzymic conversion) on the other.

39 Danielli failed to include here a crucial component of the receptor concept, i.e. the step ES → response, cf. Chapter 5).

These processes then impart to the membrane its properties as an osmotic and electrically-charged barrier separating the cell from its environment.

The progress made since can be summarized:

1. The membrane is a thin asymmetric bimolecular lipid leaflet coated and spanned by a variety of protein structures: the carbohydrate moieties of glycoproteins and glycolipids at the outer membrane face impart to the membrane recognition sites for various physiological regulatory agents and toxic substances (receptors, Chapter 5), as well as sites for cell-matrix and cell-cell adhesion. The hydrophobic portions of proteins spanning the membrane are arranged in rod-like structures forming channel-like clefts. At the cytoplasmic face, the meshwork of structural proteins (part of the cytoskeleton) contributes to the mechanical properties of the cell membrane (including cell shape and deformability). The component proteins add to the membrane a sizeable thickness, thus producing an effectively unstirred layer with ensuing complexities in the analysis of its physico-chemical properties;

2. The membrane is a dynamic structure with a considerable turnover of its components. In its character the membrane is a cell organelle;

3. The permeability barrier of the membrane permits:

(*a*) diffusion of hydrophobic substances across the lipid bilayer[40]; (*b*) diffusion of hydrophilic solutes of small molecular weight (less than 5 Å in diameter) through some 'pores' includ-

---

40 The cell physiologists make use of this phenomenon in order to introduce into cells non-dissociated (*e.g.* esterified) 'caged' substances which on intracellular hydrolysis convert into impermeant probes).

ing channels[41]; (c) transport of solutes by special[42] mechanism across the membrane using of transport proteins or pumps. (d) The translocation of functional groups across membranes, postulated by Mitchell in 1956 (cf. 1962) and later demonstrated by Kundig et al. (1967) for the hexose transferase. In Danielli's scheme, this process could be described by:

$$\text{translocase: } S_o \rightleftharpoons ES \overset{\nearrow AX}{\underset{\searrow A}{\rightleftharpoons}} ESX \rightleftharpoons SX_i$$

where AX stands for phospho*enol*pyruvate, X for the phosphate group.

In 1962, E represented the phenomenological equivalent for a variety of proteins now known to be involved in transport processes, with an underlying assumption that one protein is responsible for each of the various transport processes. As the molecular aspects of the interaction of these proteins with the transported solutes are now being unravelled, it is increasingly becoming clear that while a variety of proteins may be involved in the transport of one solute (cf. Chapter 3), an unexpected homology of proteins with differing transport characteristics is also apparent (p. 71 and Chapter 4). An additional intriguing observation simultaneously reported by the laboratories of Cunningham (Harvard) and Kabat (University of Oregon) now demonstrates that the receptor for some viruses also serves as a transport protein (cf. the review of Vile and

---

41 The notion of non-selective 'pores' is now in many cases replaced by channels of varying selectivity, this concept bridges the early discussions of the existence of pores vs. an interaction between the solute and the membrane (Brücke vs. Fick, p. 32; Ostwald vs. Nernst, p. 32).

42 In correspondence with the author, preceding the 1960 Prague Symposium, Mitchell suggested that all substrates of physiological importance may eventually be shown to be transported across membranes by special mechanisms; only cell poisons may take the direct diffusion route. The development of our knowledge in the last 30 years has shown that he was not far off the mark.

Weiss (1991). Thus, the above *functional* categories of membrane proteins provide few cues as to the nature of the protein transporters.

Many of the above pathways of solute transport across membranes have now been shown to be physiologically regulated, *e.g.* by the membrane potential and/or other physical properties (above and Chapter 8), physiological agents (*e.g.* hormones, and also ATP, *cf.* Chapter 5) or shuttles which, on stimulation, reversibly allow the retrieval of a transport system from a cellular pool into the membrane (above and Chapter 3).

4. The membrane is the prime site for the transduction of signals between cells (Chapter 5), as heralded by Sutherland's (1971) discoveries, and information on the regulation of cellular functions by hormones and other agents, including the modulation of channels by receptors for ATP. In the light of recently emerging information (*e.g.* on the role of ATP in the function of the Cl⁻ channel, see above), Danielli's simple criteria characterizing the respective transport processes may have to be refined.

5. The progress outlined above reflects the avalanche of new input of many fields of science on the unfolding perception of the molecular aspects of cell membrane structure and function. The cell membrane with its lipid, protein and carbohydrate components is now perceived to be a crucial integrative element of that basic unit of life, the cell.

6. The present studies aim at an integration of knowledge of membrane structure and function at a molecular level, including various aspects of interaction between membrane lipids and proteins. The exploration of the genetic blueprint for the synthesis and assembly of membrane components is the next major step in the understanding of membrane structure and function.

Acknowledgements

The author is greatly indebted to Drs. M.M. Civan, S. Erulkar, D.E. Goldman, B. Hille and L.L.M. Van Deenen for valuable comments on the manuscript. The advice of Drs. R.L. Barchi, W.B. Guggino, P. Mueller, L.J. Mullins, B. Pressman, L. Warren, W.F. Widdas and M.A. Zasloff on specific points was greatly appreciated.

REFERENCES

Anderson, M.P., Berger, H.A., Rich, D.P., Gregory, R.J., Smith, A.E. and Welsh, M.J. (1991) Nucleotide triphosphates are required to open the CFTR chloride channel. Cell 67, 775–784.

Armstrong, C.M. (1981) Sodium channels and gating currents. Physiol. Revs. 61, 644–683.

Armstrong, C.M. (1992) Voltage-dependent ion channels and their gating. Physiol. Revs. 72, S5–S13.

Baylor, D.A. and Hodgkin, A.L. (1973) Detection and resolution of visual stimuli by turtle photoreceptors. J. Physiol. (Lond). 234, 163–198.

Bean, R.C., Shepherd, W.C., Chan, H. and Eichner, J.T. (1970) Discrete conductance fluctuations in lipid bilayer protein membranes. J. Gen. Physiol. 53, 741–757.

Behn, V. (1897) Über wechselseitige Diffusion von Elektrolyten in verdünnten Lösungen, insbesonders über Diffusion gegen das Conzentrationsgefälle. Ann. Phys. Chem. 62, 54–67.

Bennett, V. (1989) The human erythrocyte as a model system for understanding membrane-cytoskeleton interactions. In Cell Membranes, Methods and Reviews (Elson, E., Frazier, W. and Glaser, L., eds.), Vol. 2, pp. 149–195. New York, Plenum Press.

Bernstein, J. (1902) Untersuchungen zur Thermodynamik der bioelektrischen Ströme. Pflüger's Arch. ges. Physiol. 92, 521–562.

Berridge, M.J. (1987) Inositol triphosphate and diacylglycerol: two interacting second messengers. Annu. Rev. Biochem. 56, 159–193.

Bishop, W.R. and Bell, R.M. (1988) Assembly of phospholipids into cellular membranes: biosynthesis, transmembrane movement and intracellular translocation. Annu. Rev. Cell Biol. 4, 579–610.

Blasie, J.K. and Worthington, C.R. (1969) Planar liquid-like arrangement of photopigment molecules in frog retinal receptor disk membranes. J. Mol. Biol. 39, 417–439.

Blobel, G. and Dobberstein, B. (1975) Transfer of proteins across membranes. I. Presence of proteolytically processed and unprocessed nascent immunoglobulin light chains in membrane-bound ribosomes of murine myeloma. J. Gen. Biol. 67, 835–851.

du Bois-Reymond, E. (1948) *Untersuchungen über thierische Elektricität*, Vol. I. Berlin, G. Reimer.

Brandley, B.K., Swiedler, S.J. and Robbins, P.W. (1990) Carbohydrate ligands of the LEC cell adhesion molecules. Cell 63, 861–872.

Branton, D. (1966) Fracture faces of frozen membranes. Proc. Natl. Acad. Sci. USA 55, 1048–1056.

Breckenridge, L.J. and Almers, W. (1987) Currents through the fusion pore that forms during exocytosis of a secretory vesicles. Nature (Lond). 328, 814–817.

Bretscher, M.S. (1972) Asymmetrical lipid bilayer structure for biological membranes. Nature New Biol. 236, 11–12.

Brooks, S.C. (1937) Selective accumulation with reference to ion exchange by the protoplasm. Trans. Faraday Soc. 33, 1002–1006.

Brücke, E. (1843) Beiträge zur Lehre von der Diffusion tropfbarflüssiger Körper durch poröse Scheidewände. Poggendorff's Ann. Phys. Chem. 58, 77–100.

Bütschli, O. (1894) *Investigations on Microscopic Foams and on Protoplasm*. (Minchin, E.A., trans.). London, Black Publ.

Catterall, W.A. (1986) Molecular properties of the voltage-sensitive sodium channel. Annu. Rev. Biochem. 55, 953–985.

Catterall, W.A. (1992) Cellular and molecular biology of voltage-gated sodium channels. Physiol. Revs. 72, S15–S48.

Chambers, R. (1926) The nature of the living cell as revealed by microdissection. Harvey Lectures 22, 41–58.

Chapman, D. and Leslie, R.S. (1970) Structure and function of phospholipids in membranes. In *Membranes of Mitochondria and Chloroplasts* (Racker, E., ed.) pp. 91–126, New York, Van Nostrand Reinhold Comp.

Chevalier, J., Bourguet, J. and Hugon, J.S. (1974) Membrane associated particles: Distribution in frog urinary bladder epithelium at rest and after oxytocin treatment. Cell Tissue Res. 152, 129–140.

Civan, M.M. and Shporer, M. (1989) Physical state of cell sodium. Curr. Top. Membr. Transp. 34, 1–19.

Clark, A.J. (1937) General Pharmacology. In *Handbuch d. Exp. Pharmakologie*, Vol. IV. (Heubner, W. and Schüller, J., eds.). Berlin, J. Springer.

Cohen, B.E. and Bangham, A.D. (1972) Diffusion of small non-electrolytes across liposome membranes. Nature (Lond.) 236, 173–174.

Cohn, Z.A. and Steinman, R.M. (1982) *Phagocytosis and Fluid-phase Pinocytosis*. CIBA Foundation Symp. 92, 15–28. London, Pitman Books Ltd.

Cole, K.C. (1932) Surface forces of the *Arbacia* eggs. J. cell. comp. Physiol. 1, 1–9.

Cole, K.S. (1949) Dynamic electrical characteristics of the squid axon membrane. Arch. Sci. Physiol. 3, 253–258.

Cole, K.S. (1972) *Membranes, Ions and Impulses.* Berkeley, Univ. Calif. Press.

Cole, K.S. and Cole, R.H. (1936) Electrical impedance of *Arbacia* eggs. J. Gen. Physiol. 19, 625–632.

Collander, R. (1937) The permeability of plant protoplasts to non-electrolytes. Trans. Faraday Soc. 33, 985–990.

Collander, R. and Bärlund, H. (1933) Permeabilitätsstudien an *Chara ceratophyla.* Acta Bot. Fenn. 11, 1–114.

Conolly, T.J., Carruthers, A. and Melchior, D.L. (1985) Effect of bilayer cholesterol content on reconstituted human erythrocyte sugar transporter activity. J. Biol. Chem. 260, 2617–2620.

Conway, E.J. (1947) Exchanges of K, Na and H ions between the cell and the environment. Irish J. Med. Sci. 6, 593–609; 654–680.

Cook, J.S. (ed.) (1976) *Biogenesis of Turnover of Membrane Macromolecules.* New York, Raven Press.

Cook, R.P. (1926) The antagonism of acetylcholine by methylene blue. J. Physiol. (Lond.) 62, 160–165.

Cramer, W.A., Dankert, J.R. and Uratani, Y. (1983) The membrane channel-forming bactericidal protein, colicin E1. Biochim. Biophys. Acta 737, 173–193.

Cronan, J.E. and Gelman, E.P. (1975) Physical properties of membrane lipids: Biological relevance and regulation. Bacteriol. Rev. 39, 232–256.

Danielli, J.F. (1962) Structure of the cell surface. Circulation 26, 1163–1168.

Danielli, J.F. and Davson, H. (1935) A contribution to the theory of permeability of thin films. J. Cell. Comp. Physiol. 5, 495–508.

Danielli, J.F. and Harvey, E.N. (1934) The tension at the surface of mackerel egg oil, with remarks on the nature of the cell surface. J. Cell. Comp. Physiol. 5, 483–494.

Davson, H. (1989) Biological membranes as selective barriers to diffusion of molecules. In *Membrane Transport. People and Ideas* (Tosteson, D.C., ed.) pp. 15–49. Bethesda, MD., Am. Physiol. Soc.

Davson, H. and Danielli, J.F. (1943) *The Permeability of Natural Membranes.* Cambridge, Cambridge Univ. Press.

Davson, H. and Reiner, J.M. (1942) Ionic permeability: An enzyme-like factor in the cat erythrocyte membrane. J. cell. Comp. Physiol. 20, 325–344.

De Duve, C., Berthet, J. and Beaufay, H. (1959) Gradient centrifugation of cell particles: Theory and applications. Progr. Biophys. 9, 325–369.

Deenen, L.L.M. van (1965) *Phospholipids and Biomembranes.* Oxford, Pergamon Press.

De Gier, J., Mandersloot, J.G. and Van Deenen, L.L.M. (1968) Lipid composition and permeability of liposomes. Biochim. Biophys. Acta 150, 666–675.

Devaux, P.F. (1990) The aminophospholipid translocase: A transmembrane lipid pump – physiological significance. NIPS 5, 53–58.

De Vries, H. (1884) Eine Methode zur Analyse der Turgorkraft. Jahrb. f. wiss. Bot. 14, 429–601.

Dewey, M.M. and Barr, L. (1970) Some considerations about the structure of cellular membranes. Curr. Topics Membr. Transp. 1, 1–33.

Dick, D.A.T. (1959) Osmotic properties of living cells. Internat. Rev. Cytol. 8, 388–448.

Ding, G., Franki, N., Condeelis, J. and Hays, R.M. (1991) Vasopressin depolymerizes F-actin in toad bladder epithelial cells. Am. J. Physiol. 260, C9–C16.

Donnan, F.G. (1911) Theorie der Membrangleichgewichte und Membranpotentiale bei Vorhandensein von nicht dialysierenden Elektrolyten. Z. Elektrochem. 17, 572–581.

Donnan, F.G. and Guggenheim, E.A. (1932) Die genaue Thermodynamik der Membran-gleichgewichte. Z. phys. Chem. 162, 346–360.

Dutrochet, R.J.H. (1827) Nouvelles observations sur l'endosmose et l'exosmose, et sur la cause de ce double phenomène. Ann. chim. et phys. 35, 393–400.

Edelman, G.M. (1983) Cell adhesion molecules. Science 219, 450–467.

Ehrenstein, G., Lecar, H. and Nossal, R. (1970) The nature of the negative resistance in bimolecular lipid membranes containing excitability-inducing material. J. Gen Physiol. 55, 119–133.

Ehrlich, P. (1900) On immunity with special reference to cell life. Proc. Roy. Soc. 66, 424–448.

Ehrlich, P. (1902) Über die Beziehungen von chemischer Constitution, Distribution und pharmakologischer Wirkung. In The Collected Papers of Paul Ehrlich (Dale, H.H., ed.), Vol. II, pp. 570–595. London 1960, Pergamon Press.

Eisenman, G. (1961) On the elementary atomic origin of equilibrium ionic specificity. In Membrane Transport and Metabolism (Kleinzeller, A. and Kotyk, A., eds.) pp. 163–179. New York, Academic Press.

El-Moatassim, C., Dornand, J. and Mani, J.-C. (1992) Extracellular ATP and cell signalling. Biochim. Biophys. Acta 1134, 31–45.

Evans, E.A. and Hochmuth, R.M. (1978) Mechanochemical properties of membranes. Curr. Topics Membr. Transp. 10, 1–64.

Fick, A. (1855) Über Diffusion. Poggendorff's Ann. Phys. 94, 59–86.

Finean, J.B. (1962) The nature and stability of the plasma membrane. Circulation 26, 1151–1162.

Finean, J.B. (1972) Development of ideas on membrane structure. Sub-cell. Biochem. 1, 363–373.

Finkelstein, A. (1987) Water Movement through Lipid Bilayers, Pores and Plasma Membranes. New York, Wiley-Interscience.

Frey-Wyssling, A. (1953) Submicroscopic Morphology of Protoplasm. 2nd Ed., Amsterdam, Elsevier.

Fricke, H. (1925) The electric capacity of suspensions with special reference to blood. J. Gen. Physiol. 9, 137–152.

Frye, C.D. and Edidin, M. (1970) The rapid intermixing of cell surface antigens after formation of mouse-human heterokaryons. J. Cell Sci. 7, 313–335.

Gibbs, J.W. (1876) On the equilibrium of heterogeneous substances. Trans. Conn. Acad. 3, 108–248.

Goldfarb, D. and Michaud, N. (1991) Pathways for the nuclear transport of proteins and RNA. Trends in Cell Biol. 1, 20–24.

Goldman, D.E. (1943) Potential, impedance and rectification in membranes. J. Gen. Physiol. 27, 37–60.

Gorter, E. and Grendel, F. (1925) On bimolecular layers of lipoids in the chromocytes of the blood. J. exp. Med. 41, 439–443.

Graham, T. (1861) Liquid diffusion applied to analysis. Philos. Trans. R. Soc. 151, 183–204.

Greenberg, S., Divirgilio, F., Steinberg, T.H. and Silverman, S.C. (1988) Extracellular nucleotides mediate Ca^{2+} fluxes in J774 macrophages by two distinct mechanism. J. Biol. Chem. 263, 10337–10343.

Grendel, F. (1929) Über die Lipoidschicht der Chromocyten beim Schaf. Biochem. Z. 214, 231–241.

Guharay, F. and Sachs, F. (1984) Stretch-activated single ion channel currents in tissue-cultured embryonic chick skeletal muscle. J. Physiol. 352, 685–701.

Hamburger, H.J. (1890) Die Permeabilität der rothen Blutkörperchen im Zusammenhang mit den isotonischen Coefficienten. Z. f. Biol. 26, 414–433.

Hamburger, H.J. and Bubanovic, F. (1910) La permeabilité des globules rouges, specialement vis-a-vis des cations. Arch. Int. Physiol. 10, 1–36.

Hamill, G.F., Marty, A., Neher, E., Sakmann, B. and Sigworth, F.J. (1983) Improved patch-clamp techniques for high-resolution current recording from cells and cell-free membrane patches. Pflüger's Arch. ges. Physiol. 391, 85–100.

Hardy, M.A. and DiBona, D.R. (1982) Microfilaments and the hydrosmotic action of vasopressin on toad urinary bladder. Am. J. Physiol. 243, C200–C204.

Harvey, E.N. (1911) Studies on the permeability of cells. J. Exp. Zool. 10, 507–556.

Haydon, D.A. (1979) The Organization and Permeability of Artificial Lipid Membranes. In *Membranes and Ion transport* (Bittar, E.E., ed.), pp. 64–92. London, Wiley-Interscience.

Hedin, S.G. (1897) Ueber die Permeabilität der Blutkörperchen. Pflüger's Arch. ges. Physiol. 68, 229–338.

Heinemann, S.H., Terlau, H., Stühmer, W., Imoto, K. and Numa, S. (1992) Calcium channel characteristics conferred on the sodium channel by single mutations. Nature (Lond). 356, 441–443.

Helmholtz, H. v. (1879) Studien über electrische Grenzschichten. Annln. Phys. u. Chem. 7, 337–382.

Henderson, R. (1977) The purple membrane from *Halobacterium halobium*. Annu. Rev. Biophys. Bioeng. 6, 87–109.

Hermann, L. (1899) Zur Theorie der Erregungsleitung und der elektrischen Erregung. Pflüger's Arch. ges. Physiol. 75, 574–590.

82   REFERENCES

Hertwig, O. (1893) *Die Zelle und die Gewebe*, p. 14. Jena, G. Fischer.
Hevesy, G. v., Hofer, E. and Krogh, A. (1935) The permeability of the skin of frogs to water as determined by D_2O and H_2O. Skand. Arch. Physiol. 72, 199–214.
Hewson, W. (1773) On the figure and composition of the red particles of the blood, commonly called the red globules. Philos. Trans. Roy. Soc. London 63, Part II, 303–324.
Hille, B. (1984) *Ionic Channels of Excitable Membranes*. Sunderland, MA, Sinauer Associates, Inc.
Höber, R. (1907) Beiträge zur physikalischen Chemie der Erregung und der Narkose. Pflüger's Arch. ges. Physiol. 120, 492–516.
Höber, R. (1910) Eine Methode, die elektrische Leitfähigkeit im Innern von Zellen zu messen. Pflüger's Arch. ges. Physiol. 133, 237–259.
Höber, R. (1926) *Physikalische Chemie der Zelle und Gewebe, 6.Aufl.*, 388–571; Leipzig, W. Engelmann.
Höber, R. (1936) Membrane permeability to solutes in its relations to cellular physiology. Physiol. Revs. 16, 52–102.
Hodgkin, A.L. and Huxley, A.F. (1952) A quantitative description of membrane current and its application to conduction and excitation in nerve. J. Physiol. (Lond.) 117, 500–544.
Hodgkin, A.L., Huxley, A.F. and Katz, B. (1949) Ionic currents underlying activity in the giant axon of the squid. Arch. Sci. Physiol. (Paris) 3, 129–150.
Hodgkin, A.L. and Keynes, R.D. (1955) Active transport of cations in giant axons from *Sepia* and *Loligo*. J. Physiol. (Lond.) 128, 28–60.
van 't Hoff, J.H. (1887) Die Rolle des osmotischen Druckes in der Analogie zwischen Lösungen und Gasen. Z. Phys. Chem. 1, 481–508.
Hubbell, W.C. and McConnell, H.M. (1968) Spin-label studies of the excitable membranes of nerve and muscle. Proc. Natl. Acad. Sci. USA 61, 12–16.
Jacobs, M.H. (1931) The permeability of the erythrocyte. Erg. d. Biol. 7, 1–55.
Jacobs, M.H. (1935) Diffusion processes. Erg. Biol. 12, 1–160.
Jacobs, M.H. (1962) Early osmotic history of the plasma membrane. Circulation 26, 1013–1021.
Jordan, P.C. (1987) How pore mouth charge distributions alter the permeability of transmembrane channels. Biophys. J. 51, 297–311.
Kahn, C.R. (1976) Membrane receptors for hormones and neurotransmitters. J. Cell Biol. 70, 261–286.
Kedem, O. and Katchalsky, A. (1958) Thermodynamic analysis of the permeability of biological membranes to non-electrolytes. Biochim. Biophys. Acta 27, 229–246.
Kirkwood, J.G. (1954) Transport of ions through biological membranes from the standpoint of irreversible thermodynamics. In *Ion Transport across Membranes* (Clarke, H.T., ed.) pp. 119–127. New York, Academic Press.
Kleinzeller, A. and Kotyk, A. (1961, eds.). *Membrane Transport and Metabolism*. Academic Press, N.Y.

Kleinzeller, A. and Mills, J.W. (1989) K⁺-induced swelling of the dogfish shark (*Squalus acanthias*) rectal gland cells is associated with changes in the cytoskeleton. Biochim. Biophys. Acta 1014, 40–52.

Klenk, F. and Uhlenbruck, G. (1960) Über ein neuraminsäurehaltiges Mucoproteid aus Rindererythrocytenstroma. Hoppe-Seyler's Z. f. physiol. Chem. 311, 227–233.

Klenk, F. and Uhlenbruck, G. (1960) Über neuraminsäurehaltige Mucoide aus Menschen-erythrocytenstroma, ein Beitrag zur Chemie der Agglutinogene. Hoppe-Seyler's Z. f. physiol. Chem. 319, 151–160.

Koefoed-Johnsen, V. and Ussing, H.H. (1953) The contributions of diffusion and flow to the passage of D₂O through living membranes. Acta physiol. scand. 28, 60–76.

Koeppe, H. (1897) Der osmotische Druck als Ursache des Stoffaustausches zwischen rothen Blutkörperchen und Salzlösungen. Pflüger's Arch. ges. Physiol. 67, 189–206.

Kornberg, R.D. and McConnell, H.M. (1971) Inside-outside transitions of phospholipids in vesicle membranes. Biochemistry 10, 1111–1120.

Kostyuk, P.G., Mironov, S.L. and Shuba, Ya.M. (1983) Two ion-selecting filters in the calcium channel of the somatic membrane of mollusc neurons. J. Membrane Biol. 76, 83–93.

Krnjevic, K. (1991) Cellular mechanisms of anaesthesia. Annl. New York Acad. Sci. 625, 1–16.

Kundig, W., Ghosh, S. and Roseman, S. (1967) Phosphate bound to histidine in a protein as an intermediate in a novel phosphotransferase system. Proc. Natl. Acad. Sci. USA 52, 1067–1074.

Langley, J.N. (1906) On nerve endings and on special excitable substances in cells. Proc. Roy Soc. B. 70, 170–194.

Langmuir, I. (1917) The constitution and fundamental properties of solids and liquids, II. Liquids. J. Am. Chem. Soc. 37, 1848–1915.

Langmuir, I. (1939) Molecular layers. Proc. Roy. Soc. A. 170, 1–39.

Läuger, P. (1980) Kinetic properties of ion carriers and channels. Biophys. J. 57, 163–178.

Läuger, P. (1985) Ionic channels with conformational substrates. Biophys. J. 47, 581–591.

Lawrence, A.S.C. (1961) A new type of interfacial interaction. In *Membrane Transport and Metabolism* (Kleinzeller, A. and Kotyk, A., eds.), pp. 35–44. New York, Academic Press.

Liebig, J. (1847) Über die Bestandtteile der Flüssigkeiten des Fleisches. Ann. d. Chem. u. Pharmacie 62, 257–368.

Lillie, R.S. (1916) Increase of permeability to water following normal and artificial activation in sea-urchins. Am. J. Physiol. 40, 249–266.

Ling, G.N. (1984) *In Search of the Physical Basis of Life*. New York, Plenum Publ.

Loeb, J. (1903) *The Dynamics of Living Matter.* New York, Columbia Univ. Press.

Loeb, J. and Beutner, R. (1912) Ueber die Potentialdifferenzen an der unversehrten and verletzten Oberfläche pflanzlicher und tierischer Organe. Biochem. Z. 41, 1–26.

Lowenstein, W.R. (1967) On the genesis of cellular communication. Dev. Biol. 15, 503–520.

Lowenstein, W.R. (1984) Channels in the junctions between cells. Curr. Topics Membr. Transp. 21, 221–252.

Lucké, B. and McCutcheon, M. (1932) The living cell as an osmotic system and its permeability to water. Physiol. Rev. 12, 68–139.

Luzzati, V. (1968) X-Ray diffraction studies of lipid-water systems. In Biological Membranes, Vol. 1 (Chapman, D., ed.) pp. 71–123. London, Academic Press.

MacKinnon, R. and Miller, C. (1989) Mutant potassium channels with altered binding of charybdotoxin, a pore-blocking peptide inhibitor. Science 245, 1382–1385.

Makowski, L., Caspar, D.L.D., Phillips, W.C. and Goodenough, D.A. (1977) Gap junction structures. II. Analysis of the X-ray diffraction data. J. Cell Biol. 674, 629–645.

Marmont, G. Studies on the axon membrane. I. A new method. J. Cell. Comp. Physiol. 34, 357–382, 1949.

McLaughlin, S. (1989) The electrostatic properties of membranes. Annu. Rev. Biophys. Chem. 18, 113–136.

Meyer, H. (1899) Zur Theorie der Narkose. I. Mitt. Eigenschaft der Anästhetika bedingt ihre narkotische Wirkung? Arch. f. exp. Pathol. u. Pharmakol. 42, 109–118.

Meyer, K.H. and Sievers, J.F. (1936) La permeabilité des membranes I. Theorie de la permeabilité ionique. Helv. Chim. Acta. 19, 649–664.

Michaelis, L. (1928) Investigations on molecular sieve membranes. In Colloid Symposium Monograph (Weiser, H.B., ed.) pp. 135–148. New York, Chemical Catalog Comp.

Michaelis, L. and Fujita, A. (1925) Untersuchungen über elektrische Erscheinungen und Ionendurchlässigkeit von Membranen. II. Mitt. Die Permeabilität der Apfelschale. Biochem. Z. 158, 28–37.

Mitchell, P.D. (1962) Metabolism, transport and morphogenesis: which drives which. J. Gen. Microbiol. 29, 25–37.

Mitchison, J.A. (1953) A polarized light analysis of the human red cell ghost. J. exp. Biol. 30, 397–432.

Mohl, H. von (1846) Einige Bemerkungen über den Bau der vegetabilischen Zelle. Bot. Ztg. 2, 273–277.

Mohl, H. v. (1855) Der Primordialschlauch. Bot. Ztg. 13, 689–737.

Moore, C. and Pressman, B.C. (1964) Mechanism of action of valinomycin on mitochondria. Biochem. Biophys. Res. Comm. 15, 5612–557.

Müller, J. (1838) Elements of Physiology (Baly, W., transl.), Vol. I, pp. 98–109. London, Taylor & Watson.

Mueller, P., Rudin, D.O., Tien, H.T. and Wescott, W.C. (1962) Reconstitution of cell membrane structure in vitro and its transformation into an excitable system. Nature (Lond.) 194, 979–980.

Nägeli, C. and Cramer, C. (1855) Pflanzenphysiologische Untersuchungen. 1. Heft. Zürich, F. Schulthess.

Nathanson, A. (1904) Ueber die Regulation der Aufnahme anorganischer Salze durch die Knollen von *Dahlia*. Jahrb. f. wiss. Bot. 39, 607–644.

Neher, E. and Sakmann, B. (1976) Single channel currents recorded from membranes of denervated frog muscle fibres. Nature (Lond.) 260, 799–802.

Nernst, W. (1889) Die electromotorische Wirksamkeit der Jonen. Z. phys. Chem. 4, 129–181.

Nernst, W. (1891) Verteilung eines Stoffes zwischen zwei Lösungsmitteln und zwischen Lösungsmittel und Dampfraum. Z. phys. Chem. 8, 110–139.

Nernst, W. (1899) Zur Theorie der elektrischen Reizung. Göttinger Nachrichten, Mathem.-phys. Kl., Heft 1, 104–108.

Nernst, W. (1908) Zur Theorie des elektrischen Reizes. Pflüger's Arch. ges. Physiol. 122: 275–315.

Newport, J.W. and Forbes, D.J. (1987) The nucleus: Structure, function and dynamics. Annu. Rev. Biochem. 56, 535–565.

Newton, I. (1704) *Opticks*. Reprinted 1952, pp. 215; 232, New York, Dover.

Nicolson, G.L. (1976) Transmembrane control of the receptors on normal and tumor cells. I. Cytoplasmic influence over cell surface components. Biochim. Biophys. Acta 457, 57–108.

Nishizuka, Y. (1986) Studies and perspectives of protein kinase C. Science 233, 305–312.

Noma, A. (1983) ATP-regulated K^+-channels in cardiac muscle. Nature (Lond.) 305, 147–148.

Numa, S. (1989) Molecular view of neurotransmitters receptors and ionic channels. Harvey Lect. 83, 121–165.

Ocklind, C., Forsum, U. and Öbrink, B. (1983) Cell surface localization and tissue distribution of a hepatocyte cell-cell adhesion glycoprotein (cell-CAM 105). J. Cell Biol. 96, 1168–1171.

Orbach, E. and Finkelstein, A. (1980) The nonelectrolyte permeability of planar lipid bilayers. J. Gen. Physiol. 75, 427–436.

Osterhout, W.J.V., Kamerling, J.E. and Stanley, W.M. (1934) Kinetics of penetration. VII. Molecular versus ionic transport. J. Gen. Physiol. 17, 469–480.

Ostwald, W. (1890) Elektrische Eigenschaften halbdurchlässiger Scheidewände. Z. phys. Chem. 6, 71–82.

Ottoson, D. and Svaetichin, G. (1953) The electrical activity of the retinal receptor layer. Acta Physiol. Scand. 29, 31–39.

Overton, E. (1895) Ueber die osmotischen Eigenschaften der lebenden Pflanzen und Tierzelle. Vierteljahrschr. d. Naturforsch. Ges. Zürich 40, 159–201.

Overton, E. (1896) Ueber die osmotischen Eigenschaften der Zelle in ihrer Bedeutung für die Toxikologie und Pharmakologie. Vierteljahrschr. d. Naturforsch. Ges. Zürich 41, 383–406.

Overton, E. (1899) Ueber die allgemeinen osmotischen Eigenschaften der

Zelle, ihre vermutliche Ursachen und ihre Bedeutung für die Physiologie. Vierteljahrschr. d. Naturforsch. Ges. Zürich 44, 88–114.

Overton, E. (1900) Studien über die Aufnahme der Anilinfarben durch lebende Zellen. Z. wiss. Bot. 34, 669–701.

Overton, E. (1901) *Studien über die Narkose.* Jena, G. Fischer Verl.

Overton, E. (1902) Beiträge zur allgemeinen Muskel- und Nervenphysiologie. II. Mitt. Ueber die Unentbehrlichkeit von Natrium (oder Lithium)-Ionen für den Contractionsact des Muskels. Pflüger's Arch. ges. Physiol. 92, 346–386.

Paganelli, C.V. and Solomon, A.K. (1957) The rate of exchange of tritiated water across the human red cell membrane. J. Gen. Physiol. 41, 259–277.

Palade, G.E. (1959) Functional changes in the structure of cell components. In *Subcellular Particles* (Hayaishi, T., ed.), pp. 349–408. New York, Ronald Press Co.

Pappenheimer, J.R., Renkin, E.M. and Borrero, L.M. (1951) Filtration, diffusion and molecular sieving through peripheral capillary membranes. A contribution to the pore theory of capillary permeability. Am. J. Physiol. 167, 13–46.

Parpart, A.K. and Dziemian, A.J. (1940) The chemical composition of the red cell membrane. Cold Spring Harbor Symp. 8, 17–22.

Parsegian, V.A. (1969) Energy of an ion crossing a low dielectric membrane: solution to four relevant electrostatic problems. Nature (Lond.). 221, 844–846.

Pasternak, A.C. (1975) Biosynthesis of membranes. In *Biological Membranes* (Parsons, D.S., ed.), pp. 55–67, Oxford, Clarendon Press.

Pethica, B.A. (1984) Introduction. In *Cell Fusion*, Ciba Foundation Symp. 103 (Evered, D. and Whelan, eds.), pp. 1–7. London, The Pitman Press.

Pfeffer, W.F.P. (1877) Osmotische Untersuchungen. Leipzig, W. Engelmann.

Planck, M. (1890) Über die Erregung von Electricität und Wärme in Electrolyten. Annln. Phys. u. Chem., NF 39, 161–186.

Pressman, B.C. (1985) The discovery of ionophores: An historical account. In *Metal Ions in Biological Systems* (Sigel, H., ed.). New York, M. Dekker, Inc.

Preston, G.M., Carroll, T.P., Guggino, W.B. and Agre, P. (1992) Appearance of water channels in *Xenopus* oocytes expressing red cell CHIP 28 protein. Science 256, 385–387.

Pringsheim, N. (1854) *Untersuchungen über den Bau und die Bildung der Pflanzenzelle.* Berlin, Hirschwald.

Quaranta, V. and Jones, C.R. (1991) The internal affaires of an integrin. Trends Cell Biol. 1, 2–4.

Quincke, G. (1888) Über periodische Ausbreitung an Flüssigkeitsoberflächen und dadurch hervorgerufene Bewegungserscheinungen. Ann. Phys. u. Chem. NF 35, 580–642.

Redman, C.M. and Hokin, L.E. (1959) Phospholipid turnover in microsomal membranes of the pancreas during enzyme secretion. J. Biophys. Biochem. Cytol. 6, 207–214.

Rideal, E.K. (1937) Factors in membrane permeability. Trans. Faraday Soc. 33, 1081–1085.

Riordan, J.R., Rommens, J.M., Kerem, B., Alon, N., Rozmahel, R., Grzelczak, Z., Zielinski, M.J., Lok, S., Plavcic, N., Chou, J.-L., Drumm, M.L., Iannuzzi, M.S., Collins, F.S. and Tsui, L.-C. (1989) Identification of the cystic fibrosis gene: cloning and characterization of complementary DNA. Science 245, 1066–1073.

Robertson, J.D. (1960) The molecular structure and contact relationships of cell membranes. Progr. Biophys. Chem. 10, 344–418.

Rosenberg, P.A. and Finkelstein, A. (1978) Water permeability of gramicidin A-treated lipid bilayer membranes. J. Gen. Physiol. 72, 341–350.

Rothman, J.E. and Lenard, J. (1977) Membrane asymmetry. The nature of membrane asymmetry provides clues to the puzzle of how membranes are assembled. Science 195, 743–753.

Rottem, S. (1982) Transbilayer distribution of lipids in microbial membranes. Curr. Topics Membr. Transp. 17, 235–261.

Ruhland, W. and Hoffmann, C. (1925) Die Permeabilität von Beggiatoa mirabilis. Ein Beitrag zur Ultrafiltrationstheorie des Plasmas. Planta 1, 1–83.

Sachs, F. (1992) Stretch-sensitive ion channels: An update. In Sensory Transduction (Corey, D.P. and Roper, S.D., eds.), pp. 241–260. New York, The Rockefeller University Press.

Sachs, J. (1865) Handbuch der Experimental-Physiologie der Pflanzen. Leipzig, W. Engelmann.

Schlögl, R. (1964) Stofftransport durch Membranen. Darmstadt, D. Steinkopf Verl.

Schmid, G. (1950) Elektrochemie feinporiger Kapillarsysteme. I. Übersicht. Z. Electrochem. 54, 424–429.

Schmidt, W.J. (1935) Doppelbrechung und Feinbau der Markscheide der Nervenfasern. Z. Zellf. u. mikr. Anat. 23, 667–676.

Schmitt, F.O. and Bear, A. (1939) The ultrastructure of the nerve axon sheath. Biol. Rev. 14, 27–50.

Schulman, J.H. (1937) Structure in relation to living biological function. Trans. Faraday Soc. 33, 1116–1125.

Schultz, C.H. (1836) Das System der Cirkulation, in seiner Entwicklung durch die Thierreihe und im Menschen. Stuttgart, Cotta.

Schwann, T. (1847) Microscopical Researches into the Accordance of the Structure and Growth of Animals and Plants (Smith, H., transl.), London, Sydenham Soc.

Seddon, J.M. (1990) Structure of the inverted hexagonal (H_{II}) phase, and non-lamellar phase transitions of lipids. Biochim. Biophys. Acta 1031, 1–69.

Simon, S.M., Peskin, C.S. and Oster, G.F. (1992) What drives the translocation of proteins? Proc. Natl. Acad. Sci. USA 89, 3770–3774.

Singer, S.J. and Nicolson, G.L. (1972) The fluid mosaic model of the structure of cell membranes. Science 175, 720–731.

Sjöstrand, F.S. (1956) The ultrastructure of cells as revealed by the electron microscope. Int. Rev. Cytol. 5, 455–533.

Smith, H.W. (1962) The plasma membrane, with notes on the history of botany. Circulation 26, 987–1012.

Sollner, K., Dray, S., Grim, E. and Neihof, R. (1954) Electrochemical studies with model membranes. In *Ion Transport Across Membranes* (Clarke, H.T., ed.), pp. 144–188. New York, Academic Press.

Staverman, A.J. (1952) Non-equilibrium thermodynamics of membrane processes. Trans. Faraday Soc. 48, 176–185.

Steck, T.L. (1974) The organization of proteins in the human red blood cell membrane. J. Cell Biol. 62, 1–19.

Stein, W.D. (1967) *The Movement of Molecules across Cell Membranes*. New York, Academic Press.

Stein, W.D. and Danielli, J.F. (1956) Structure and function in red cell permeability. Disc. Faraday Soc. 21, 238–251.

Steinman, R.M., Mellman, I.S., Muller, W.A. and Cohn, Z.A. (1983) Endocytosis and the recycling of plasma membrane. J. Cell Biol. 96, 1–27.

Stoeckenius, W. (1962) Structure of the plasma membrane. An electron-microscopic study. Circulation 26, 1066–1069.

Straub, F.B. (1953) Über die Akkumulation der Kaliumionen durch menschliche Blutkörperchen. Acta Physiol. Akad. Sci. Hung. 4, 235–240.

Stühmer, W., Ruppersberg, J.P., Schröter, K.H., Sakmann, B., Stocker, M., Giese, K.P., Perschke, A., Baumann, A. and Pongs, O. (1989) Molecular basis of functional diversity of voltage-gated potassium channels in mammalian brain. EMBO J. 8, 3235–3244.

Sutherland, E.W. (1962) The biological role of adenosine-3′,5′-phosphate. The Harvey Lectures 57, 17–33.

Sutherland, E.W. (1971) Cyclic AMP and hormone action. In *Cyclic AMP* (Robison, G.A., Butcher, R.W. and Sutherland, E.W.), pp. 17–48, New York, Academic Press.

Teorell, T. (1935) Studies on the 'diffusion effect' upon ionic distribution. I. Some theoretical considerations. Proc. Natl. Acad. Sci. USA 21, 152–161.

Teorell, T. (1953) Transport processes and electrical phenomena in ionic membranes. Progr. Biophys. Biophys. Chem. 3, 305–369.

Traube, M. (1867) Experimente zur Theorie der Zellbildung und Endosmose. Arch. f. Anat., Physiol. u. wiss. Med. pp. 87–165.

Troshin, A.S. (1976) *Problems of cell permeability*. (Hall, M.G. transl.). London, Pergamon Press.

Tschopp, J., Müller-Eberhard, H.J. and Podack, E.R. (1982) Formation of transmembrane tubules by spontaneous polymerization of the hydrophilic complement protein C9. Nature (Lond). 298, 534–538.

Unwin, P.N.T. and Ennis, P.D. (1984) Two configurations of a channel-forming membrane protein. Nature (Lond). 307, 609–613.

Ussing, H.H. (1949) The distinction by means of tracers between active transport and diffusion. Acta physiol. scand. 19, 43–56.

Ussing, H.H. (1954) Ion transport across biological membranes. In *Ion Transport across Membranes* (Clarke, H.T., ed.) pp. 3–22. New York, Academic Press.

Van Slyke, D.D. (1926) *Factors affecting the Distribution of Electrolytes, Water and Gases in the Animal Body.* Philadelphia, Lippincott.

Verkleij, A.J., Zwaal, R.F.A., Roelofsen, B., Comfurius, P., Kastelijn, D. and van Deenen, L.L.M. (1973) The asymmetric distribution of phospholipids in the human red cell membrane. A combined study using phospholipases and freeze-etch electron microscopy. Biochim. Biophys. Acta 323, 178–193.

Verkman, A.S. (1989) Mechanisms and regulation of water permeability in renal epithelia. Am. J. Physiol. 257, C837–850.

Vile, R.G. and Weiss, R.A. (1991) Virus receptors as permeases. Nature (Lond.) 352, 666–667.

Voelker, D.R. (1985) Lipid assembly into cell membranes. In *Biochemistry of Lipids and Membranes* (Vance, D.E. and Vance, J.E., eds.) pp. 474–502. Menlo Park, CA, Benjamin/Cummings Publ.

Voelker, D.R. (1991) Organelle biosynthesis and intracellular lipid transport in eukaryotes. Microbiol. Rev. 55, 543–560.

Wade, J.B., Stetson, D.L. and Lewis, S.A. (1981) ADH action: Evidence for a membrane shuttle mechanism. Ann. New York Acad. Sci. 372, 106–117.

Wienhues, U. and Neupert, W. (1992) Protein translocation across mitochondrial membranes. Bioessays 14, 17–23.

Wilbrandt, W. (1935) The significance of the structure of a membrane for its selective permeability. J. Gen. Physiol. 18, 933–9654.

Wilbrandt, W. (1938) Die Permeabilität der Zelle. Erg. Physiol. 204–291.

Winzler, R.J. (1969) A glycoprotein in human erythrocyte membranes. In *Red Cell Membrane, Structure and Function* (Jamieson, G.A. and Greenwalt, T.J., eds) pp. 157–171. Philadelphia, J. Lippincott.

Wirtz, K.W.A. and Zilversmit, D.B. (1968) Exchange of phospholipids between liver mitochondria and microsomes in vitro. J. Biol. Chem. 243, 3596–3602.

Yoda, S. and Yoda, A. (1987) Phosphorylated intermediates of Na-K-ATPase proteoliposomes controlled by bilayer cholesterol. Interaction with cardiac steroid. J. Biol. Chem. 262, 103–109.

Young, J.D-E., Peterson, C.G.B., Venge, P. and Cohn, Z.A. (1986) Mechanism of membrane damage mediated by human eosinophil cationic protein. Nature (Lond.) 321, 613–615.

Yu, J., Fischman, D.A. and Steck, T.L. Selective solubilization of proteins and phospholipids from red blood cell membranes by non-ionic detergents. J. Supramol. Struct. 1, 233–248.

Yu, J. and Steck, T.L. (1975) Associations of band 3, the predominant polypeptide of the human erythrocyte membrane. J. Biol. Chem. 250, 9176–9184.

Zwaal, R.F.A., Demel, R.A., Roelofsen, B. and van Deenen, L.M. (1976) The
lipid bilayer concept of cell membranes. Trends Biochem. Sci. 1, 112–114.

A. Kleinzeller (Editor) A History of Biochemistry: Exploring the Cell Membrane.
(Comprehensive Biochemistry Vol. 39) © 1995 Elsevier Science B.V.

Chapter 3

The Concept of a Membrane Transport Carrier

A. KLEINZELLER

Department of Physiology, School of Medicine, The University of
Pennsylvania, Philadelphia, PA, U.S.A.

Introduction

Studies of membrane phenomena (Chapter 2) started with the assumption that solutes entered (or exited) across cell membranes by the physical process of diffusion. Quantitative appraisals of this assumption by measurements of the permeability coefficient (amount of solute crossing a unit membrane area in unit time under the influence of a unit concentration gradient) demonstrated inconsistencies, which eventually led to the postulate of separate transport pathways involving interaction of the solutes with specific transport proteins (carriers). Hence, the carrier concept reflects the kinetic era of studies of membrane phenomena.

The work of Osterhout (1933) on the mechanism of electrolyte distribution in marine algae, and his model experiments led to an early postulate that ions may interact with a hypothetical, lipid-soluble membrane component which would then shuttle the solute across the membrane as a non-ionized complex; a more detailed model of a hypothetical carrier of cations operating in the cell membrane was put forward by Lundegårdh (1940). The reviews of Jacobs (1931), Davson and

Danielli (1943), Wilbrandt and Rosenberg (1961), LeFevre (1975) and Widdas (1988) reflect milestones with regard to the development of ideas that many ions and non-electrolytes may interact with a membrane component in the course of their transport across cell membranes. To a great extent, the development of the carrier concept was fostered by the advent of radioactive tracers as labels for the measurement of unidirectional hydrophilic solute fluxes (*cf.* Ussing, 1952).

I. The postulate of membrane permeability pathways involving mechanisms other than simple diffusion

Overton's concept of a lipid layer separating the cell from its environment (*cf.* Chapter 2) raised questions as to the pathway by which hydrophilic solutes (salts, and other physiological substrates) passed this surface layer (membrane). Overton (1895) realized this point by stating:

> 'In many of these processes (i.e. uptake or release of substances by cells) the protoplasm participates actively and transports substances often in the opposite direction than would occur if only the laws of diffusion were responsible. ... One might even say that organisms avoid in their metabolism such compounds which can rapidly pass through the plasma membrane.'

This view implied that for a solute to pass a cell membrane, two possibilities had to be considered: *1.* Diffusion (a) through a lipid membrane; this included the possibility of hydrophilic substrates being transiently made lipophilic by interaction with some other (membrane) component, or (b) through aqueous 'pores' (*cf.* the mosaic and molecular sieve theories of cell membranes, Chapter 2). Solute transport by both these pathways had to be consistent with predictions stemming from Fick's laws of diffusion (*i.e.* thermal agitation as the driving force). *2.* Other pathways, equivalent to Höber's (1899) term 'physiological permeability', as opposed to the passive 'physical permeability'[1]. Höber understood that physiological permeability might involve the transfer of a solute as a chemical unit (*i.e.* not changed metabolically), but this point was firmly established only later. Gradually, a variety of observations inconsistent with simple diffusion shifted the emphasis of investigations towards 'physiological permeability'.

1. The temperature dependence (Q_{10}) of the permeability coefficient was higher than that predicted for simple diffusion at physiological temperatures (*i.e.* exceeding the value of 1.4, corresponding to an activation energy of some 4000 cal/mol); thus, for glycerol, Jacobs et al. (1935) reported Q_{10} values of 3-4, characteristic for chemical reactions. Although this point might be explained on the basis of potential energy barriers in the membrane for the diffusion of solutes (*cf.* Davson and Danielli, 1943), differences in the Q_{10} for glycerol in different animal species were more difficult to deal with.

2. The structural specificity of solute uptake represented another stumbling block. Nagano (1902) noticed that the absorption of hexoses from dog intestine showed some structural specificity (galactose being more rapidly absorbed than glucose); such observations were later extended to the intestinal absorption of sugars in rats (Cori, 1925) and other species. Diffusion of substances of the same molecular diameter and polarity should proceed at the same rate. The observation of the stereo-specificity of intestinal sugar absorption (Westenbrink, 1937) certainly was not consistent with a diffusion process. The first studies on the structural specificity of the uptake of sugars were impressed with similarities between uptake and hexokinase activity, and ideas were put forward that sugar phosphorylation in the membrane (hexokinase activity)

---

1 'One has to assume that the physiological import and export is a complex, not analyzed process ...based on the cell surface (plasma skin). One can not but think that a complex organization is required for such action of the plasma membrane' (Höber, 1911).

might be responsible for the observations. A direct examination of such views (*e.g.* Blakeley and Boyer, 1955) demonstrated the need for a transport step preceding metabolic reactions.

3. Diffusion through a lipid (or mosaic) membrane might be expected to show similar properties in cells of different species. However, Kozawa (1914) in Höber's laboratory established major differences in the erythrocyte permeability for sugars: while the human (and ape) erythrocyte was permeable to glucose and a number of other hexoses, that of rodents, carnivores, and other mammalian species was impermeable to these sugars. Similar differences were observed for the species permeability of erythrocytes to amino acids (Kozawa and Miyamoto, 1921) and glycerol (Jacobs, 1931).

4. Diffusion kinetics predicts a linear relationship between the rate of solute entry and its concentration gradient (Chapter 2). However, Ege (1919) noted that sugars entered red blood cells faster from lower than from higher concentrations. Wilbrandt et al. (1947) then extended these studies by quantitative measurements, establishing the saturability of the transport process. Danielli's calculations (1954) demonstrated that some solutes (*e.g.* urea, glycerol or glucose) penetrated the cell membrane considerably faster than predicted on the basis of diffusion through a strictly homogeneous lipid layer, and this was directly demonstrated by Wood et al. (1968).

5. Diffusion should not be affected by the presence of other solutes. Contrary to such prediction, marked effects of various substances on solute entry were reported: Jacobs and Corson (1934) found that the uptake of glycerol by erythrocytes was inhibited by traces[2] (10^{-7} M) of Cu^{2+}; this heavy metal also inhibits the Na^+ entry into cat red blood cells (Davson and Reiner, 1942). Effects of heavy metals on various

---

2 Jacobs later calculated that at 6.10^{-7} M Cu^{2+}, at most 1% of the total cell surface could have been covered by copper; thus, active patches of a small fraction of the red blood cell membrane were the site of a specific mechanism for glycerol penetration.

transport processes were then amply documented (cf. Passow et al., 1961). Wilbrandt and Rosenberg (1961) summarized extensive evidence for a selective inhibition of membrane permeability for many solutes by a variety of agents.

Competition of aldoses for the entry mechanism into various cells: Cori (1926) may have been the first to demonstrate this phenomenon for the absorption of hexoses by the intestine, and analogous phenomena were then observed for sugar reabsorption in the kidney by Shannon (1938) and in red blood cells by LeFevre and Davies (1951) and by Widdas (1954) (cf. LeFevre, 1975). Clear evidence for competition for a *membrane* entry step was provided by Blakeley and Boyer (1955) who found in yeast a competitive inhibition of glucose uptake by the non-metabolizable 6-deoxy-6-fluoro-D-glucose at concentrations which were ineffective on glucose fermentation by an extract from the same cells.

The effect of insulin on sugar transport is an instance of effectors on transport processes: On the basis of Kozawa's (1914) experiments, Höber (1914) suggested the possibility[3] that this hormone might affect the sugar entry into cells; Häusler and Loewi (1925) arrived at a similar conclusion. The first more direct evidence on this point was provided by Lundsgaard (1939): He demonstrated in perfused limbs of animals (a) a saturation of glucose uptake by the muscles of perfused hind limbs of cats on increasing the glucose concentration of perfusing plasma; and (b) insulin increased the disappearance of glucose from the plasma into the muscle while the apparent glucose concentration in muscle cells was nil. These observations were not consistent with simple diffusion of glucose across the cell membrane, but suggested an 'active' process of glucose transfer, stimulated by insulin. Direct evidence for the primary action of insulin on glucose entry into

---

3 'It might be that given an abnormal constitution of the membrane, simple sugars cannot reach the normal site of their degradation; ... a membrane theory of diabetes might thus replace today's theories on diabetes.'

muscle was then reported by Levine et al. (1950) (at that time, this report was considered to being highly unorthodox, *cf.* Chapter 1) and by Park et al. (1956).

6. The unequal distribution of electrolytes between cells and their environment also invited speculations about specific mechanisms of entry. An analysis of the entry of Na^+ and K^+ into cat red blood cells led Davson and Reiner (1942) to the conclusion that an enzyme-like membrane component reduced the activation energy barrier of the cell membrane for the sodium ion; and they postulated a similar system for the fast, Cu^{2+}-inhibitable entry of glycerol.

7. The discovery by Monod and his school (Cohen and Monod, 1957) of the uphill, genetically controlled uptake of sugars by bacteria provided evidence for an enzyme-like mechanism of the process; at the same time, the genetic approach to transport phenomenology represented a novel experimental tool.

8. The observations of Lundsgaard (1939) on the effect of insulin (see above) raised questions concerning the relationship between a putative transport process and cellular metabolism. Given Overton's views that most metabolic substrates crossed the cell membranes by pathways different from diffusion, the rate of such solute transport across the membrane may become limiting for the subsequent cellular metabolism dependent on the kinetic parameters of both transport and metabolic processes. Subsequent studies provided ample evidence for transport as the rate-limiting step for metabolism and raised questions as to the mechanism(s) by which transport processes may be regulated (*cf.* Park et al., 1956; Randle and Smith, 1958; Kipnis, 1959; Horrecker et al., 1960; Kleinzeller and Kotyk, 1963; Kornberg, 1973).

All these observations were difficult to reconcile with a simple diffusion process as the mechanism of transfer of the investigated solutes across cell membranes. On the other hand, the described properties of the systems recalled those of enzymes. Thus, the concept was put forward that hypothetical (enzyme-like) membrane components[4], carriers, are involved in the transfer of solutes across cell membranes (cf. Davson and Reiner, 1942; Hodgkin et al., 1949; Ussing, 1952; Rosenberg and Wilbrandt, 1954). As opposed to the then well-defined concept of enzyme action, the carrier would bring about a translocation of the substrate across the membrane, but would not catalyze a chemical transformation of the substrate to the product. Some alternative explanations of the above observations were considered: (a) A kinetic analysis would not distinguish between a carrier-mediated translocation of a solute from its intracellular adsorption; such possibility appeared to be eliminated by the counterflow phenomenon (see below). (b) The idea of an enzyme-catalyzed group translocation was put forward by Mitchell (1957), and was indeed later demonstrated (Roseman, 1969). (c) A role of metabolic enzymes acting as carriers in the membrane appeared to be remote in the light of evidence above (cf. also Winkler and Wilson, 1966). (d) Particularly for ions, the possibility of intracellular binding to a protoplasmic polyelectrolyte system (cf. Ling, 1984) has found rather few adherents. Thus, the essentially kinetic concept of a specific, carrier-mediated transport process became acceptable. Two essentially equivalent terms for the observed phenomenology were put forward: Ussing in 1947 (cf. 1952) suggested the term 'exchange diffusion' for the phenomenon where a membrane component B interacts with the impermeant solute A to form the transient compound AB capable of ferrying the solute across the membrane.

Danielli (1954) classified existing information by differentiating phenomenologically between:

(a) carrier-mediated equilibrating systems (facilitated diffusion [or facilitated transport], with thermal agitation as the driving force); 'the system differs from diffusion in that the

---

4 It should be noted that in this respect, pharmacologists (cf. Clark, 1937) were well ahead of the transport field in suggesting an enzyme-like interaction between drugs and their membrane receptors (cf. Chapter 5).

rate at which molecules diffuse is strongly influenced by structural and steric factors'; and

(b) *active* (net uphill; i.e. away from equilibrium) solute transport requiring an input of metabolic energy.

At the same time, it was realized that cell membranes are characterized by the operation of several transport systems operating in parallel:

> 'If no other information is available, a given flux may be the resultant of any combination of three processes, namely diffusion, exchange diffusion, and active transport' (Ussing, 1952).

The possibility that active transport represented an energized carrier-mediated process was early recognized[5].

This chapter deals with the broad concept of carrier-mediated phenomena. The phenomenology of active transport is discussed in Chapter 4.

II. The phenomenology of carrier-mediated transport: kinetics and models

Increasing knowledge of kinetics of carrier-mediated fluxes permitted the modelling of the process in terms of its components, *i.e.* (a) interaction of the carrier in the membrane with the external solute to form the carrier-solute complex at one face of the membrane; (b) reorientation (or translocation) of the complex to face the other side of the membrane; and (c) dissociation of the complex, releasing the solute to the other side. Such a concept reflected available knowledge on the mechanism of enzyme reactions.

---

5 Mitchell expressed this since 1963 (*cf.* 1992): 'The idea that all transport processes in biology are mechanistically dependent on facilitated diffusion, whether metabolically coupled or not, provides the basis for the general concept of a specific vectorial ligand conduction.

1. CRITERIA OF CARRIER-MEDIATED TRANSPORT PROCESSES

Experimental observations (see above) defined essential criteria for the participation of a carrier in a transport process not involving chemical transformation of the solute (cf. Danielli, 1954; Wilbrandt and Rosenberg, 1961; Stein, 1964; LeFevre, 1975; Widdas, 1988):
1. Saturation kinetics of solute transport across a cell membrane, including non-linearity with the gradient, competition of solutes for transfer, high uni-directional fluxes (exceeding those for simple diffusion); 2. Structural specificity of solute transport; 3. Temperature dependence associated with a chemical reaction; 4. Susceptibility to enzyme inhibitors, protein reagents, hormones. 5. Flux coupling phenomena, including cotransport and countertransport (counterflow); (see below); 6. Genetic control of the transport process (control of the carrier protein).

Most (but not all) of these criteria have to be met in order to characterize a transport process as carrier-mediated; the effects of ionophores (Chapter 2) are consistent with some (cf. Hladky, 1979). Thus defined, the carrier has to be viewed as essentially passive in nature (Widdas, 1952); Widdas realized that

'where the membrane has a potential difference across it, however, more work may be done on carriers crossing in one direction than in the other. Thus 'against the gradient' transfer may result, although the carriers themselves may still be regarded as acting passively'.

However, it should be pointed out here that the above criteria, with the possible exception of counterflow, do not permit an easy discrimination between a carrier-mediated process from an internalization of a solute via an endocytotic mechanism.

2. KINETICS AND MODELS FOR THE CARRIER-MEDIATED TRANSPORT OF NON-ELECTROLYTES

Given the available experimental techniques at the time, the study of kinetics of sugar transport in erythrocytes proved at first to be most rewarding: As opposed to the red blood cells of most mammalian species, in the cells of some species (*e.g.* human, or rabbit) the rate of sugar (*e.g.* glucose) transport exceeds by some 10^3 that of its subsequent intracellular phosphorylation (Wood et al., 1968), and free glucose is found in the cells; hence, the transport is an equilibrating system, and metabolism does not greatly affect the interpretation of flux analyses. The introduction of labelled substrates further facilitated experimentation, as did the use of nonmetabolizable analogues (*e.g.* 3-O-methyl-glucose; Csáky, 1942). The reviews of Crane (1977) and Widdas (1988) summarize the development of ideas.

The kinetic analysis of a transport phenomenon allows predictions as to the molecular properties of the involved carrier (*i.e.* 'models'). The first such model, in rather simple terms, was put forward by Shannon (1939) to describe the renal reabsorption of sugars, and this model was subsequently extended by LeFevre and LeFevre (1952): the hypothesis visualized that the membrane provided a fixed sugar binding entity with access to both interfaces. The now classical model of Widdas (1952) (a symmetric, mobile carrier in adsorption equilibrium with the solute at both interfaces, and a rate-limiting translocating step brought about by thermal agitation[6]) was quanti-

---

[6] A hypothetical carrier mechanism specific for Na^+ was proposed by Hodgkin et al. (1949).

tatively consistent with available information (saturation kinetics[7]; competitive phenomena) and permitted testable predictions: one such major prediction concerned the situation
where two substrates with differing affinities, separated by a
membrane, competed for the same carrier; under these conditions, the downhill flux for one sugar was predicted to drive a
(transient) uphill flux of the competing sugar[8]. This prediction
was then experimentally verified by Park et al. (1956). The
phenomenon of counterflow (countertransport), characterized
by a transient 'overshoot' of one of the substrates against an
apparent concentration gradient also became an important
criterion of facilitated diffusion (cf. Rosenberg and Wilbrandt,
1957). Subsequently, the still relatively simple model of Widdas was modified to meet results of further enquiry. Kinetic
evidence for an intrinsic asymmetry of the human erythrocyte
glucose carrier[9], signalled by Wilbrandt (1954), was demonstrated by Geck (1971) and in Widdas' laboratory (cf. Widdas,
1980). The model of Regen and Morgan, but particularly that
of Regen and Tarpley (1974) removed some restrictive assumptions and added additional elements, e.g. an unstirred
layer on both interfaces of the membrane.
The substrate-carrier interaction. This rapid process reflects
the capability of the carrier to select the substrate, defining its
specificity. Hydrogen bonding between the erythrocyte glucose carrier and the glucose molecule appears to occur at two

---

7 Widdas at first expressed his data in terms of Langmuir's adsorption isotherm;
after presenting his data at the Physiological Society meeting in 1951, R.B. Fisher
and D.S. Parsons (1953) mentioned to him their data showing that a Michaelis-Men-
ten relationship applied for sugar absorption by the intestine. Subsequently, Widdas
(personal communication) employed this biologically more relevant approach.
8 'In the competitive system, the change in chemical potential of the one species
transferred from an area of greater to an area of less concentration provides the
energy which enables the second species to be transferred against the gradient, and
there is, as may be expected, no circumvention of the Second Law of Thermodynamics.
However, metabolic activity could supply the necessary energy for a continuing process based on either mechanism to work in a biological system.'
9 This asymmetry does not appear to be universal: erythrocytes from some species,
and also some other cells, do not show kinetic asymmetry (cf. Carruthers, 1990).

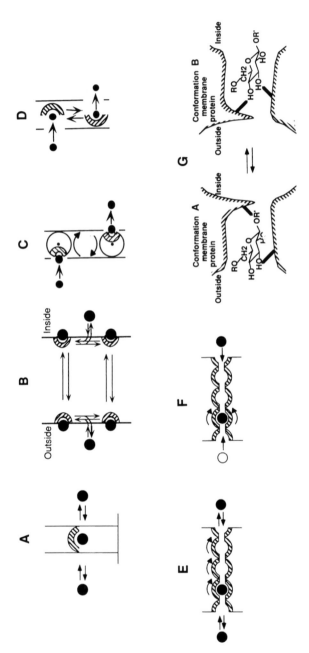

Fig. 1. Development of carrier models. Black and white balls represent substrate molecules. A. The fixed-site concept of a carrier (Shannon, 1939; LeFevre and LeFevre, 1952). B. and C. The mobile carrier (Widdas, 1952); B. Diffusible model. C. Rotating model. D. Conformational model (Patlak, 1957). E. The pore model (Stein and Danielli, 1956). F. The lattice-membrane model (Vidaver, 1966; Naftalin, 1970); α: substrate-specific interaction sites; m: translocation process for substrate from one site to another; during counterflow, two substrates can interchange between adjoining sites. G. Alternating conformational model (Barnett et al., 1975); reprinted with permission of the Biochemical Society and Portland Press.

sites (Barnett et al., 1975) while that in the brush border of
intestinal or renal epithelial cells appears to involve most (or
all) oxygens of the hexose molecule (Barnett et al., 1968;
Kleinzeller et al., 1980).
The translocating step. The nature of the translocating mechanism was the subject of extensive debates (Fig. 1). The fixed-
site model was at first superseded by the concept of a mobile
carrier, and this idea was considered to be clinched by the
discovery of the counterflow phenomenon (consistent with a
diffusing (Fig. 1B) or rotating (Fig. 1C) model[10]; counterflow
implied *some* movement within the system. Alternative views
have been presented: The conformational model (Patlak, 1957)
suggested a system gated by conformational changes, similar
to that of Mitchell (1957) (Fig. 1D). The Stein and Danielli
(1956) model of a protein-lined pore (Fig. 1E) with transient
hydrogen bonding of substrates[11] represented an alternative
to the mobile carrier model, but did not allow an easy explanation of the counterflow phenomenon until Vidaver (1966) and
Naftalin (1970) provided indications for this in their models
(Fig. 1F); movement within such system did not necessarily
involve a whole molecule – the swinging of a couple of hydrogen bonds within a region of the carrier might suffice (Vidaver,
1966). Such a model (now designated as an alternating conformation model, Fig. 1G) was then presented by Barnett et al.
(1975), the conformational changes of the carrier protein in
the course of the translocation step may be related to thermal
energy, or, possibly, also to the interaction of substrate and

---

10 The unquestioning belief in a mobile carrier actually delayed the exploration of
alternative models.
11 Mitchell (1956) was critical of such model: 'I suggest that specific permeation
reactions of high specificity may occur by a process quite analogous to that of enzyme
reactions, cause a change of configuration or position of a protein carrier in the membrane such that the accessibility of its adsorbed passenger changes from one side of
the membrane to the other. I think that for membrane reactions of high specificity,
this type of model has the merit of conforming more closely to what is known of the
kinetics of biochemical reactions than the simple pore model of Stein and Danielli'.

protein. This model accounts for (a) substrate specificities of the transport that differ at the cytoplasmic and external surfaces of the membrane; (b) the countertransport phenomenon; (c) the apparent slower rate of outward movement of the unloaded carrier as compared with the inward solute transfer; the model also predicts that at any moment, there is only one binding site for the substrate (this prediction was then verified in the laboratory of Lienhard (Appleman and Lienhard, 1985).

The nature of the carrier. By analogy to enzymes, the inhibitory effect of many agents suggested that the carriers are proteins (Wilbrandt and Rosenberg, 1961; LeFevre, 1975). Furthermore, the assumption of the genetically-controlled mechanism of bacterial sugar accumulation as an enzyme-like system (Cohen and Monod, 1957) started efforts to isolate the involved protein (Fox and Kennedy, 1965). The availability of specific, high-affinity inhibitors of transport processes proved to be crucial for the characterization, purification and isolation of carrier proteins (*cf.* Tanner, 1979); for such an inhibitor to be a useful tool, its affinity for the transport protein should exceed that of the its substrates by at least 3 orders of magnitude. Hence the effort of investigators to explore the properties of inhibiting agents.

Heterogeneity of carriers. The carrier concept was first developed with the view that one carrier was responsible for one class of solutes. This idea had to be abandoned in favor of some heterogeneity. For sugar transport, the difference between equilibrating vs. active (uphill) transport *per se* was not sufficient to postulate different carriers, particularly in view of the experiments of Winkler and Wilson (1966). Differences in cell sensitivity to activation by insulin invited discussions on the topic of heterogeneity. However, major phenomenological differences between equilibrating and active sugar transport as to substrate specificity, cationic requirement, and pharmacology became crucial. The equilibrating system (in red blood cells, muscle, or adipocytes) did not require C_2-OH in the *gluco*-configuration or cations as co-substrates, and was selectively inhibited by phloretin[12]; on the other hand, the active transport system did require C_2-OH in the *gluco*-configuration, was cation-dependent and phlorizin was a selective inhibitor. Thus, the possibility of carrier heterogeneity for one solute could not be avoided.

The development of the phenomenology of amino acid transport followed a similar trend. Gale (1947) first drew attention to the capability of micro-organisms of accumulating amino acids against large concentration gradients. It was Heinz (1954) who demonstrated some phenomenology of carrier-mediated transport for the glycine uptake by Ehrlich ascites tumor cells (saturability, high Q_{10}, and acceleration of uptake of the labelled compound by preloading the cells with the 'cold' amino acid). As compared with progress in unravelling the carrier-mediated transport of sugars, the development of information of the mediation of carrier(s) in the uptake of amino acids was slowed by several factors: First, as opposed to sugars, the charged nature of amino acids cannot be neglected[13]. Second, Christensen's discovery of a multiplicity of carriers with overlapping specificity (Christensen, 1979, 1984), and the absence of selective inhibitors made studies more complex. In order to deal with the heterogeneity of carriers for amino acids, Christensen introduced numerous non-metabolizable amino acid analogues for transport studies and made use of the competitive phenomena by developing the strategy of inhibition analysis for pinpointing individual transport pathways (Christensen, 1979): Two substrates share one transport pathway, *a*, only if they *mutually* compete for the flux; a competitive inhibition of substrate A uptake by B may not reflect a shared pathway for both substrates, since B may also use an

---

12 The aglycone of phlorizin
13 'Amino acid transport is ion transport' (Christensen, 1984).

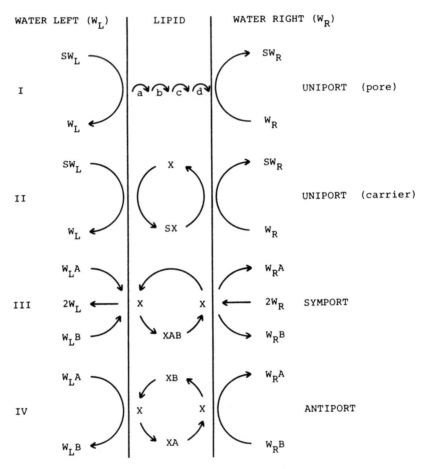

Fig. 2. Mechanisms of carrier-mediated translocation of solutes across a membrane (cf. Mitchell, 1962). Carrier: X; Solutes S, A, B; The solutes exist on both aqueous phases of the membrane as hydrates SW, AW and WB. Left and right aqueous phases are denoted by suffixes L and R. I. Translocation through a pore; II, translocation by a carrier-mediated process; III and IV, coupling of fluxes by one carrier system, symport (III) or antiport (IV). Reproduced with permission of the Biochemical Society and Portland Press.

alternative pathway *b*, and hence, A does not necessarily compete with B for this second pathway.

3. CARRIER-MEDIATED COUPLING OF FLUXES

The demonstration of the counterflow phenomenon, with the carrier acting as a revolving door for two fluxes in opposite directions, an antiport in Mitchell's (1962) terminology, opened the door for a reappraisal of some existing observations, as well as new investigations. Fig. 2 shows Mitchell's (1963) concept of the phenomenology of carrier-mediated transport systems. From the energetic point of view, the carrier-mediated coupling of solute fluxes may be treated as an energy transducer with the function of converting the free energy stored in the electrochemical gradient of one solute to that of the other solute (Rosenberg, 1954; Turner, 1983). The analysis of carriers in ion-exchange systems proved to be particularly laborious in view of the variety of driving forces involved, *i.e.* the membrane electrical potential for electrogenic ionic transfers, the coupling mechanisms, etc. Kinetically, two models for a tightly coupled exchange system may be visualized (Segel, 1975): (a) The sequential (ping-pong) model postulates that at any given time, only one substrate binds to the exchanger; (b) in the simultaneous model, two different substrates are bound to the protein at the same time, one at each face of the membrane, and both substrates are transported simultaneously, each in the opposite direction. These models may have to be reflected in the architecture of the transport protein. The pharmacology of the transport system became an invaluable tool in their identification; in fact, often it was the pharmacology of a given physiological function which eventually led to the identification of a given transport process and its carrier.

Here, the development of ideas on a few of such processes will be mentioned.

3.1. The Na^+-dependent transport of organic solutes

Christensen and Riggs (1952) noticed the cationic dependence of amino acids accumulation by cells. Subsequently, Crane (1960, 1961) proposed a hypothesis to explain the cationic dependence of sugar accumulation in intestinal mucosal cells, suggesting that the uphill sugar entry into the cells was driven by a downhill transport of Na^+; the active accumulation of the sugar in the cells was explained by a continuous sink in the cells for Na^+ produced by an extrusion of the cations. Thus, particularly in the model of Schultz and Zalusky (1964) the active, transcellular sugar transport system was essentially a passive, carrier-mediated (phlorhizin-inhibitable) cotransport (symport in Mitchell's, 1962, terminology) of sugar + Na^+ at the apical face of the intestinal cells, coupled to the energy-requiring process (sodium pump) at the basolateral cell membrane. For a more extensive appraisal of the concepts of secondary active transport systems of non-electrolytes, see Chapter 4.

Such a cotransport system may also be driven by some other cations: based on studies of the inhibition of the bacterial 'permease'-induced accumulation of sugars by dinitrophenol, Mitchell (1962) suggested a proton-sugar cotransport system as the mechanism of the phenomenon, and direct evidence for this process was established (West, 1970). Subsequent studies led to the belief that in bacteria, proton-organic solute cotransport is the predominant mechanism (Kaback, 1986). However, Na^+ (and even Li^+) may also be the cotransport cation (Chen et al., 1985).

The concept of cation-organic solute cotransport was extensively explored and substantiated in many laboratories and with different solutes, including many amino acids and small peptides, organic acids, etc. (Schultz and Curran, 1970; Heinz, 1972; Crane, 1977; Eddy, 1978). Direct evidence for the carrier-mediated (passive) cotransport concept was forthcoming when the technique of membrane vesicles was perfected (thus eliminating the possible contribution of metabolic processes)

and a transient ('overshoot') sugar (and amino acid) accumulation in the vesicles was produced by external Na^+ (Kinne,
1976); on dissipation of the sodium gradient, the sugar or
amino acid re-equilibrated between the vesicle and the external medium.

The concept of interaction of the carrier with a solute and an
ionic species raised several questions: First, in the case of a
sugar and Na^+ or H^+, the loaded carrier would become charged;
hence the translocating process would be electrogenic, and be
dependent on the pertaining electrical potential gradient
across the membrane. This point was experimentally established by the elegant experiment of Beck and Sacktor (1975)
using membrane vesicles. In the case of amino acids, the zwit-
terionic nature of the solute makes the analysis more difficult.
However, the electrogenic nature of (at least some) the ion-
organic solute cotransport system raised additional questions
pertaining to charge transfer, and, possibly, the charged nature of the carrier itself. The second major point related to the
stoichiometry of the interacting molecules. Vidaver had already demonstrated a ratio of two Na^+ per glycine molecule
translocated across the pigeon erythrocyte membrane. The
stoichiometry of the process is a determinant of the energetic
aspects of solute transfer, defining the maximal accumulation
ratio achievable in a given system (Eddy, 1978; Aronson,
1981). Third, the order of solute interactions with, and dissociation of, the carrier become of interest. Thus, while the order
of interaction may be random in some co- and countertransport systems (Stein, 1986), for the intestinal glucose-Na^+
cotransport system it was suggested that Na^+ interacted with
the carrier first at one face, while at the other face, glucose
interacted first (Hopfer and Grosclose, 1980). A similar, four-
state model for the Na^+-alanine cotransport system was postulated (Jauch and Läuger, 1988): while at the outer face of the
membrane, Na^+ interacts with the carrier only after alanine
was bound, the reverse order holds for the cytoplasmic side; in
the alanine-carrier-Na^+ state, alanine exchanges only with

the cytoplasmic side, Na^+ only with the extracellular medium. The modelling of this process in molecular terms will have to take account of the revealed kinetic information.

3.2. Cation/cation exchange systems

An exchange mechanism between external cations and cell H^+ was suggested by several pioneering observations. Goto (1918) described a loss of muscle K^+ on acidosis. In his studies on gastric acid secretion, Teorell (1939) suggested an exchange diffusion mechanism of H^+ against alkali ions[14]. In baker's yeast, Rothstein and Enns (1946) and Conway and O'Maley (1946) independently demonstrated that the presence of glucose produced an acidification of the medium associated with an uptake of K^+. Ussing in 1947 (cf. 1952) described an exchange diffusion system which facilitates the exchange of labelled Na^+ between the medium and muscle. In the kidney, Pitts and Alexander (1948) suggested an exchange between urine Na^+ and cell H^+ as the mechanism of urine acidification; these authors also noted that the experimental conditions did not allow a distinction between an exchange of Na^+ for H^+ and a movement of Na^+ together with OH^-. Subsequently, more direct evidence for the existence of cation exchange mechanisms in the membranes of a variety of systems was reported, e.g. in the red blood cell membrane, Na^+/H^+ and K^+/H^+ exchangers (Harris and Pressman, 1967); in mitochondria, specific electroneutral exchange systems H^+/Na^+ and H^+/K^+ (Mitchell and Moyle, 1967); in the frog skin Na^+/H^+ and Cl^-/HCO_3^- exchange carriers, Garcia-Romeu (1971). Studies of these systems were greatly stimulated by the recognition of their involvement in the regulation of intracellular pH (cf. Roos and Boron, 1981).

---

14 'The electrolyte concentration changes in the stomach content are pictured as an exchange diffusion – for instance, the hydrogen ions of an acid are exchanged against alkali ions of the mucosa or blood'.

The sodium pump in the sodium exchange mode, *i.e.* when the extracellular [K^+] is low or zero (Glynn and Karlish, 1975) may well be responsible for a major portion of the exchange diffusion system of Ussing. Direct evidence for the presence of a membrane exchange carrier was provided by Murer et al. (1976), using brush border membrane vesicles from epithelial cells: the proton gradient inside → outside produced a Na^+ gradient directed from the medium into the vesicles. This transport system also satisfied other criteria for a carrier-mediated process, and amiloride (as well as its derivatives) was found to be a reliable inhibitor of the process. With a stoichiometry of 1 Na^+ : 1 H^+, the process is electroneutral, and this mode of operation is found in mammalian cells under most experimental conditions. However, electrogenic Na^+/H^+ exchange systems were demonstrated particularly in bacteria with a stoichiometry varying from about 2H^+ : 1 Na^+ to 2 Na^+ : 1 H^+ depending on species and experimental conditions (Krulwich, 1983), and have since been demonstrated in a variety of cells, including eukaryotes. The K^+/H^+ exchanger, described in some cells, appears to be but a mode of the Na^+/H^+ antiporter (Cala, 1986).

Antiports similar to, or related to that of the Na^+/H^+ system are known: the transport of organic cations appears to be coupled to protons (Sokol et al., 1988). The Na^+/Ca^{2+} exchange system may be more complex, being electrogenic, with a stoichiometry of 3 (or more?) Na^+ per Ca^{2+} (Blaustein and Nelson, 1982). The suggested divalent cation/H^+ exchange in mitochondria (Chappel et al., 1963) appears to be the resultant of a coupling between a Na^+/Ca^{2+} exchange system with a Na^+/H^+ antiport (Crompton, 1985).

3.3. The anion exchange systems

Nasse (1878) first described an exchange of Cl^- by HCO_3^- in erythrocytes (see the 'Hamburger effect', Chapter 4). The elegant studies of Jacobs and Stewart (1942) led at first to the postulate of an OH^-/Cl^- exchange system in erythrocytes, which was subsequently modified to a HCO_3^-/Cl^- exchange. Later, indications of a tightly coupled anion exchange system as an important pathway of anion transport across cell membranes were found in a variety of cells, e.g. erythrocytes (Harris and Pressman, 1967) and the frog skin epithelium (Garcia-Romeu, 1971). Kinetic studies of the anion exchange system in erythrocytes revealed major differences between the permeabilities of various ionic species: (a) The anion/cation selectivity ratio was more than six orders of magnitude in spite of the similarities in size of Cl^- and K^+; while this phenomenon might be due to Cl^- crossing the membrane through a pore in a membrane with a cationic matrix (fixed charge model, see Chapter 2), the possibility of the involvement of a carrier-mediated process was also considered in the light of major differences in the rates of halide transport (Tosteson, 1959). The pH-dependence of the chloride transport system led to the postulate of a titratable (amino) group of the putative carrier (Gunn, 1972). The high anion selectivity of the membrane made it possible to study the anion fluxes independently from those of cations. (b) The mechanism for the exchange of labeled Cl^- was some 200,000-fold faster than that of net Cl^- flux (Passow, 1969). (c) Fluxes of anions displayed a temperature dependence, saturability, and competitive inhibition between anionic spacies (Dalmark and Wieth, 1972; Gunn et al., 1973). (d) The unexpected inhibition of the Cl^- exchange flux at higher Cl^- concentrations (Case and Dalmark, 1973) led to the possibility that the transport site might be characterized by an additional Cl^-- binding site capable of modifying the transport system (modifier site). (e) The availability of selective, high-affinity inhibitors of anion transport (particularly sulfonated stilbene derivatives SITS and DIDS (Cabantchik and Rothstein, 1972) greatly contributed to the exploration of this transport system. (f) Indications of an obligatory one-to-one exchange mechanism were also forthcoming (Harris and

Pressman, 1967). In the light of the quantitative prevalence of Cl^- and HCO_3^- in biological systems, and similar transport rates, the Cl^-/HCO_3^- exchange system appears to be functionally crucial from the point of view of cell physiology[15]. The scholarly reviews of Knauf (1979) and Passow (1986) provide an insight into the development of experimental data and ideas on the anion antiporter.

The association of the anion exchange system with a specific membrane component also played a major role in the development of ideas about this transport system. In the early 70's two major polypeptides were found in erythrocyte membranes, (cf. Steck, 1978). One of these proteins, designed band 3, was associated with the anion transport system, particularly in the light of a linear relationship between the binding of stilbene sulfonate agents to membrane proteins and the inhibition of anion translocation (Cabantchik and Rothstein, 1974). Since the band-3 glycoprotein is a dimer spanning the membrane, this information had a bearing on ideas of how anions might pass through the cleft between the two parts of this protein (see below).

The anion antiport appears to have a relatively broad spectrum of specificity (Knauf, 1979). In particular, this system might also be the vehicle by which anionic metabolic substrates (e.g. pyruvate or lactate, [cf. Halestrap, 1976]) might enter (or exit) the cell. However, more specific anion/anion antiports, e.g. between the monovalent p-aminohippuric acid and dicarboxylic acids, were characterized (Ullrich et al., 1987).

The study of anion-exchange systems in bacteria allowed a resolution and reconstitution of the exchange proteins, including the phosphate-sugar phosphate antiport (Maloney, 1990).

---

15 Available techniques do not allow the distinction between two possibilities concerning the HCO_3^- transport: (a) translocation of the bicarbonate ion; or (b) translocation of OH^- with a simultaneous release of CO_2.

3.4. Cation / anion symport systems

The discovery of these carrier-mediated transport systems stemmed from detailed investigations of anion transport: phosphate absorption by the kidney has been well established. The elucidation of the actual mechanism had to await the development of the membrane vesicle technique which permitted dissociation of the actual transport mechanism from phosphate metabolism. Using brush border membrane vesicles, Hoffmann et al. (1976) established an electroneutral Na^+- inorganic phosphate cotransport system; a proton-phosphate cotransport system has been previously indicated in mitochondria (Mitchell and Moyle, 1969).

The recognition of cotransport systems involving Cl^- developed from studies of the osmotic behavior of fishes. (1) An active chloride extrusion by fish gills against sizeable osmotic gradients was described in Krogh's laboratory by Keyes (1931) and such a process was then convincingly demonstrated in a variety of tissues by Zadunaisky (1979); an involvement of Na^+ in this process was also seen. (2) The apparent concentration of Cl^- in a variety of cells was substantially higher than that corresponding to a passive (Nernst) distribution. Various suggestions were put forward to explain these sets of phenomena, including a Na^+/Cl^- cotransporter, until Geck et al. (1980) documented an electroneutral cation-anion cotransport system characterized by a stoichiometry of 1 Na^+ : 1 K^+ : 2 Cl^-. Given the properties of most cells, the downhill flux of Na^+ drives the uphill flux of K^+ and 2 Cl^-. The so-called loop-diuretics (*e.g.* furosemide and bumetanide) proved to be relatively selective inhibitors of this transport system. The carrier-mediated, essentially passive character of this quaternary transport system has been amply documented in a variety of cells (Geck and Heinz, 1985) and using membrane vesicles (Kinne et al., 1985). Its coupling with the operation of the sodium pump is then responsible for the phenomenology of active chloride extrusion (or absorption (Chapter 4)). Some modes of operation of this transport system have been described; thus, some anions may replace chloride (*e.g.* taurine or γ-aminobutyrate), and some interchange between the cations also appears feasible (Geck and Heinz, 1985).

Some variations on the theme, including simpler cation-anion symport systems have been identified, *e.g.* a Na^+/Cl^- transporter (Stokes et al., 1984), with thiazides as selective inhibitors; a K^+/Cl^- symport (Kregenow and Caryk, 1979), sensitive to loop diuretics, but with a somewhat different pharmacology; and an electrogenic Na^+/HCO_3^- system, first described by Boron and Boulpaep (1983); this last transport system may be actually more complex, with a stoichiometry of $nNa^+ : CO_3^{2-} : HCO_3^-$ (Soleimani et al., 1991) or $2\,HCO_3^- : 1\,Na^+$ (Deitmer and Schlue, 1989). The now recognized Na^+-dependent Cl^-/HCO_3^- exchange system (Russell et al., 1983) may also belong to this category. Heinz (1984) would consider such systems as modes of the Na^+-K^+, $2Cl^-$ transporter.

3.5. The carrier-mediated coupling of flux systems: tertiary coupling

In analogy to the known chemical coupling of enzyme systems in metabolic chains (with the product of one reaction serving in a cyclic process as substrate for the subsequent one (Green et al., 1934), the developing knowledge of symport and antiport systems led to suggestions (Heinz, 1974) of their possible indirect (tertiary) coupling. Although originally applied to the specific case of the linkage of the carrier system for amino acids to an energy-yielding metabolic reaction, the concept has broader implications. The possibility of such coupling was first demonstrated by Thomas (1977) for the Na^+/H^+ and the HCO_3^-/Cl^- antiports (with the intermediate participation of the $HCO_3^- + H^+ \rightleftharpoons H_2CO_3 \rightleftharpoons CO_2 + H_2O$ reactions), by Cala (1983) when analyzing the KCl efflux from erythrocytes, showing that this process is actually the result of a linkage between a

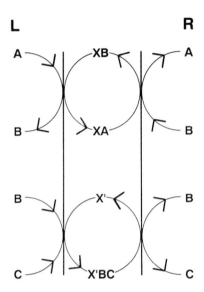

Fig. 3. *Coupling between two carrier systems (antiport A/B and symport B/C), carriers X and X'. Because of this coupling, the phenomenology of the system is that of a A/C symport. However, as opposed to a A/C symport, the system will also display a saturable requirement for B, and will be sensitive to known inhibitors of antiports of the A/B type, or symports of the B/C type.*

K^+/H^+ antiport and a Cl^-/HCO_3^- antiport (Fig. 3), and the driving force was an imposed osmotic gradient; the participation of the anion/anion antiport was demonstrated using the selective stilbene sulfonate reagents. Another, somewhat differing type to coupling between two flux systems has been identified by Schild et al. (1990). Here, the Na^+/H^+ antiport is linked to the anion exchange system by catalytic amounts of formate, which readily passes the membrane as the undissociated formic acid, while the formate anion is the substrate for the $Cl^-/$ anion exchange system: the resultant of this linkage is a unidirectional Na^+/Cl^- flux. In a similar way, a variety of other ionic fluxes might be dissected as a coupling between two (or several) simpler systems, using appropriate inhibitors. Ultimately, the decision between a phenomenologically described sym- or antiport, and an actual coupling between two sym- or antiport systems may have to await isolation of the appropriate protein(s).

III. Towards the molecular mechanisms of carrier-mediated processes

Studies of the D-glucose transporter provided the conceptual approach to its purification, and reconstitution in lipid vesicles (Kasahara and Hinckle, 1977) and its detailed description in molecular terms (Mueckler et al., 1985). Such efforts marked the transition from the kinetic era to a study of the molecular aspects of carrier function. This new chapter in the exploration of membrane carriers was made possible by the availability of high-affinity inhibitors (for the glucose carrier, particularly cytochalasin B), the use of antibodies to the purified transporter, and the application of methods of protein chemistry and molecular biology. The result of this effort in several laboratories led to the first description (Mueckler et al., 1985) of the glucose transporter as an integral membrane glycopeptide of 55 kDa consisting of 492 amino acids, with 12 membrane-spanning intramembranous, alpha-helical domains, each consisting of 21 amino acids. This progress required a reappraisal of existing concepts, and raised new questions, *e.g.*: (a) Is there a heterogeneity of glucose transporters with differing properties (and amino acid sequences); (b)What is the spatial arrangement of the intramembranous domains, and how can the structure of the transporter be reconciled with models derived by kinetic studies; (c) What is the molecular nature of hydrogen bonding between the transported substrate (sugar) and a particular amino acid (or acids) of the intramembranous portion of the peptide; (d) For the cation-dependent sugar transporters, which moiety (acidic amino acid) of the peptide is responsible for such interaction; (e) How can the function of the transporter be modulated (in the light of the activating effects of experimental conditions such as cell activity, hormones, and second messengers).

Heterogeneity of glucose transporters. At first, investigators
(Henderson and Maiden, 1990) were impressed by major homologies of sugar transport proteins (amino acid, as well as
DNA sequences) from a variety of mammalian tissues and bacteria, both for facilitated transport and for some proton-linked
sugar transporters from *E. coli*; hence, the conclusion was
warranted that these sugar transport systems evolved rather
early in the evolutionary process. On the other hand, some
bacterial sugar transport systems differed widely from this
model. Subsequent detailed comparisons of kinetic properties
of glucose transporters from different tissues, as well as their
encoding in the DNAs and RNAs, suggested significant heterogeneity. By 1990, it was recognized that the facilitative glucose transporters represented an expanding family[16] (Gould
and Bell, 1990) in spite of the fact that there are several regions of the molecules in which the sequence of amino acids is
identical. Of particular interest was the finding from several
laboratories (Zorzano et al., 1989) that insulin-sensitive tissues express an unique transport protein.

The tetrameric transport protein for the Na^+/glucose
cotransport (Hediger et al., 1987) displays no homology with
either the mammalian facilitated glucose carrier or the bacterial sugar transport proteins; hence, the mammalian Na^+-
driven transporter has no evolutionary relationship to the
other sugar transporters.

Spatial arrangement of the transporter and kinetic models. In
order to satisfy the models developed on the basis of kinetics
(see above, section 3.2), these domains had to be arranged in a
way to form a cleft capable of interacting with, and passage of,
the substrate with appropriate conformational changes (Fig.
1G). The first such possible model has now been presented
(Widdas and Baker, 1991), with a cleft formed by nine domains

---

16 At present consisting of 7 proteins.

in the open form, and six in the closed form; the rocking of the domains in the course of function would be brought about by only 3 (out of 12) domains. This model also allows for the assumed partial slippage of the carrier, i.e. with conformational changes even in the absence of substrate as visualized in model Fig. 1B.

Hydrogen bonding of substrate and peptide. For the hydrogen bonding between sugar hydroxyls and peptides, aspartate, asparagine, glutamate and glutamine residues appear to play a major role (Quiocho, 1990), at least in the periplasmic sugar-binding protein in bacteria. As to the possible interaction of the carrier with the cation, several anionic groups have been identified in the Na^+-glucose cotransport peptide.

Regulation of the carrier activity. With advancing knowledge on the glucose carriers, the mechanism by which insulin stimulates sugar entry into cells (see Section I) could be analyzed in greater detail (Carruthers, 1990): In 1980 two laboratories (Cushman and Wardzala, 1980; Suzuki and Kono, 1980) reported evidence for an increase of carrier molecules in the insulin-stimulated plasma membranes of adipocytes, and suggested a translocation of intracellular sequestered carrier molecules into membranes[17] (Fig. 4). This novel concept of the hormone's action was extensively explored kinetically: at first, in adipocytes the increased number of glucose carrier molecules in the plasma membrane could not quantitatively account for the increased transport activity. A major argument in favor of the recruitment hypothesis was based on the observation that adipocyte plasma membranes contained two insu-

---

17 As with the reversible translocation of water through specific channels (aggrephores, Chapter 2, Section III.1), an involvement of the cytoskeleton in the translocating process is likely. The concept of the translocation of a carrier between the cytoplasm and the membrane also raises the question of the mechanism by which the hydrophobic portion of the carrier is correctly imbedded within the membrane to be functional. Singer and Yaffe (1990) suggest that this insertion (of large molecules, see Chapter 2) does not occur directly, but requires energy and an organized system of proteins.

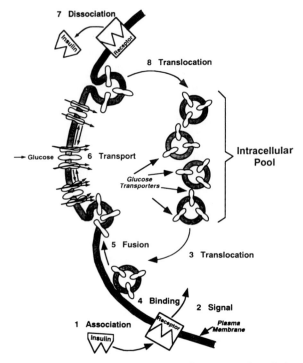

Fig. 4. Carrier translocation between a cytoplasmic pool and the membrane as a regulatory mechanism for solute transport (Karnielli et al., 1981). Reproduced with permission of the authors.

lin-stimulated glucose carriers. While the activity of the erythrocyte-type of glucose carrier was to a minor degree enhanced by insulin, the newly characterized insulin-responsive glucose transporter (Zorzano et al., 1989; James et al., 1989) was the major glucose transporter. On insulin stimulation, this transporter can be translocated from an intracellular, low-density microsomal fraction into the plasma membrane and the operation of this transporter was consistent with the insulin-enhanced rate of glucose transport. However, arguments in favor of an intrinsic effect of insulin on the carrier activity still persist (Baly and Horuk, 1988). The actual mechanism by which insulin stimulates this transporter is now being explored, and the possibility was raised that phosphorylation of a portion of the transporter may be involved. These findings vindicate the imaginative suggestions of Höber (1914) and Häusler and Loewi (1925) that diabetes is, after all, at least in part a transport-related malfunction.

A redistribution of glucose transporters between the plasma membrane and intracellular storage sites may well be the mechanism by which cells react to a variety of stresses, e.g. anaerobiosis, temperature changes or virus infection (Widnell et al., 1990).

So far, little progress has been made on the molecular mechanisms involved in hormone effects on other carrier-mediated processes, e.g. phosphate transport or the Na^+/H^+ symporter (Sacktor and Kinsella, 1985). A variety of ion transport systems is affected (stimulated or inhibited) by cyclic AMP (Kim, 1991).

IV. Questions, questions

The now mushrooming new knowledge about the carriers, their amino acid sequence, as well as their organization in cell membranes begins to answer some old question and raise new ones. An increasing number of carrier proteins has been isolated, their amino acid sequence determined and the systems were reconstituted. A better understanding of the molecular mechanisms involved in the carrier function will now require an elucidation of the three-dimensional structure of the transport proteins in a membrane. Such studies should also reveal the intrinsic aspects differentiating between the structural arrangements of the hydrophobic portions of peptides responsible for either an ionic channel, the cleft responsible for facilitated diffusion or active transport; the possibility can not be excluded that it is the nature (or magnitude) of the conformational changes occurring in the course of solute-protein interaction which is a determinant here. A recent analysis (Rini et

al., 1992) demonstrated an induced conformational arrangement in the course of antibody-antigen recognition, while in
other protein-protein (protein-solute?) interactions the respective conformational changes may be rather small.

One of the basic pertinent questions relates to the nature of
the driving forces for a simple carrier system at the molecular
level. Mitchell (1962) visualized a role of the intrinsic asymmetry of carrier as the basis for the difference between the
scalar nature of enzymic reactions, and the vectorial flow of
solutes catalyzed by membrane carriers. Widdas and Baker
(1991) have just raised the possibility of the surface energy of
water as a driving force. New information on the nature and
role of the asymmetry of transport proteins will be revealing.

Crane (1986) raised a number of pertinent questions concerning the nature of the Na-dependence of the Na^+-organic
solute cotransport systems. As time and techniques progress,
details as to the (hydrogen) bonding between the transported
solute and specific amino acid groups of the transporter are
emerging; this implies also further knowledge about the respective sites of interaction of solute and cotransported cation.
At the same time, new questions can be raised, e.g. the nature
of interaction between the transporter and the embedding
lipid membrane (and its possible role in the function of the
transporter, cf. Chapter 2). The exploration of the mechanism
of regulation of transport phenomena at the level of carriers
continues to be pertinent. The rather complex phenomenology
of some cotransport systems, e.g. the quaternary $Na^+/K^+/2 Cl^-$
system (section 3.4) invites the inquiry into the possibility
that the process may be dissected into simpler components,
e.g. as a sequence of events on a single protein molecule (Stein,
1986).

The molecular approach to the carrier concept (and appropriate techniques) heralds a new era which will be reflected by
a new chapter on this topic. Indications of such new emerging
facets are the observations from two laboratories (Kim et al.,
1991; Wang et al., 1991) that the mammalian cell membrane

receptor for a virus is actually the membrane transporter of cationic amino acids. Increasingly detailed information on transport proteins also raises questions as to the simple categorization of transport processes as visualized by Danielli Chapter 2, p. 47). Thus, the peptide of the chloride channel requires ATP hydrolysis for the maintenance of its activity (Chapter 2): the actual transport process still appears to be Cl⁻ diffusion through the channel in spite of the (indirect) involvement of metabolic energy.

Acknowledgements

The author is indebted to Drs. H.N. Christensen, S.W. Cushman, S. Erulkar, P.D. Mitchell, R.L. Post and W.F. Widdas for their valuable comments and help.

REFERENCES

Appleman, J.R. and Lienhard, G.E. (1985) Rapid kinetics of the glucose transporter from human erythrocytes. Detection and measurement of a half-turnover of the purified transporter. J. Biol. Chem. 260, 4575–4578.
Aronson, P.S. (1981) Identifying secondary active solute transport in epithelia. Am. J. Physiol. 240, F1–F11.
Barnett, J.E.G., Ralph, A. and Munday, K.A. (1970) The mechanism of active intestinal transport of sugars. Biochem. J. 116, 537–538.
Barnett, J.E.G., Holman, G.D., Chalkley, R.A. and Munday, K.A. (1975) Evidence for two asymmetric conformational states in the human erythrocyte sugar transport system. Biochem. J. 145, 417–429.
Beck, J.C. and Sacktor, B. (1975) Energetics of the Na⁺-dependent transport of D-glucose in renal brush border membrane vesicles. J. Biol. Chem. 250, 8674–8680.
Blakely, K.R. and Boyer, P.D. (1955) The effect of 6-deoxy-6-fluoro-D-glucose on yeast fermentation and hexokinase. Biochim. Biophys. Acta 16, 576–582.
Blaustein, M.P. and Nelson, M.T. (1982) Sodium-calcium exchange: its role in the regulation of cell calcium. In *Membrane Transport of Calcium* (Carafoli, E., ed.), pp. 217–236. London, Academic Press.
Boron, W. and Boulpaep, E.J. (1983) Intracellular pH regulation in the renal proximal tubule of the Salamander. J. Gen. Physiol. 81, 53–94.
Cabantchik, Z.I. and Rothstein, A. (1972) The nature of the membrane sites

controlling anion permeability of human red blood cells as determined by studies with disulfonic stilbene derivatives. J. Membrane Biol. 10, 311–320.

Cala, P.M. (1986) Volume-sensitive alkali metal-H transporter in *Amphiuma* red blood cells. Curr. Top. Membr. Transp. 26, 79–99.

Carruthers, A. (1990) Facilitated diffusion of glucose. Physiol. Rev. 70, 1135–1176.

Cass, A. and Dalmark, M. (1973) Equilibrium dialysis of ions in nystatin-treated red cells. Nature (Lond.), New Biol. 244, 47–49.

Chappell, J.B., Cohn, M. and Greville, G.D. (1963) The accumulation of divalent ions by isolated mitochondria. In *Energy-linked Functions of Mitochondria* (Chance, B., ed.) pp. 219–231. New York, Academic Press.

Chen, C.-C., Tsuchiya, T., Yamane, Y., Wood, J.M. and Wilson, T.H. (1985) Na⁺ (Li⁺)-proline cotransport in *Escherichia coli*. J. Membrane Biol. 84, 157–164.

Christensen, H.N. (1969) Some special kinetic problems of transport. Adv. Enzymol. 32, 1–20.

Christensen, H.N. (1979) Exploiting amino acid structure to learn about membrane transport. Adv. Enzymol. 49, 41–101.

Christensen, H.N. (1984) Organic ion transport during seven decades: The amino acids. Biochim. Biophys. Acta 779, 255–269.

Christensen, H.N. and Riggs, T.R. (1952) Concentrative uptake of amino acids by the Ehrlich mouse ascites carcinoma cells. J. Biol. Chem. 194, 57–68.

Clark, A.J. (1937) *General Pharmacology*. Handb. d. exp. Pharmakol., Vol. IV (Heubner, W. and Schüller, eds.). Berlin, J. Springer.

Cohen, G.N. and Monod, J. (1957) Bacterial permeases. Bacteriol. Rev. 21, 169–194.

Conway, E.J. and O'Malley, E. (1946) The nature of the cation exchanges during yeast fermentation, with formation of 0.01 N H-ion. Biochem. J. 40, 59–67.

Cori, C.F. (1925) The fate of sugar in the animal body. I. The rate of absorption of hexoses and pentoses from the intestinal tract. J. Biol. Chem. 66, 691–715.

Cori, C.F. (1926) Competition between sugars for sugar absorption in the intestine. Proc. Soc. exp. Biol. N.Y. 23, 290–291.

Crane, R.K. (1960) Intestinal absorption of sugars. Physiol. Rev. 40, 789–825.

Crane, R.K., Miller, D. and Bihler, I. (1961) The restrictions on the possible mechanism of intestinal active transport of sugars. In *Membrane Transport and Metabolism* (Kleinzeller, A. and Kotyk, A., eds.), pp. 439–449. New York, Academic Press.

Crane, R.K. (1977) The gradient hypothesis and other models of carrier-mediated active transport. Rev. Physiol. Biochem. Pharmacol. 78, 101–159.

Crompton, M. (1985) The regulation of mitochondrial Ca^{2+} transport in heart. Curr. Top. Membr. Transp. 25, 231–276.

Csáky, T. (1942) Über die Rolle der Struktur des Glucosemoleküls bei der Resorption aus dem Dünndarm. Hoppe-Seyler's Z. physiol. Chem. 277, 47–57.

Dalmark, M. (1975) Chloride and water distribution in human red cells. J. Physiol. (Lond.). 250, 65–84.

Dalmark, M. and Wieth, J. (1972) Temperature dependence of chloride, bromide, iodide, thiocyanate, and salicylate transport in human red cells. J. Physiol. (Lond.) 244, 583–610.

Daly, D.L. and Horuk, R. (1988) The biology and biochemistry of the glucose transporter. Biochim. Biophys. Acta 947, 571–590.

Danielli, J.F. (1954) The present position in the field of facilitated diffusion and selective active transport. Colston Symp. 7, 1–14.

Davson, H. and Danielli, J.F. (1943) *The Permeability of Natural Membranes.* Cambridge, University Press.

Davson, H. and Reiner, J.M. (1942) Ionic permeability: An enzyme-like factor in the cat erythrocyte membrane. J. Cell Comp. Physiol. 20, 325–344.

Deitmer, J.W. and Schlue, W.-R. (1989) An inversely directed electrogenic sodium-bicarbonate cotransport in leech glial cells. J. Physiol. (Lond.) 411, 179–194.

Eddy, A.A. (1978) Proton-dependent solute transport in microorganisms. Curr. Top. Membr. Transp. 10, 280–360.

Ege, R. (1919) Thesis, Copenhagen. cf. Bang, O. and Ørskov, S.L. (1937) Variations in the permeability of the red blood cells in man, with particular reference to the conditions obtaining in pernicious anemia. J. Clin. Invest. 16, 279–286.

Fisher, R.B. and Parsons, D.S. (1953) Glucose movements across the wall of the small intestine. J. Physiol. (Lond.). 119, 210–224.

Fox, C.F. and Kennedy, E.P. (1965) Specific labelling and partial purification of the M protein, a component of the β-galactoside transport system in *Escherichia coli.* Proc. Natl. Acad. Sci. USA 54, 891–899.

Garcia-Romeu, F. (1971) Anionic and cationic exchange mechanism in the skin of anurans, with special reference to *Leptodactylidae* in vivo. Phil. Trans. R. Soc. Lond. B. 262, 163–174.

Geck, P. (1971) Eigenschaften eines asymmetrischen Carrier Modells für den Zuckertransport am menschlichen Erythrozyten. Biochim. Biophys. Acta 241, 462–472.

Geck, P. and Heinz, E. (1985) The Na-K-2Cl cotransport system. J. Membrane Biol. 91, 97–105.

Geck, P., Pietrzyk, C., Burckhardt, B.-C., Pfeiffer, B. and Heinz, E. (1980) Electrically silent cotransport of Na$^+$, K$^+$, 2Cl$^-$ in Ehrlich cells. Biochim. Biophys. Acta 600, 432–447.

Glynn, I.M. and Karlish, S.J.D. (1975) Different approaches to the mechanism of the sodium pump. Ciba Fndn. Symp. 31, 205–220.

Goto, K. (1918) Mineral metabolism in experimental acidosis. J. Biol. Chem. 36, 355–376.

Gould, G.W. and Bell, G.I. (1990) Facilitative glucose transporters: an expanding family. Trends Biochem. Sci. 15, 18–25.

Green, D.E., Stickland, L.H. and Tarr, H.L. (1934) Studies on reversible dehydrogenase systems III. Carrier-linked reactions between isolated dehydrogenases. Biochem. J. 28, 1812–1824.

Gunn, R.B. (1972) A titratable carrier model for both mono- and divalent anion transport in human red blood cells. In *Oxygen Affinity of Hemoglobin and Red Cell Acid-Base Status* (Rorth, M. and Astrup, P., eds.). pp. 823–827. Copenhagen, Munksgaard.

Gunn, R.B., Dalmark, M., Tosteson, D.C. and Wieth, J. (1973) Characteristics of chloride transport in human red blood cells. J. Gen. Physiol. 61, 185–2.

Halestrap, A.P. (1976) Transport of pyruvate and lactate in human erythrocytes. Evidence for the involvement of the chloride carrier and a chloride-independent carrier. Biochem. J. 156, 193–207.

Harris, E.J. and Pressman, B.C. (1967) Obligate cation exchanges in red cells. Nature (Lond.) 216, 918–920.

Häusler, H. and Loewi, O. (1925) Zur Frage der Wirkungsweise des Insulins. I. Mitt. Insulin und die Glukoseverteilung zwischen flüssigen und nicht- flüssigen Systemen. Pflüger's Arch. ges. Physiol. 210, 238–279.

Hediger, M.A., Coady, M.J., Ikeda, T.S. and Wright, E.M. (1987) Expression cloning and cDNA sequencing of the Na^+/glucose co-transporter. Nature (Lond.). 330, 379–381.

Heinz, E. (1954) Kinetic studies on the influx of glycine-1-C^{14} into the Ehrlich mouse ascites carcinoma cell. J. Biol. Chem. 211, 781–790.

Heinz, E. (Ed.). (1972) *Na-linked Transport of Organic Solutes*. Berlin, Springer Verl.

Heinz, E. (1974) Coupling and energy transfer in active amino acid transport. Curr. Top. Membr. Transp. 5, 137–159.

Henderson, P.J.F. and Maiden, M.C.J. (1990) Homologous sugar transport proteins in *Escherichia coli* and their relatives in both prokaryotes and eukaryotes. Phil. Trans. R. Soc. Lond. B. 326, 391–410.

Hladky, S.B. (1979) The carrier mechanism. Curr. Top. Membr. Transp. 14, 332–425.

Höber, R. (1899) Über Resorption im Dünndarm. Pflüger's Arch. ges. Physiol. 74, 246–271.

Höber, R. (1911) *Physikalische Chemie der Zelle und Gewebe*. 3. Aufl. Leipzig, Engelmann, pp. 264.

Höber, R. (1914) Nachwort. Biochem. Z. 60, 253–256.

Hodgkin, A.L., Huxley, A.F. and Katz, B. (1949) Ionic currents underlying activity in the giant axon of the squid. Arch. Sci. Physiol. (Paris) 3, 129–150.

Hoffmann, N., Thees, M. and Kinne, R. (1976) Phosphate transport by isolated renal brush border vesicles. Pflüger's Arch. ges. Physiol. 362, 147–156.

Hopfer, U. and Grosclose, R. (1980) The mechanism of Na⁺-dependent-D-glucose transport. J. Biol. Chem. 255, 4453–4462.

Jacobs, M.H. (1931) The permeability of the erythrocyte. Erg. d. Biol. 7, 1–55.

Jacobs, M.H. and Corson, S.A. (1934) The influence of minute traces of copper on certain hemolytic processes. Biol. Bull. 67, 325–326.

Jacobs, M.H., Glassman, H.N. and Parpart, A.K. (1935) Osmotic properties of the erythrocyte. VII. The temperature coefficients of certain hemolytic processes. J. Cell. Comp. Physiol. 7, 197–225.

Jacobs, M.H. and Stewart, D.R. (1942) The role of carbonic anhydrase in certain ionic exchanges involving the erythrocyte. J. Gen. Physiol. 25, 539–552.

James, D.E., Strube, M. and Mueckler, M. (1989) Molecular cloning and characterization of an insulin-regulatable glucose transporter. Nature (Lond.) 338, 83–87.

Jauch, P. and Läuger, P. (1988) Kinetics of the Na⁺/alanine cotransporter in pancreatic acinar cells. Biochim. Biophys. Acta 939, 179–188.

Kaback, H.R. (1986) Active transport in *Escherichia coli*: passage to permease. Annu. Rev. Biophys. 15, 279–319.

Karnielli, E., Zarnowski, M.S., Hissin, P.J., Simpson, I.A., Salans, L.B. and Cushman, S.W. (1981) Insulin-stimulated translocation of glucose transport systems in the isolated rat adipocyte cell. J. Biol. Chem. 256, 4772–4777.

Keys, A.B. (1931) Chloride and water secretion and absorption by the gills of the eel. Z. vergl. Physiol. 15, 364–388.

Kim, H.D. (1991) Ion transport and adenylyl cyclase system in red blood cells. Curr. Top. Membr. 39, 181–225.

Kim, J.W., Closs, E.I., Albritton, L.M. and Cunningham, J.M. (1991) Transport of cationic amino acids by the mouse ecotropic retrovirus receptor. Nature (Lond.) 352, 725–728.

Kinne, R.K. (1976) Properties of the glucose transport system in the renal brush border membrane. Curr. Top. Membr. Transp. 8, 209–267.

Kinne, R., Koenig, B., Hannafin, J., Kinne-Saffran, E., Scott, D.M. and Zierold, K. (1985) The use of membrane vesicles to study the NaCl/KCl cotransporter involved in active transepithelial chloride transport. Pflüger's Arch. ges. Physiol. 405 (Suppl. 1), 101–105.

Kipnis, D.M. (1959) Regulation of glucose uptake by muscle: Functional significance of permeability and phosphorylating activity. Ann. N.Y. Acad. Sci. 82, 354–365.

Kleinzeller, A. and Kotyk, A. (1963) Transport of sugars into yeast cells and the regulation of metabolism. In: *Colloque Internat. sur les Méchanismes de Régulation*, pp. 373–377. Marseille, CNRS.

Kleinzeller, A., McAvoy, E.M. and McKibbin, R.D. (1980) Active renal hexose transport. Structural requirements. Biochim. Biophys. Acta 600, 513–529.

Knauf, P.A. (1979) Erythrocyte anion exchange and the band 3 protein: Transport kinetics and molecular structure. Curr. Top. Membr. Transp. 12, 251–263.

Kornberg, H.L. (1973) Fine control of sugar uptake by *Escherichia coli*. Symp. Soc. exp. Biol. 27, 175–193.

Kozawa, S. (1914) Beiträge zum arteigenen Verhalten der roten Blutkörperchen. III. Artdifferenzen in der Durchlässigkeit der roten Blutkörperchen. Biochem. Z. 60, 231–253.

Kozawa, S. and Miyamoto, N. (1921) Note on the permeability of the red corpuscles for amino acids. Biochem. J. 15, 167–170.

Kregenow, F.M. and Caryk, T. (1979) Cotransport of cations and Cl⁻ during volume regulatory responses of duck erythrocytes. The Physiologist 22, 73.

Krulwich, T.A. (1983) Na⁺/H⁺ antiporters. Biochim. Biophys. Acta 726, 245–264.

LeFevre, P.G. (1948) Evidence for active transfer of certain non-electrolytes across the human red cell membrane. J. Gen. Physiol. 31, 505–527.

LeFevre, P.G. and Davies, R.I. (1951) Active transport into human erythrocytes. Evidence from comparative kinetics and competition among monosaccharides. J. gen. Physiol. 34, 515–524.

LeFevre, P.G. and LeFevre, M.E. (1952) The mechanism of glucose transfer into and out of the human red cell. J. Gen. Physiol. 35, 891–906.

LeFevre, P.G. (1975) The present state of the carrier hypothesis. Curr. Top. Membr. Transp. 7, 109–215.

Levine, R., Goldstein, M.S., Huddlestun, B. and Klein, S.P. (1950) Action of insulin on the 'permeability' of cells to free hexoses, as studied by its effect on the distribution of galactose. Amer. J. Physiol. 163, 70–76.

Ling, G.N. (1984) *In Search of the Physical Basis of Life*. New York and London, Plenum Press.

Lundegårdh, H. (1940) Investigations as to the absorption and accumulation of inorganic ions. Ann. Agric. Coll. Sweden 8, 234–404.

Lundsgaard, E. (1939) On the mode of action of insulin. Uppsala Läksareforen Förh. 45, 143–152.

Maloney, P.C. (1990) Resolution and reconstitution of anion exchange reactions. Phil. Trans. R. Soc. Lond. B. 326, 437–454.

Mitchell, P.D. (1956) Faraday Soc. Discussion of Membrane Phenomena 21, p. 279.

Mitchell, P. (1957) A general theory of membrane transport from studies in bacteria. Nature (Lond.) 180, 134–136.

Mitchell, P.D. (1962) Molecule, group, and electron translocation through natural membranes. Biochem. Soc. Symp. 22, 142–168.

Mitchell, P.D. (1992) Foundations of vectorial metabolism and osmochemistry. Bioscience Reports 11, 297–344.

Mitchell, P.D. and Moyle, J. (1959) Translocation of some anions, cations and acids in rat liver mitochondria. Eur. J. Biochem. 9, 149–155.

Mitchell, P.D. and Moyle, J. (1967) Respiration-driven proton translocation in rat liver mitochondria. Biochem. J. 105, 1147–1162.

Mueckler, M., Caruso, C., Baldwin, S.A., Panico, M., Blench, I., Morris, H.R., Allard, W.J., Lienhard, G.E. and Lodish, H.F. (1985) Sequence and structure of a human glucose transporter. Science 229, 941–945.

Murer, H., Hopfer, U. and Kinne, R. (1976) Sodium/proton antiport in brush-border-membrane vesicles isolated from rat small intestine and kidney. Biochem. J. 154, 597–604.

Naftalin, R.J. (1970) A model for sugar transport across red cell membranes without carriers. Biochim. Biophys. Acta 211, 65–78.

Nagano, J. (1902) Zur Kenntniss der Resorption einfacher, im besonderen stereoisomerer Zucker im Dünndarm. Pflüger's Arch. ges. Physiol. 90, 389–404.

Nasse, H. (1878) Untersuchungen über den Austritt und Eintritt von Stoffen (Transsudation und Diffusion) durch die Wand der Haargefässe. Arch. ges. Physiol. 16, 604–634.

Osterhout, W.J.V. (1933) Permeability in large plant cells and in models. Erg. Physiol. 35, 967–1021.

Overton, E. (1895) Uber die osmotischen Eigenschaften der lebenden Pflanzen und Tierzelle. Vierteljahrsschr. d. Naturf. Ges. Zürich 40, 159–201.

Park, C.R., Bornstein, J. and Post, R.L. (1955) Effect of insulin on free glucose content of rat diaphragm in vitro. Am. J. Physiol. 182, 12–16.

Park, C.R., Post, R.L., Kalman, C.F., Wright, J.H. Jr., Johnson, L.H. and Morgan, H.E. (1956) The transport of glucose and other sugars across cell membranes and the effect of insulin. Ciba Found. Colloq. Endocrinol. 9, 240–256.

Passow, H. (1969) Passive ion permeability of the erythrocyte membrane. Progr. Biophys. Mol. Biol. 19, 425–467.

Passow, H. (1986) Molecular aspects of band 3 protein-mediated anion transport across the red blood cell membrane. Rev. Physiol. Biochem. Pharmacol. 103, 61–223.

Passow, H., Rothstein, A. and Clarkson, T.H. (1961) The general pharmacology of the heavy metals. Pharmacol. Rev. 13, 185–224.

Patlak, C.S. (1957) Contributions to the theory of active transport: II. The gate type non-carrier mechanism and generalizations concerning tracer flow, efficiency, and measurement of energy expenditure. Bull. Math. Biophys. 19, 209–235.

Pitts, R.F. and Alexander, R.S. (1948) The nature of the renal tubular mechanism of acidifying the urine. Am. J. Physiol. 144, 239–254.

Quiocho, F.A. (1990) Atomic structures of periplasmic binding proteins and the high affinity active transport systems in bacteria. Phil. Trans. R. Soc. Lond. B 326, 341–351.

Randle, P.J. and Smith, G.H. (1958) Regulation of glucose uptake by muscle:

The effects of insulin, anaerobiosis and cell poisons on the penetration of isolated rat diaphragm by sugars. Biochem. J. 70, 501–508.

Regen, D.M. and Tarpley, H.L. (1974) Anomalous transport kinetics and the glucose carrier hypothesis. Biochim. Biophys. Acta 339, 218–233.

Rini, J.M., Schulze-Gahmen, U. and Wilson, I.A. (1992) Structural evidence for induced fit as a mechanism for antibody-antigen recognition. Science 255, 959–965.

Roos, A. and Boron, W.F. (1981) Intracellular pH. Physiol. Rev. 61: 296–434.

Roseman, S. (1969) The transport of carbohydrate by a bacterial phosphotransferase system. J. Gen. Physiol. 54, 138s–180s.

Rosenberg, T. (1954) The concept and definition of active transport. Symp. Soc. exp. Biol. 8, 27–41.

Rosenberg, T. and Wilbrandt, W. (1954) Enzymic processes in cell membrane penetration. Symp. Soc. exp. Biol. 8, 65–92.

Rosenberg, T. and Wilbrandt, W. (1957) Uphill transport induced by counterflow. J. Gen. Physiol. 41, 289–296.

Rothstein, A. and Enns, L.H. (1946) The relationship of potassium and carbohydrate metabolism in baker's yeast. J. Cell. Comp. Physiol. 28, 231–252.

Russell, J.M., Boron, W.F. and Brodwick, M.S. (1983) Intracellular pH and ion fluxes in barnacle muscle with evidence for reversal of the ionic mechanism by intracellular pH regulation. J. Gen. Physiol. 82, 47–78.

Sacktor, B. and Kinsella, J.L. (1985) Hormonal effects on sodium cotransport systems. Ann. N.Y. Acad. Sci. 456, 438–444.

Schild, L., Aronson, P.S. and Giebisch, G. (1990) Effects of apical membrane Cl^--formate exchange on cell volume in rabbit proximal tubule. Am. J. Physiol. 258, F530–F536.

Schultz, S.G. and Curran, P.F. (1970) Coupled transport of sodium and organic solutes. Physiol. Rev. 50, 637–718.

Segel, I.H. (1975) Enzyme Kinetics. New York, J. Wiley.

Shannon, J.A. (1938) The tubular reabsorption of xylose in the normal dog. Am. J. Physiol. 122, 775–781.

Singer, S.J. and Yaffe, M.P. (1990) Embedded or not? Hydrophobic sequences and membranes. Trends Biochem. Sci. 15, 369–373.

Sokol, P.P., Holohan, P.D., Grass, S.M. and Ross, C.R. (1988) Proton-coupled organic cation transport in renal brush border membrane vesicles. Biochim. Biophys. Acta 940, 209–218.

Soleimani, M., Lesoine, G.A., Bergman, J.A. and Aronson, P.S. (1991) Cation specificity and modes of the $Na^+ : CO_3^{2-} : HCO_3^-$ cotransporter in renal basolateral membrane vesicles. J. Biol. Chem. 266, 8705–8710.

Steck, T.L. (1978) The band 3 protein of the human red cell membrane. A review. J. Supramol. Struct. 8, 311–324.

Stein, W.D. (1964) Facilitated diffusion. Recent Progr. in Surface Science 1, 200–337.

Stein, W.D. (1986) *Transport and Diffusion across Cell Membranes*. New York, Academic Press.

Stein, W.D. and Danielli, J.F. (1956) Structure and function in red cell permeability. Discuss. Faraday Soc. 21, 238–251.

Stein, W.D. (1986) Intrinsic, apparent, and effective affinities of co- and countertransport systems. Am. J. Physiol. 250, C523–C533.

Stokes, J.B., Lee, I. and D'Amico, M. (1984) Sodium chloride absorption by the urinary bladder of the winter flounder. A thiazide-sensitive, electrically neutral transport system. J. Clin. Invest. 74, 7–16.

Suzuki, K. and Kono, T. (1980) Evidence that insulin causes translocation of glucose transport activity to the plasma membrane from an intracellular storage site. Proc. Natl. Acad. Sci. USA 77, 2542–2545.

Tanner, M.J.A. (1979) Isolation of integral membrane proteins and criteria for identifying carrier proteins. Curr. Top. Membr. Transp. 12, 1–51.

Teorell, T. (1939) On the permeability of the stomach mucosa for acids and some other substances. J. Gen. Physiol. 23, 263–274.

Thomas, R.C. (1977) The role of bicarbonate, chloride and sodium in the regulation of intracellular pH in snail neurones. J. Physiol. (Lond.) 273, 317–338.

Tosteson, D.C. (1959) Halide transport in red blood cells. Acta physiol. scand. 48, 19–41.

Turner, R.J. (1983) Quantitative studies of cotransport systems: Models and vesicles. J. Membrane Biol. 76, 1–15.

Ullrich, K.J., Rumrich, G., Fritzsch, G. and Klöss, S. (1987) Contraluminal paraaminohippurate (PAH) transport in the proximal tubule of the rat kidney. II. Specificity: aliphatic dicarboxylic acids. Pflüger's Arch. ges. Physiol. 408, 38–45.

Ussing, H.H. (1952) Some aspects of the application of tracers in permeability studies. Adv. Enzymol. 13, 63–93.

Vidaver, G.A. (1964) Some tests of the hypothesis that the sodium-ion gradient furnishes the energy for glycine-active transport by pigeon red cells. Biochemistry 3, 803–808.

Vidaver, G.A. (1966) Inhibition of parallel flux and augmentation of counter flux shown by transport models not involving a mobile carrier. J. Theoret. Biol. 10, 301–306.

Wang, H., Kavanaugh, M.P., North, R.A. and Kabat, D. (1991) Cell-surface receptor for ecotropic murine retroviruses is a basic amino acid transporter. Nature (Lond.) 352, 729–731.

West, I.C. (1970) Lactose transport coupled to proton movements. Biochem. Biophys. Res. Commun. 41, 655–661.

Westenbrink, H.G.K. (1937) Über die Spezifizität der Resorption einiger Monosen aus dem Darme der Ratte und der Taube. Arch. Néerl. Physiol. 21, 433–454.

Widdas, W.F. (1952) Inability of diffusion to account for placental glucose transfer in the sheep and considerations of the kinetics of a possible carrier transfer. J. Physiol. (Lond.) 118, 23–39.

Widdas, W.F. (1954) Facilitative transfer of hexoses across the human erythrocyte membrane, J. Physiol. (Lond.) 125, 163–180.

Widdas, W.F. (1980) The asymmetry of the hexose transfer system in the human red cell membrane. Curr. Top. Membr. Transp. 14, 165–223.

Widdas, W.F. (1989) Old and new concepts of the membrane transport for glucose in cells. Biochim. Biophys. Acta 947, 385–404.

Widdas, W.F. and Baker, G.F. (1991) The role of the surface energy of water in the conformational changes of the human erythrocyte glucose transporter. Cytobios 66, 179–204.

Wilbrandt, W. (1954) Secretion and transport of non-electrolytes. Symp. Soc. Exp. Biol. 8, 136–162.4.

Wilbrandt, W., Guensberg, E. and Lauener, H. (1957) Der Glukoseeintritt durch die Erythrocytenmembran. Helv. Physiol. Pharmacol. Acta 5, C20–C22.

Wilbrandt, W. and Rosenberg, T. (1961) The concept of carrier transport and its corollaries in pharmacology. Pharmacol. Rev. 13, 109–183.

Wood, R.E., Wirth, E.P. Jr. and Morgan, H.E. (1968) Glucose permeability of lipid bilayer membranes. Biochim. Biophys. Acta 163, 171–178.

Zadunaisky, J.A. (1979) Characteristics of chloride secretion in some intestinal epithelia. In Mechanisms of Intestinal Secretion (Binder, H.J., ed.) pp. 53–64. New York, Alan Liss.

Zorzano, A., Wilkinson, W., Kotliar, N., Thoidis, G., Wadzinski, B.E., Ruoho, A.E. and Pilch, P.F. (1989) Insulin-regulated glucose uptake in rat adipocytes is mediated by two transporter isoforms present in at least two vesicle populations. J. Biol. Chem. 264, 12358–12363.

A. Kleinzeller (Editor) A History of Biochemistry: Exploring the Cell Membrane.
(Comprehensive Biochemistry Vol. 39) © 1995 Elsevier Science B.V.

Chapter 4

The Concept of a Solute Pump

A. KLEINZELLER

Department of Physiology, School of Medicine, The University of
Pennsylvania, Philadelphia, PA, U.S.A.

I. The perception of the phenomenon

The concept of 'active transport' (or a pump) in animal cells developed on the basis of spirited discussions between physiologists in the period 1860–1900 on the mechanism of absorptive and secretory processes[1]. Although other possible indications of such process, *i.e.* the unequal distribution of electrolytes between cells and their immediate environment was already known around 1850, they became the focus of interest only some 100 years later.

Ludwig (1861) in his *Textbook of Human Physiology* sum-

---

1 As so often, physiologists working with animal cells disregarded work on plants. Hales (1727) demonstrated that plant roots generated a flow of sap against considerable pressure gradients. Hales' ideas on that phenomenon: 'Tho' vegetables have not an engine, which, by its alternate dilatations and contractions, does forcibly drive blood through the arteries and veins; yet has nature wonderfully contrived other means, most powerfully to raise and keep in motion the sap' (p. 76 *ff.*). And he then quotes I. Newton's observation that 'by the same (mechanical) principle, a sponge sucks in water, and the glands....suck in various juices from the blood'. This concept was then pushed further by plant physiologists: Nägeli in 1855 suggested that sap flow might be due to a polarity of osmotic properties of plant cell 'membranes' (see Chapter 2) with an involvement of cell metabolism, and later (Sachs, 1865) postulated sap formation and flow as a transcellular flow of water and salts, produced by diffusional forces due to the existing high concentration of salts in the cells, and unequal resistance to flow (we would now say permeability) at both poles of the absorbing rootlet cells. Sachs clearly stated that this phenomenon reflects a steady state of living cells, sap flow being abolished by cell death.

marized the results of his studies on the salivary gland which demonstrated that the driving force for the nerve-stimulated saliva secretion could not be the blood pressure (*i.e.* filtration of water and solutes from the blood into the glandular lumen); rather, the gland itself developed a driving force which exceeded the capillary pressure and was independent of the same. The participation of tissue oxidative processes in the secretion was also suggested by Cl. Bernard's observation of differences in the blood color flowing out of the resting and stimulated glands. Since 1874, Heidenhain and his school emphasized that the active role of cells involved in secretory (kidney, 1874) and absorptive (intestine, 1894) processes characterized all glandular tissues. As opposed to a vitalistic view, he considered hitherto unknown physical and chemical changes in cells as the basis for phenomena such as mobility or secretion. His views and experimental results contested simple physical concepts of intestinal absorption by diffusion: Under in vivo conditions with preserved blood circulation, intestinal loops absorbed serum (of the same osmotic pressure as the plasma of the experimental animal) and also hypotonic solutions of NaCl. Both these absorptive processes proceeded against the prevailing osmotic gradients. Heidenhain's student Röhmann also demonstrated the complete absorption of glucose solutions from the intestinal loop. Such observations were not consistent with predictions of simple diffusion. Overton (1896) clearly postulated the role of metabolic processes in epithelial cells for the up-hill transport of solutes in the kidney and intestine. Another clear indication of the selective role of cells in the intestinal absorptive process was provided by the little known report of Tappeiner (1878) who demonstrated a highly selective absorption of bile salts (cholate, glycocholate and taurocholate) in different sections of the small intestine. Heidenhain's views were vigorously contested by Hamburger who invoked intraabdominal pressure as the physical driving force for absorption. However, further support for the participation of intestinal mucosal cells was provided by Cohnheim

(1898) who showed that while in living dogs the intestine absorbed water, glucose, and salts, the dead intestine behaved like any dead membrane with fluid movement following established laws of diffusion and osmosis. This was true, whether the epithelial layer died on its own (24 h) or by heat treatment. With a rare insight, Cohnheim postulated that resorption is produced by an easy permeability of the intestinal epithelium in one direction, and the absence of permeability in the opposite direction; in effect, this would be analogous to a concept of a mechanistic valve[2]. Three years later, Cohnheim (1901) provided still more convincing evidence for 'active water reabsorption' by demonstrating that the net absorption of sea water from intestinal sacs of the sea cucumber was blocked by inhibition of cell metabolism by anoxia, or poisons such as chloroform or NaF.

Anglo-Saxon investigators are occasionally teased about their neglect of 'foreign' scientific literature. However, 100 years ago, the pioneering work of the English physiologist E.W. Reid was for a long time not acknowledged by the then dominant German school of physiology. Reid (1890, 1892) used an in vitro approach to study absorption (and secretion), having devised an apparatus in which he clamped a strip of intestine or frog skin to separate two isotonic bathing solutions of identical composition; this was a respectable forerunner of the well-known Ussing chamber (*cf.* Chapter 7, Fig. 1A). By measuring volume changes he demonstrated a net transfer of fluid from the 'luminal' to the 'serosal' side, corresponding to absorption[3]. A reversal of this process was obtained when the tissue was poisoned with pilocarpine: here, a net fluid transfer from the outer to the inner side was observed, resembling in-

---

2 Such concept (Klappenvorrichtung) was suggested even earlier by Pfeffer (1877) and then by Overton (1899) when discussing the unilateral transport of water in tadpoles.

3 Reid was also the first to demonstrate that the absorption of saline from the pond side of frog skin is inhibited by the cardiac glycoside, digitalin. This observation later became the most powerful tool to study the sodium pump (see p. 153).

testinal secretion. Neither fluid transfer can be accounted for by osmosis, and hence Reid's conclusion that 'positive evidence was obtained of a vital absorptive action in the intestine'. The qualitative description of these active absorptive and secretory processes also revealed that physical processes (diffusion, filtration, etc.) cannot be neglected as components. In this criticism of Höber's view postulating intestinal absorption of salts solely by an intercellular pathway, Reid (1900) concluded that

> 'one is inclined to think that the cement (between epithelial cells) is permeable to sodium chloride in both directions, but only to a slight degree, while it is a peculiarity of living cells to be permeable in only one direction'.

Thus, future investigators had to shoulder the additional burden of dissecting the quantitative contributions of the individual transfer pathways comprising the net absorptive (or secretory) process.

Similar indications of an active role of cells were drawn from studies of the concentrative capacity of the kidney, and the excretion of urea. Because of the complexity of this tissue, and the greater analytical difficulties, progress was relatively slow, although already in 1874 Heidenhain observed an apparent secretion of a dye into the lumen of the convoluted proximal tubule.

In all these studies the term *active* denoted the cellular contribution to the accumulative transport process, in order to distinguish physiological absorption or secretion from simple physical processes. The next step in the recognition of the phenomenon was to provide more definitive evidence that metabolic energy is the basis for the observed osmotic work. Brodie and Vogt (1910) established that intestinal absorption of water and salts increased O_2 uptake and CO_2 formation, implicating epithelial cells in this process. When Höber (1926) reviewed the state of knowledge and spoke of active physiological (as opposed to physical) permeability, he based his appraisal on two sets of data: first, it became clear that many solutes are transferred across epithelial membranes without chemical modification. While self-evident for non-metaboliza-ble solutes (*e.g.* salts), it was proven that e.g. sugars are ab-sorbed unchanged by the kidney (Clark, 1922); this absorption was blocked by metabolic inhibitors such as $HgCl_2$. Van Slyke and Meyer (1913) had already shown that amino acids absorbed by the intestine enter the circulation, are then accu-mulated in various tissues (evidently against their concentra-tion gradient[4]: 'that amino acids should pass ... uphill ... requires another explanation than mere osmosis') and meta-bolized particularly in the liver.

Second, the metabolic dependence of the absorption of sol-utes (against the concentration gradient) was better substan-tiated. Höber was most impressed by observations originated on mammalian cells by Heidenhain (1874) and on plant cells by Collander (1921), showing that the accumulation of lipid-insoluble vital stains could be blocked by narcotics. Such stud-ies were then extended in Höber's laboratory. Hence, the dependence on metabolism was set as a criterion for active physiological permeability.

Wilbrandt (1938) sought to distinguish between equilibrat-ing and accumulative transport: by his definition, an accumu-lative process is coupled to cell metabolism and leads to a dis-tribution which (in the case of ions) cannot be reconciled with equilibria produced by selective membrane permeability; in the experiments of Donnan (1911) and Teorell (1933), ionic gradients of up to 10^3 can be achieved by model equilibrating systems. Wilbrandt's criterion reflected the results of the labo-ratories of Osterhout (1933), Collander (1921) and Lunde-

---

4 Christensen (1993) points out that this fundamental observation was actually not emphasized in textbooks of physiology and biochemistry, and thus a generation of students was not aware of an early recognition of active transport of a nutrient

138   A. KLEINZELLER

gårdh (1937) where an accumulation of salts in plant cells was demonstrated[5].

An equilibrating system could not explain why the giant sea alga *Valonia* accumulated K^+ against a gradient of 41–48, whereas the ratio of Na^+ concentrations inside and outside was 0.07–0.18. In his study of gastric acid secretion, Teorell (1933) emphasized that model systems could not suffice to produce a H^+ gradient of 10^7, and called for a theory which would deal with transport phenomena as a steady state, rather than an equilibrating system. Huf (1935,1936) used the term *active transport* to describe the metabolic dependence of the net NaCl transport across the frog skin, with lactate or pyruvate serving as metabolic substrate: with rare insight, he suggested that 'active' or 'vital driving' forces denoted only the unknown chemical reactions necessary to maintain an unknown physical mechanism; chemical free energy, generated by an oxidative process, might be converted into work of absorption by an electrical process. Huf also postulated that such active transport system might be the mechanism by which osmoregulatory phenomena are brought about. Krogh (1938) adopted this term. Shannon (1939) was well ahead of his time when characterizing secretory and absorptive phenomena in the kidney as processes deriving their energy from local cellular metabolism; he visualized each of the reactions of the transfer mechanism to be made up of two reactants, *i.e.* the transported substrate and a cellular component. This cellular component appears to be specific for a given solute (or class of solutes), imparting saturability (indicated by maximal rates), and substrate competition. His views were summarized by the stoichiometric scheme:

---

5 In fact, the unequal distribution of salts between the environment and plant cells was already known in the middle of the 19th century (*cf.* Sachs, 1865) and used as an explanation for the absorption of water and salts by rootlets and the formation of plant sap.

$$A + B \rightleftharpoons AB \longrightarrow B + T_S$$

(A: transported solute; B: cellular component; T_S: transferred solute); in this scheme, the reaction $A + B \longrightarrow AB$ appears to be that requiring an input of metabolic energy.

The introduction of radioactive labels from 1936 onwards catalyzed new concepts on active transport. This was largely reflected by reports by Brooks, Hoagland and Osterhout at the memorable Cold Spring Harbor Symposium in 1940, and by the studies of the Copenhagen group headed by Hevesy. The concept of a cationic pump was introduced at the Cold Spring Harbor Symposium in 1940 by Steinbach and by Dean, providing a rational explanation for the unequal ionic distribution in red blood cells and muscle.

1. ELECTROLYTE PUMPS

1.1. Ionic pumps in plant cells

The membrane of plant cells was long considered to be impermeable (or, at least, poorly permeable) for inorganic ions in the light of early studies by Collander (1921), Osterhout (1933), Hoagland (1937), S.C. Brooks and M.M. Brooks (1941), and others, demonstrating only small losses of salts when cells and plant tissues were placed in distilled water. The slow reversibility of plasmolysis produced by inorganic salts pointed in the same direction. In the period 1920–1940, giant algal cells were a choice object for such studies owing to their size and hence the availability of material from single cells for the determination of individual ions. When it became clear that a selective cellular accumulation of ions (particularly of K^+, Cl^- and phosphate against gradients of up to two orders of magnitude, especially in fresh-water algae) took place, the view gradually emerged that specific mechanisms of ion accumulation, dependent on metabolism, were localized in the cell protoplasm,

and the accumulated ions then flowed passively into the vacu-
olar sap. This view was substantiated by the early use of radioisotopes (Brooks, 1938), who demonstrated the entrance of
radioactive Na^+, K^+, Rb^+, and phosphate into marine egg cells
and algae (cf. 1941). An ion accumulation by plant storage
tissue, e.g. potato tubers, was found by Steward. Brooks (1941)
appears to be the first to postulate in 1935 that the unequal
ionic distribution between plant cells and their environment
represent a dynamic equilibrium (steady state), rather than a
static system. Two early hypotheses were put forward to explain ion accumulation: Osterhout visualized a molecular
mechanism by modelling active transport with 'lipid-soluble
carriers': in this model system K^+ interacted with a carrier
anion (OH^- or X^-) faster than Na^+ and passed through the
membrane as the non-dissociated molecule; Brooks postulated
an 'ion exchange mechanism'.

Plant roots represented another important material for
study, being able to take up salts from very dilute solutions,
accumulate them in the cells, and secrete particularly K^+ and
NO_3^- into the xylem. The technique introduced by Lundegårdh
in 1943 (cf. 1945) permitted direct measurements of the secretory process[6], showing that the epidermal cells of plant roots
pumped H_2O (and electrolytes) against an osmotic equivalent
of 4 atm.; this process required aeration in order to persist.
The   anion   respiration   theory   of   active   absorption
involved the interaction of an anion (in nature: NO_3^-) with a
hypothetical membrane (carrier) cation (e.g. Fe of the cytochrome system) and subsequent release of the anion at the
inner membrane face, with the oxidation-reduction cycle of the
cytochrome system as the driving force.

---

6 Lundegårdh's method actually owed much to the pioneering study of Hales (1727)

1.2. The cation pump in erythrocytes and muscle

Early studies on muscle (Liebig[7], 1847) and red blood cells (Schmidt, 1850) showed that the predominant intracellular cation is K^+, while Na^+ is the main cation in the immediate cellular environment. When questions began to be raised as to the mechanisms of this phenomenon, the main arguments were that (a) the cell surface (membrane) is impermeable to both these cations; or (b) the cytoplasm selectively binds potassium.

The mammalian erythrocyte membrane has been known to be highly permeable to anions such as Cl^- and HCO_3^- particularly on the basis of studies of Hamburger (1891) and Koeppe[8] (1897). The phenomenon of the chloride shift is usually associated with Hamburger's name, who rediscovered and extended the observation of Nasse (1878) that on bubbling CO_2 through blood, Cl^- passed from the plasma into cells, while the titratable plasma alkalinity increased. This observation was interpreted by suggesting that on entry of CO_2 into the cells, carbonic acid combined with cellular bases and the formed HCO_3^- exchanged for plasma chloride; such process had to be associated with some cation movements across the cell membrane. On the other hand, until 1940 the membrane of red blood cells was taken to be impermeable to K^+ and Na^+ under normal conditions; abnormal conditions (e.g. presence of poisons) increased their permeability (cf. Jacobs, 1931; Ponder, 1934). A respectable amount of evidence demonstrating that

---

7 Liebig pointed out that a variety of cells was characterized by high K, low Na, e.g. sea weed and algae; he hypothesized that such differences in ionic distribution might reflect different permeabilities, and also could be the basis of bioelectric phenomena. Schmidt's careful analysis clinched the concept that the intracellular composition of inorganic components greatly differs from that of their environment, as opposed to the views of Prevost and Dumas; Schmidt was also the first to point out that, as opposed to the red blood cells of humans and some other species, those of the dog, cat and sheep are characterized by high intracellular Na, low K.

8 a practicing physician who carried out his beautiful studies as a hobby.

erythrocytes could leak K^+ and take up Na^+, beginning with Hamburger's experiments (1891,1910) were dismissed with arguments that the authors failed to correct for volume changes. An irreverent thought might occur that the concept of an erythrocyte membrane permeable to anions and cations would pose problems for the explanation of the unequal cation distribution between cells and their environment. Several factors contributed to the change of mind:

First, various pressures at the end of the thirties (economic in the U.S.A., military in Europe) prompted greater interest in the use of preserved human blood for transfusion (*cf.* DeGowin et al., 1940; Dowman et al., 1940). However, for more than a decade observations had accumulated to show that storage of blood at 0°C produced a continuous loss of cell K^+ into plasma, subsequently causing toxic K^+ effects in patients receiving transfusions; the loss of K^+ could be slowed by adding glucose to the blood. These experiences then prompted the now classic studies of Danowski (1941), and independently, of Harris (1941), a member of DeGowin's team. Their experiments demonstrated that the view of the erythrocyte membrane impermeable to Na^+ and K^+ was untenable. Human red blood cells suspended in isotonic saline at 0°C lost K^+ and took up Na^+ in what appeared to be a reciprocal flux; this process was reversed at 37°C in the presence of glucose as substrate. Inhibition of glucose utilization by F^- also blocked the metabolically dependent cation transport. The view was put forward that the cation concentration in the protoplasm is maintained by the metabolic functions of the cells.

Secondly, the introduction of radioisotopes for studies of transport phenomena made it possible to directly measure unidirectional fluxes and thus define the transport processes in greater detail. Hahn et al. (1939) demonstrated in vivo a slow exchange of $^{42}K^+$, added to the plasma, with K^+ present in the erythrocytes (and muscle) of rats. In the high-Na^+, low-K^+ erythrocytes, of the dog and cat, the membrane permeability to Na^+ and K^+ was established: Cohn and Cohn (1939) demonstrated that ^{24}Na$^+$ readily entered the cells; Davson (1940) showed that the membrane of cat red blood cells was permeable to K$^+$ and Na$^+$ when appropriate gradients of these cations were imposed. It might be (and was) argued that these cells qualitatively differ from the red blood cells of most mammalian species, and thus results obtained might not be general.

In a clear case of 'to have seen what everybody has seen, but have thought what nobody has thought' (Szent-Györgyi, 1957), discussions at the Cold Spring Harbor Symposium in 1940 by Steinbach (1940b) and by Dean (cf. 1941) put forward a different interpretation of existing information. Steinbach (1940b) proposed the term sodium pump:

> 'In order to account for the high concentration of potassium in the cells and for the permeability to sodium indicated by much older work.. there must be some mechanism present for pumping out the sodium that wanders into protoplasm'.

At the same 1940 symposium, in a discussion with Davson, Dean (cf. 1941) suggested (in analogy to HCl secretion by the stomach) that

> 'the cat red blood cell might be excreting Na$^+$ rather inefficiently, whereas other cells that have a high K$^+$ have a more efficient method for excreting sodium and therefore only K$^+$ is left in the cells'.

Subsequently, the full paper of Dean et al. (1941) extended the concept of a steady state of intracellular cations put forward by Brooks and Brooks (1941, p.5):

> 'If a cell is permeable to a given ion, and does not change its content of that ion over a period of time, we can say that the numbers of K$^+$ ions crossing its membrane in each direction in unit time are equal'.

Thus, the static concept of the electrolyte distribution in red blood cells gave way to a dynamic concept with the cells being in a steady state resulting from a balance between influx and efflux of ionic species across the membrane, one of these processes being active. It may not be fortuitous that this recognition coincided with the introduction of the concept of the dynamic state of cell metabolism by Schönheimer (1941); as with membrane phenomena, the idea of a dynamic state of the cells became convincing by the use of isotopes as an experimental tool.

The development of the concept of the sodium pump in muscle cells followed a similar pattern, as indicated by the chronology[9]:

1902: Overton proposes an exchange of Na^+ for K^+ during muscle contraction; Bernstein presents a theory of the resting potential in muscle and nerve, implying membranes permeable to K^+ (and Cl^-);

1928–1934: in Höber's laboratory, Mond and his collaborators demonstrate a metabolism-dependent uptake of K^+ from perfusion fluid or plasma into muscle and suggest a K^+/H^+ exchange (cf. Netter, 1934);

but

1931: Hill categorically states that muscle cells normally are impermeable to Na^+, K^+ and Cl^-;

1936: Fenn acknowledges the permeability of the muscle membrane for K^+; the membrane is impermeable for Na^+ and Cl^-;

1940: Fenn recognizes that Na^+ can enter muscle cells; Steinbach (1940a) demonstrates an exchange of Na^+ for K^+;

1941: Boyle and Conway demonstrate the permeability of the muscle membrane for Cl^- (and K^+)[10];

By 1946, the concept that an ionic pump (or pumps) is in-

---

9 Two points may have contributed to the delay in the recognition that the muscle membrane is permeable to Na^+, K^+ and Cl^-: a) as compared with studies in erythrocytes, the analysis of ion exchanges in muscle is complicated by the sizeable extracellular space, as already realized by Overton (1902); and (b) the resting muscle cell was assumed to differ fundamentally from the excited one: the contractility of the preparation being taken as the crucial criterion of the preparation's viability, results obtained on non-contracting muscles were considered to be unphysiological.

10 The accumulation of K^+ in muscle cells (deemed to be impermeable to Na^+) is explained by an electrostatic interaction of this cation with intracellular non-diffusible anions, and hence is a passive phenomenon..

volved in the transport of bulk electrolytes in animal cells was embraced by Krogh (1946). The acceptance of the pump concept put the enquiry on a new level, raising essential questions as to the molecular mechanism of the process, viz. (a) which ion is pumped; (b) how does a pump recognize the pumped solute(s); (c) how does the energy supply (chemical or physical) required for the pump action interact with the pumping mechanism; and (d) how is the solute actually propelled across the membrane. A number of possible mechanism was considered at that time.

Franck and Mayer (1947) put forward the provocative idea that active transport might be produced by a sequence of two events at separate membrane sites: a catalytically mediated interaction of a solute or solvent with a cellular molecule and a subsequent reversal of this reaction, driven by some other chemical exergonic reaction[11]. Although at the time this concept was applied to the possibility of a water pump[12], in essence it was later specified in terms of the operation of the sodium pump. Goldacre (1952) suggested that the folding and unfolding of protein molecules could be the cellular basis for performing osmotic work. The most provocative was Mitchell's (1957) concept[13] of a pump (Fig. 1), suggesting an interaction of the 'carrier' with the transported substrate and the subsequent substrate translocation across the membrane by a conformational change of the carrier, followed by a dissociation of the solute and the carrier at the other side of the membrane; a

---

11 The original model of Osterhout (cf. 1933) with guaiacol as a lipid-soluble carrier for cations may have played a role here
12 The possibility of a water pump was explored further when Opie (1949) found that a modest hypertonicity of external NaCl solutions was required to prevent cell swelling; hence, the cells maintained an osmotic water gradient. The suggestion of Robinson (1950) of a specific water pump became untenable by the experiments of Mudge (1951) showing that a metabolically-dependent movement of cations, associated with an osmotically-obliged water flux, could satisfactorily explain the phenomenon later pursued as the mechanism of cell volume regulation.
13 This was an extension of Danielli's (1954) analysis of possible mechanisms of membrane transport.

146   A. KLEINZELLER

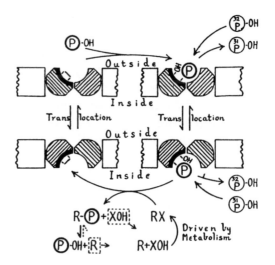

Fig. 1. Model for the uphill translocation of phosphate (Mitchell, 1957).
Reproduced with permission of Nature (Lond.).

'Phosphorylated carriers are present in the membrane, the phosphoryl group of which, due to thermal movements, become alternately accessible to the media on either side where a specific exchange may occur between the phosphoryl group on the carriers and phosphate ions without loss of the bond-energy of the carrier-phosphoryl compound'. 'The displacement of analogues that also occurs during the accumulation of amino acids and sugars (Cohen and Monod, 1956; Rickenberg et al., 1956) ... can be explained by the uptake of the amino acids into the membrane as covalent translocator (that is, activated) compounds'.

direct involvement of metabolic energy (ATP) as the donor of the required energy was envisioned.

2. ACTIVE TRANSPORT OF NON-DISSOCIATED SOLUTES

The active absorption of sugars and amino acids from the intestine and the renal tubule was recognized very early (see above). Cori (1925) demonstrated the structural specificity of intestinal sugar absorption, and Verzár's laboratory in Basel established the metabolic dependence of this process. The understanding of the process was greatly enhanced when it was

shown beginning with the work of Clark (1922), and beauti-
fully extended by Wearn and Richards (1924) that the phlo-
rizin-induced renal excretion of glucose (phlorizin diabetes)
was the result of a rather selective inhibition of glucose reabsorption by the kidney; subsequently, the selective inhibition
of glucose absorption was also shown for the intestine beginning with the studies of Nakazawa, and in a more definitive
way by Lundsgaard (1933). Thus, a powerful tool for the study
of the active sugar transport system was established; unfortunately, no such selective inhibitor of active amino acid transport is known. Further progress was hampered by the rapid
intracellular metabolism of studies substrates, *e.g.* glucose.
This point was resolved by detailed studies of substrate specificity showing that several non-metabolizable substrates were
also absorbed, *e.g.* the important synthetic glucose analog, 3-
O-methyl glucose (Csáky and Wilson, 1956). The site of the
actual absorption process was conclusively shown to be the
epithelial cells (with an important role of the luminal (brush-
border) cell membrane) and a role of the ionic environment
(particularly of Na^+, but also of K^+) for solute absorption was
found by Ricklis and Quastel (1958), and in more detail in
Ussing's laboratory by Csáky and Thale (1960); Crane et al.
(1961) established the essential phenomenology of intestinal
active transport of non-dissociated solutes on the cellular level
by demonstrating the role of the Na^+-gradient (*cf.* Crane,
1977). In epithelial membranes, Schultz and Zalusky (1964)
added the measurement of the short-circuit current as an important tool for the study of the Na-dependent sugar transport, having shown that the intestinal transport of sugars
across the cells was associated with an increase in the net
transcellular flux of Na^+.

   In bacteria, Gale, in 1947, discovered a massive accumulation of amino acids, particularly of glutamate. At some stage,
this energy requiring process was considered to differ from
diffusion by proceeding against the electrostatic and concentration gradient across the membrane (Gale and Mitchell,

1947). Influenced by the views of his illustrious preceptor, M. Stephenson, (*cf.* Gale, 1971), he proposed in 1954 that upon passing into cells, glutamate entered into some metabolic process which resulted in the formation of a non-diffusible form. Rickenberg et al. (1956) in a classical paper put forward the concept that the observed accumulation of the non-metabolizable methyl-β-thiogalactoside in *E. coli* against steep concentration gradients was brought about by an enzyme-like process localized in the cell membrane, distinct from metabolic enzymes; this process, catalyzed by a 'permease' (not the best possible term), was characterized by saturation kinetics, metabolic dependence and stereo-specificity; it could be induced by the presence of substrate, and was shown to be genetically controlled (Cohen and Monod, 1957). The first reports on the uptake of amino acids by ascites cells against major concentration gradients were by Christensen and his group, beginning in 1952 (Christensen, 1975), and the dependence of this process on the ionic environment (Riggs et al., 1958) was quickly recognized. Since all these cellular accumulative processes were dependent on cell metabolism, they fell into the category of active transport. The above discoveries opened the door for a detailed study of the transport phenomenology, leading to the understanding of the involved mechanism(s).

II. The phenomenology of active transport

1. CRITERIA FOR ACTIVE TRANSPORT

The search for a comprehensive definition of a phenomenon represents a crucial step in its analysis and eventual elucidation. This was clearly stated by Rosenberg (1954):

'Fundamental definitions have varying functions in the study of a group of allied phenomena. In the early stages they facilitate the collection of the necessary empirical data by differentiating between superficially similar phenomena. In later stages they assist in the formulation of models and in comparing of analogous groups of phenomena and thus in the understanding of the basal mechanisms. In the final stages they render possible the immediate recognition and classification of relevant phenomena in the study of new systems and the complete cataloguing of the whole group of phenomena.'

Once it was recognized that active transport is a physiological reality, Rosenberg (1948) defined active transport as

'a transfer of chemical matter from a lower to a higher chemical (in the case of charged components: electrochemical) potential. Thus, work (osmotic, and/or electrical) has to be done to actively transfer solutes across membranes; it is generally assumed to take place by coupling to intracellular chemical processes'.

A thermodynamic analysis then led to the postulate of a mediating (coupling) link in the membrane. This definition was thermodynamic in nature, applying to the process of solute transport against its (electro)chemical gradient[14]; metabolic dependence of a process per se was not a sufficient criterion, including, e.g. regulatory phenomena.

Ussing (1949), applying the flux ratio equation derived for diffusion (see Chapter 2) to transport phenomena, emphasized that deviation from the equation

'indicates that the solute (ion) in question does not diffuse in a free state only, but ... as a component of some other moving particle in the membrane; metabolic energy then provides the energy necessry for the deviation from simple diffusion'.

The physicochemical treatment of the phenomenon clearly im-

---

14 This concept of active transport was not concerned about the actual coupling mechanism between the transported solute and the driving metabolic process, and thus appeared to be 'mechanistically mysterious' (Mitchell, 1992): 'The mystery disappears as soon as one recognizes that the so-called active transport of the species S must actually occur by the facilitated diffusion of one or more thermodynamically mobile components that are derivatives (i.e. products, compounds, and/or complexes) of S'.

plied that the same molecular species is present on both sides of the membrane. This essentially phenomenological criterion proved to be eminently useful, particularly for the analysis of epithelial transport since the determination of both fluxes, J, and concentrations, c, (more correctly, activities) in the media separated by the epithelial membrane) could be readily assessed. On the other hand, when applied to transport across a cell membrane, caution had to be used in view of major assumptions necessary when assessing the intracellular concentrations (activities) of a solute. Another problem proved to be the basic assumption in Ussing's criterion, *i.e.* the independence of the fluxes of the studied ionic species from other ionic fluxes (the independence principle). Hodgkin and Keynes (1955b) were the first to show that this assumption is not generally valid, and a correctional exponential term n had to be applied, i.e.

$$J_i/J_o = [c_i/c_o \cdot \exp (zEF/RT)]^n \tag{1}$$

The value of n may vary between 0.5 to 2.5 (*cf.* also Schagina et al., 1983). Thus, some limitations to Ussing's criterion were recognized for the single-file diffusion considered by Hodgkin and Keynes, and for counterflow (see below).

Rosenberg (1954) recognized that

'the actually observed phenomenon represented only part of a (then) unknown transport complex; the rest is transported "downhill" and thus furnishes, owing to coupling, the energy necessary for uphill transport[15]. The total transport of the whole complex is never active'.

---

15 Conceptually, the term uphill, or accumulative, transport is strictly phenomenological; there is little reason to exclude the possibility that under some specific conditions, metabolic energy might be used to drive a process downhill; it would then be rather difficult to distinguish such event from a simple passive downhill flux. This point has been repeatedly discussed (but not solved) (Wilbrandt, 1961: 'Should a car running down-hill be called a bicycle?'; Christensen, 1976).

The salient properties of active transport were defined as: (1) saturability; (2) competition by chemically similar substances; (3) a temperature coefficient of the order of chemical reactions; (4) structural specificity; (5) dependence on metabolism; (6) effects of enzyme activators and inhibitors. It will be recognized that excepting (5), the above criteria are identical with those put forward for a carrier-mediated transport system (Chapter 3). The advantage of this definition of active transport lies in the more molecular approach to the same problem. Moreover, this analysis was extended by Wilbrandt and Rosenberg (1961) by showing that Ussing's criterion applied also to the (transient) uphill transport produced by counterflow (Chapter 3), *i.e.* without actual input of *metabolic* energy: here the carrier was actually assumed to be the mobile particle. In effect, in some cases an assumed active transport system for an amino acid proved to be counterflow, driven by the downhill flux of an endogenous amino acid sharing the carrier. In a phenomenological way, application of irreversible thermodynamics to active transport phenomena (Kedem, 1961) described the need for a coupling mechanism in the membrane (expressed as the cross-coefficient R_{ir}) between a solute flux and metabolism, leaving open the possibility that the vectorial flux of the actively driven solute might be coupled either to a scalar process (*e.g.* a chemical reaction) or to another vectorial process (*e.g.* a solute flux and/or electrical potential). It should be recognized that it is the (direct or indirect) dependence on metabolism which distinguishes active transport from simpler transport mechanisms such as diffusion, facilitated diffusion, etc.

The above criteria provided a solid basis for the study of the active transport of any solute, in spite of some shortcomings found in every attempt at defining a concept (Danielli, 1954: the thermodynamic approach emphasizes differences between initial and final states, to the exclusion of the processes intervening between these states). However, for a long time the term was still used in the old sense, *i.e.* denoting any transport process involving cellular activity as *active*. Thus, the equilibrating transport of sugars across the red blood cell membrane was termed to be active in 1954. Still more surprising, the accumulation of Ca^{2+} by mitochondria was considered to be active transport because of 'metabolic dependence', in spite of evidence that the intramitochondrial Ca was trapped as an insoluble phosphate. Such inconsistencies prompted an acerbic remark by A. Katchalsky when he told this author in 1968:

'I just returned from a 5-week stay with a mitochondrial biochemist and I had to give him a good brainwash about the physicochemical principles underlying active transport processes'.

2. WHICH ION IS PUMPED?

While in plants a case for active anion transport was made by Lundegårdh (1937), in mammalian cells both Na^+ extrusion and/or K^+ accumulation had to be considered. At the same time, existing knowledge required a more stringent appraisal and definition of the conceptual framework of active transport and it's testable criteria. Such statements would have to point out ways how to dissect the implied multiplicity of pathways by which a solute can cross the membrane in both directions, *i.e.* for a cation driven by the pump across the membrane and lost (or reentered) through a leak.

The concept of a Na^+-pump as envisioned by both Dean and Steinbach visualized Na^+ extrusion by an electrogenic process; K^+ was the accumulated in the cells due to electrostatic forces. Krogh's observations (1946) that in frog skin the active transport of Na^+ did not require external K^+ was consistent with such view, as was Conway's suggestion of a muscle membrane effectively impermeable for Na^+. Conclusive evidence for the operation of the Na^+ pump in the frog skin was provided by Ussing and Zerahn (1951) by demonstrating an equality between the measured short-circuit current and the net flux of

Na^+ through the cells. The active transport of Na^+ produced the electrical potential across this epithelial membrane. The technique of measuring the short-circuit current across a membrane became a standard tool for the study of ion transport against its electrochemical gradient.

The resolution of the question as to the nature of the pumped cation, and its coupling to metabolism was advanced by studies in erythrocytes which appeared to be the simplest model. The whole spectrum of possibilities was explored: On the basis of kinetics, Solomon (1952) proposed a model in which both Na^+ and K^+ are actively transported across the membrane by separate mechanisms, the only link being the source of metabolic energy; for both cations a 'ferry'-type of mechanism was visualized, imparting saturation kinetics for the transport, and each undergoing a (thermodynamically irreversible) cyclic process in the course of crossing the membrane. A major breakthrough was provided by Schatzmann's (1953) discovery: using the protocol of Harris (1941) for following cation transport, a selective inhibitory effect of cardiac glycosides was found. He envisaged these inhibitors to displace the cation(s) from a carrier which brings about an equivalent uphill exchange of Na^+ and K^+. Hodgkin and Keynes (1955a) put forward a model for the squid axon in which metabolic energy drives a cyclic process effecting an absorption of K^+ and an extrusion of Na^+. Continuing the careful kinetic studies started by Shaw, Glynn's model (1956) suggested in erythrocytes a carrier effecting a coupled Na^+ and K^+ transport driven by glycolysis. Conceptually, the mechanisms envisaged by Solomon, Schatzmann and Glynn were electroneutral, representing an exchange of two positively charged cationic species; Hodgkin and Keynes (1955a) were aware of this question, having observed that in the nerve[16] the efflux of Na^+ exceeded the

---

16 Electrogenicity of the Na-pump was also indicated when Ritchie and Straub (1957) demonstrated a post-tetanic hyperpolarization of the nerve and this process was inhibitable by ouabain, absence of K^+ and inhibitors of oxidative phosphorylation.

absorbed K⁺. Once Post and Jolly (1957) established a stoichiometry of 3 Na^+ extruded for 2 K^+ accumulated, the coupled active transport of Na^+ and K^+ was indeed electrogenic and could explain the establishment of electrical potential gradients across the membrane. The authors also recognized the advantages of the direct coupling of both cations in the active transport process for the physiology of the cell (control of cell volume).

The search for defining the specificity of individual ionic pumps tended to detract from an exploration of functional similarities; these aspects have come to the fore only more recently (see below).

3. THE NATURE OF THE COUPLING BETWEEN SOLUTE FLUXES AND METABOLISM

The elucidation of the driving force required for the active transport of a given solute was essential for a better understanding of the phenomenology of the transport process. As pointed out above, an uphill flux of a solute might be driven by a direct coupling to an enzymic reaction which would provide the required energy, as known at the time for a variety of enzymic systems. Alternatively, an indirect coupling might take place, the solute moving uphill at the expense of the flux of another solute moving down its electrochemical gradient, corresponding to the counterflow mechanism (Chapter 3). In both these processes, the nature of the solute is identical on either side of the membrane. Mitchell (1957) suggested still a third mechanism, group translocation: here, as later shown for the phosphotransferase system (see Chapter 2), the solute (sugar) is phosphorylated in the course of its transport across the membrane by a 'translocase'. Strictly speaking, this process does not constitute active transport as defined above.

3.1. Direct coupling

At the 1960 Prague Symposium (Kleinzeller and Kotyk, 1961), several concepts of the coupling between ion transport and metabolism were discussed. Conway presented his concept of a redox pump, another version of Lundegårdh's (1945) anion pump where the electron transfer system was assumed to have anionic group capable of binding a cation. If such a system were located in a membrane, it could bring about a transfer of electrolytes across the membrane. The scheme should meet two predictions: As pointed out by Robertson in 1948 (*cf.* Briggs et al., 1961), the stoichiometry between O_2 used and cation transported should be 4, and the process should proceed predominantly under aerobic conditions; anaerobic organisms would have difficulties making use of such mechanism. The scheme proposed by Hokin and Hokin (1960) suggested phosphatidic acid as the carrier for cations, permitting the transport of 2 Na^+ per one ATP-ADP cycle. And finally, several investigators suggested a more direct coupling via the newly discovered Na^+-activated, K^+-stimulated ATPase in crab nerve[17] (Skou, 1957). The concept of a direct coupling of cation transport to ATP hydrolysis was actually indicated earlier. Thus, in an apparently little appreciated paper, Clarkson and Maizels (1952) demonstrated an ATPase activity in the erythrocyte membrane and suggested:

'The purpose of this enzyme activity is obscure; but if the uptake of Na and discharge of K at the inner face of the cell membrane depends on phosphorylation, it is possible that return of Na and uptake of K depend on phosphorolysis'.

---

17 This major discovery resulted from Skou's (*cf.* 1989) investigation of a membrane ATPase from crab nerves. At first, a slight, variable stimulation of enzyme activity by Na^+ was found, and only subsequent studies disclosed a stimulating role of both Na^+ and K^+. The relationship of these observations to the operation of the Na^+-pump was recognized only later.

The Hungarian school introduced the use of permeabilized erythrocytes for the study of the problem. The experiments of Straub (1953) indicated the role of macroergic bonds as the driving force for the accumulation of K^+ against its concentration gradient, and his student Gárdos (1954) then proved by more direct experiments that the uptake of K^+ by erythrocyte ghosts can be driven by ATP. Hodgkin and Keynes (1955a) showed that the extrusion of Na^+ (and uptake of K^+) from nerve was inhibited by agents known to block oxidative phosphorylation in mitochondria (*e.g.* dinitrophenol, azide and cyanide); Ritchie and Straub (1957) arrived at a similar conclusion.

The discovery of the Na^+-activated, K^+-stimulated ATPase by Skou offered a plausible explanation for the observed stoichiometry between Na^+ extrusion and metabolic parameters. The involvement of the ATP–ADP cycle in the active cation transport satisfied both the requirement of being operative also under anaerobic conditions, and a better stoichiometry between O_2 utilization and Na^+ transported: 18 Na^+/O_2 was found by Zerahn in the frog skin and by Leaf for the toad bladder, corresponding to predictions based on evidence obtained in studies of oxidative phosphorylations ($6 \sim P/O$, *i.e.* 3 $Na^+/ \sim P$). These studies established the $(Na^+$-$K^+)$-ATPase as a component of the sodium pump.

3.2. Indirect coupling

For non-electrolytes, the analysis of the coupling between transport and metabolism proved to be more complex as long as physiological (metabolizable) substrates were used as models. On the suggestion of Verzár, Wilbrandt and Laszt proposed a scheme for the intestinal glucose absorption by which glucose was phosphorylated at the luminal cell membrane, and dephosphorylated at the basal membrane, thus producing transcellular glucose transport driven by metabolism. The discovery of active transport of non-metabolizable analogues by Csáky falsified this hypothesis.

At the Prague symposium the general nature of the role of cations in the active accumulation of non-dissociated solutes was recognized. Mitchell developed concepts by which spatially anisotropic enzymes in the membrane might produce a vectorial component of metabolism, and Crane et al. (1961); (cf. also Crane, 1977; Mitchell, 1979) proposed a new mechanism of coupling active transport with metabolism. Starting with the known phenomenology of sugar transport in the intestine, i.e. saturability, structural specificity, dependence on metabolism, and requirement of cations in the medium, Crane postulated a mechanism by which the downhill flow of Na^+ in intestinal cells was coupled by a specific carrier with the uphill flux of a sugar, the free energy of the downhill translocation of Na^+ thus driving the uphill flux of the sugar. The downhill gradient of Na^+ was maintained by the operation of the sodium pump. This gradient hypothesis thus visualized the indirect coupling of an equilibrating transport system with the active extrusion of another solute; this mechanism was later denoted by Heinz (1978) to be a secondary active system. Such process is characterized by the specific properties of both transport components; thus, the Na^+-dependent sugar transport is inhibited by phlorizin (specific for the transport protein interacting with Na^+ and glucose) and by cardiac glycosides (selective for the Na^+-pump). The reversibility of the Na^+-solute cotransport system has been demonstrated more recently (Nelson and Blaustein, 1982).

Crane's concept was rather close to that of Mitchell: whereas Crane postulated a role of the Na^+ pump as the direct driving force, Mitchell visualized a role of the electron transport chain for generating the H^+ flux. Subsequently, several possibilities were explored (Mitchell, 1971), i.e. (a) convective coupling between solute and ion fluxes; (b) a role of the intracellular Na^+ concentration and (c) coupling produced by a carrier mechanisch which involved an interaction with both solute and ion. Later studies demonstrated the viability of this last concept by broadening it in several respects: the coupling between substrates (*e.g.* sugars, amino acids) and H^+ transport, predicted by Mitchell's chemiosmotic hypothesis, was found by West (1970), thus providing a mechanism for the accumulative process of sugars in bacteria. At present, the mechanism of solute accumulation in bacteria appears to parallel that in animal cells in that an equilibrating proton-solute symport is coupled to the extrusion of protons ($\Delta\mu_H^+$ inside negative and alkaline), generated by respiration and a proton-ATPase (Kaback, 1986). A cation can interact in the transport of its co-substrate in several ways (*cf.* Crane, 1986): Crane's 1961 model implied an interaction of both Na^+ and substrate with a common carrier; flux coupling by this mode is essentially mechanical (the flux of Na^+ pulling or propelling that of glucose). The other possibility, *i.e.* a chemical interaction of the cation with the co-substrate, thus facilitating the flow via the carrier, now appears to be remote at least for sugars in the light of our knowledge that the stoichiometry between Na^+ and hexoses may vary (see below); however, such a possibility cannot be dismissed for substrates carrying a charge, *e.g.* amino acids (Christensen, 1984).

The value of a hypothesis can be measured by the number of new questions asked, and new testable predictions it stimulates. The original gradient hypothesis implied two conceptual points: (a) the coupling of the uphill sugar transport to the downhill chemical gradient of Na^+; and (b), a simple stoichiometry of 1 sugar accumulated per 1 Na^+ transported inward; the sugar being a non-dissociated molecule, a net transport of one charge across the membrane might be assumed, *i.e.* the process could be electrogenic. This prediction was verified particularly in the laboratories of Armstrong and of Schultz (*cf.* Schultz, 1977) by measurements of changes of the electrical potential gradient across the mucosal membrane on addition of sugars or amino acids. Hence, the uphill solute transport is coupled to the electrochemical gradient of the cation. The com-

petition of various, structurally unrelated solutes for uphill
transport was first hypothesized as a competition for the existing Na^+ gradient (Semenza, 1971); in the light of more recent
knowledge, it appears to be a competition for the driving force,
i.e. the electrochemical gradient of Na^+ (Murer and Hopfer,
1974).

The 1:1 stoichiometry withstood many tests in several laboratories, although rather early a dissonant note was sounded
when Vidaver showed that the accumulation of the neutral
amino acid glycine in pigeon erythrocytes was coupled to the
entry of two Na^+. The question of stoichiometry became particularly important when the energetic aspects of the secondary
active transport were examined (see, *e.g.* Turner, 1983): on
thermodynamic grounds it is rather obvious that the energy
required for pushing *e.g.* a sugar S against its concentration
gradient cannot exceed the driving force (the electrochemical
potential) of the downhill flow of the coupled cation A; this
statement can be expressed by the equation

$$\ln([S_i]/[S_o]) = \ln([A_o]/[A_i] + FE/RT) \tag{3}$$

where the subscripts i and o describe the inner and outer compartments; thus, both the concentration gradient of the cation
and the electrical potential E^{18} across the membrane are determinants of substrate accumulation.

When Kimmich and Randles (1980) found that the accumulation ratio of a sugar in intestinal cells far exceeded the established electrochemical gradient, the concept of a 1:1 stoichiometry had to be modified; a stoichiometry of 1 sugar:2 Na^+
easily satisfied the thermodynamic requirements; eq. (4) then
changes to:

---

18 Fleckenstein (1948) appears to be the first to suggest that the electrical potential
gradient across membranes might drive other processes.

$$\ln([S_i]/[S_o]) = n\{\ln([A_o]/[A_i]\} + FE/RT) \tag{3}$$

where n is the number of cations coupled to the transfer of one S. For the stoichiometry of 2:1, substrate accumulation ratios of more than 400 are thermodynamically feasible. It has now become clear that cells and tissues vary in this stoichiometry: the simplest explanation is to assume that at least two transport systems for a sugar exist, differing in the coupling stoichiometry. Physiologically, $n > 1$ signifies a more effective process in that secretion or absorption can proceed against steeper gradients. Thus, the accumulation of amino acids in the central nervous system against gradients of 3–4 orders of magnitude is effected by a coupling of 2 Na^+ per 1 amino acid (*cf.* Erecinska, 1987).

The concept of secondary active transport is now well established (*cf.* Heinz, 1989): a number of such transport systems have been described, and the criteria for distinguishing this from primary active transport (directly linked to reactions providing the required energy, *i.e.* ATP hydrolysis) have been put forward by Aronson (1981). The discovery of the active chloride transport may serve as an example (*cf.* Zadunaisky, 1979): Krogh first suggested the possibility of an active chloride transport across cell membranes. The introduction of isotopic and electrophysiologic methods permitted characterization of the process in detail. On the basis of experiments on intestinal transport, Nellans et al. (1973) first suggested the coupling of an electroneutral, carrier-mediated cotransport of Na^+ and Cl^- with the operation of the sodium pump as the primary driving force. Later, the electroneutral component was defined as a coupling of Na^+, K^+, and Cl^--fluxes in the proportion of 1:1:2 (Geck et al., 1980). However, a chloride transport directly coupled to ATP hydrolysis also appears to be operative in some systems (see below).

Such results place on the shoulders of investigators the additional burden of dissecting an active transport phenomenon in terms of the effective driving forces, thus discriminating between direct and indirect coupling. At the same time, the student may be relieved of the need to search for separate directly coupled active transport systems.

The above development of ideas did not preclude the search for other coupling mechanisms. Again based on Mitchell's concepts, Roseman (1969) discovered in bacteria another mechanism for sugar transport by demonstrating sugar phosphorylation in the membrane, and subsequent transfer of the sugar phosphate into the cells (the phosphotransferase system). Of course, this mechanism does not meet the strict criteria of active transport, as defined above.

4. CRITIQUE OF THE ACTIVE TRANSPORT CONCEPT

The concepts presented above were based on the hypothesis of specific membrane phenomena. Such views were and are vigorously opposed by several groups of investigators. The Prague Symposium witnessed a major clash of ideas on active transport: The sorption theory, represented by Troshin (*cf.* 1966) and Ling's (*cf.* 1984) association-induction hypothesis maintain that the unequal distribution of electrolytes between cells and their environment, as well as associated electrical phenomena, may be explained by an adsorption particularly of cations on intracellular proteins; Ling's more recent views postulate an involvement of cell ATP (and the (Na^+-K^+)-ATPase) in these adsorption-desorption processes; he concedes the existence of active transport across epithelial cells[19] but visualizes this as the result of a sequence of adsorptive and desorptive processes at two sites in series.

It would appear that at this stage, Ling is one of few outspoken opponents of the membrane concept of active transport (although the respective models are not necessarily mutually

---

19 As did Troshin's teacher Nasonov, in a discussion with this author in 1967.

exclusive). As also discussed in Chapter 2, the association-induction hypothesis of Ling's does not account for information
forthcoming in the last decade *e.g.* the reconstitution of pump
action in liposomes using purified ATPases, and the expression of membrane function by single genes (see below). Rus-
sian investigators now back the membrane theory (Dr. A. Lev,
personal communcations).

III. Mechanism of active transport

The membrane theory of transport phenomena actually implies a new criterion for active transport, i.e. the localization of
the transport mechanism in the membrane, and hence the
possibility of working with such membrane preparations in
vitro. The clarified phenomenology of the investigated process
permits focusing on its mechanism. The definitive demonstration of an active transport mechanism requires that the process be dissected into a sequence of individual steps (reactions),
the respective protein(s) be identified, isolated, characterized
and eventually the process be reconstituted in defined lipid
bilayer membranes. The nature of the questions asked also
requires a change in the employed experimental approaches,
from predominantly kinetic studies towards a more biochemical analysis (enzymic properties of the transporter; its protein
chemistry, etc.), with kinetics employed as an experimental
tool. This development will be traced particularly for two active transport systems, *i.e.* the Na$^+$-pump, and the secondary
active transport of glucose.

1. TOWARDS THE MECHANISM OF THE NA$^+$-PUMP

The recognition of the (Na$^+$-K$^+$)-ATPase as a component of the
Na$^+$-pump raised the question of correlation of the pump with
the enzyme. The closeness of the characteristics of the (Na$^+$-

K^+)-ATPase and the pump in a variety of cells was demonstrated by Post et al., and by Dunham and Glynn in the early 60s and subsequent studies have substantiated this view, leading to the concept that the Na^+-pump and the (Na^+-K^+)-ATPase are a single entity (see *e.g.* Hoffman, 1962 and Bonting, 1971). The use of specific inhibitors of the enzyme, cardiac glycosides, as probes proved to be essential in the course of establishing its localization in the cell membrane, and in its preparation and purification. An effect of a specific antibody-antigen reaction on the (Na^+-K^+)-ATPase of sheep erythrocytes (Ellory and Tucker, 1969) provided further support for the identity of the enzyme and the pump. Careful studies in several laboratories (*cf.* Bonting, 1971) then demonstrated the presence of the (Na^+-K^+)-ATPase in most living cells (animal, plant, and bacterial): the operation of the Na^+-pump thus was established as the crucial mechanism by which the unequal distribution of the bulk cations between cells and their environment is brought about.

The study of the mechanism of the Na^+-pump aims at answering the question how the energy from the hydrolysis of ATP is used to perform osmotic and electrical work by transporting ions against their electrochemical gradients.

Studies of the biochemical mechanism of the (Na^+-K^+)-ATPase required work with membrane preparations in which the vectorial nature of the ion transport process is lost, and the hydrolysis of ATP is scalar as in bulk solution (*cf.* Mitchell, 1979). The asymmetric nature of the enzyme was clearly recognized by Whittam (1962). Skou (1960) visualized a phosphorylated intermediate, comprising the enzyme, ATP, Mg^{2+}, Na^+ and K^+. Subsequently, several groups of investigators (*cf.* Post, 1974) succeeded in isolating an intermediate which satisfied criteria of competence and eventually allowed the description of the enzyme action in terms of a cyclic process (the Albers-Post reaction scheme) involving two conformational states with a distinct asymmetry as to interaction with ATP and Na^+ at the inner face, and K^+ (and the cardiac glycoside ouabain) on the outer face of the membrane. A stoichiometry of 3 Na^+:2 K^+:1 ATP for the enzyme preparation was established, identical with that found on the cellular level. An important further step was the demonstration in several laboratories of the reversibility of the individual reactions of the enzyme, resulting in the formation of ATP when the ionic gradients were reversed. Several modes of the operation of the sodium pump were described (*cf.* Glynn and Karlish, 1978). From the fact that a net transfer of a charge across the membrane takes place, it could be deduced that the operation of the sodium pump would be affected by the membrane potential, and such prediction could be verified (*cf.* De Weer et al., 1988). Thus, while important details of the mechanism of the (Na^+-K^+)-ATPase are still being elucidated (*cf.* Hoffman and Forbush, 1983), the broad outlines of this process have been sketched by 1978. Any model of a transport ATPase requires: (1) a cytosolic site for bonding and hydrolysis of ATP (plus Mg^{2+}); (2) a site(s) for the extruded ion; (3) phosphorylation of the protein after binding of ATP and the extruded ion, associated with conformational changes[20] of the protein; at the same time, a barrier must exist which prevents a back-leak of the extruded ion while the ATP is not bound, thus permitting a vectorial flow of the extruded ion inside → out; (4) an occluded portion of the pump molecule (possibly its hydrophobic portion) separating the inner and outer side of the molecule; (5) an outer site at which the extruded ion is released and the inflowing ion binds. The definite establishment of the mechanism requires a reconstitution of the transport system after imbedding the asymmetric protein in a lipid bilayer.

Once the putative Na^+-pump (the [Na^+-K^+]-ATPase) was isolated in sufficient purity, the next crucial step in establishing the mechanism could be taken: In several laboratories the pump was reconstituted by incorporation of the purified en-

---

20 This term is now used to denote the actual propelling function of the pump.

zyme in liposomes and demonstrating an electrogenic, uphill transport of Na^+ and K^+ in this system (cf. Dixon and Hokin, 1980). These studies also provided a substantive link of the enzyme preparation to the function of the pump in cells.

1.1. Other ionic pumps

The success of the analytical approach which led to the description, isolation and characterization (in biochemical terms) of the Na^+-pump was paralleled in studies of other ionic pumps (cf. Läuger, 1991), particularly:

The Ca^{2+}-ATPase from the sarcoplasmic reticulum (Hasselbach and Makinose, 1961) and the erythrocyte membrane (described by Schatzmann in 1966, cf. Schatzmann, 1982);

The K^+-ATPase e.g. from insect midgut, first clearly established by Harvey and Needegard (1963);

The $(H^+$-$K^+)$-ATPase from the gastric mucosa, recognized by Ganser and Forte (1973) as a K^+-stimulated ATPase and later characterized further (cf. Sachs, 1987);

The H^+-ATPase e.g. from plants and fungi (cf. Rea and Pook, 1985; Serrano, 1988);

The H^+-ATPase, e.g. from phosphorylating membranes, such as the proton pump coupled to cytochrome-c oxidase in mitochondria by Wikström (1977).

In addition to the above cation pumps, an anion-stimulated ATPase has been described by Kasbekar and Durbin (1965) and such anion pumps have been demonstrated in various cells. An ATP-driven pump (the multidrug resistance P-glycoprotein) extruding a variety of drugs has also been found (Horio et al., 1988); the protein of this pump shows major homology with a variety of functionally differing transport proteins (cf. Chapter 2, see also Chen et al., 1986). As pointed out in Section II.1, ATP hydrolysis associated with a transport process does not necessarily constitute evidence for a pump: The protein of the Cl^--channel requires ATP hydrolysis for the maintenance of its conformation in the open state (Anderson et al., 1991), while the actual Cl⁻-transport does not proceed against its electrochemical gradient.

Pedersen and Carafoli (1987) group known ion-motive ATPases into three major categories: The 'P' class forms covalent phosphorylated intermediates as part of the reaction cycle (*e.g.* Na⁺-K⁺-ATPase; Ca²⁺-ATPase; H⁺-ATPase of plasma membranes of eukaryotes; K⁺-ATPases of bacteria). The ion-motive 'V' ATPases are associated with membrane organelles other than mitochondria or sarcoplasmic reticulum, and are found e.g. in vacuoles, lysosomes, secretory granules, etc. The 'F' category encompasses enzymes of the F_0F_1 type found in mitochondria, chloroplasts and bacteria. Other ion-motive ATPases may be still be discovered; thus, the possibility of an electrogenic, ATP-dependent Cl⁻ transport in the intestinal cells of some species is now being explored. A relationship between individual enzymes may eventually be revealed by comparisons of their amino acid sequences: efforts in this direction (*cf.* Jørgensen and Andersen, 1988) have shown a great homology of the catalytic subunits of the (Na⁺-K⁺)-ATPase, Ca²⁺-ATPase and the gastric (H⁺-K⁺)-ATPase, and thus it is not surprising that several of their enzymic functions also show major similarities (*cf.* Polvani and Blostein, 1988; Rabon and Reuben, 1990) in that the at first described ionic specificities of pumps might be less definite than originally thought. Thus, protons may to some extent substitute for Na⁺ or K⁺ in the sodium-pump, and Na⁺ may act as a substrate for the gastric (H⁺-K⁺)-ATPase. Such information could add further to the unifying concept of pumps.

Ionic pumps can be driven by a variety of forces: in halobacteria, the proton (*cf.* Stoeckenius and Bogomolny, 1982) and chloride (Schobert and Lanyi, 1982) pumps are driven by light. The redox energy (*cf.* earlier views of Lundegårdh and Conway, p. 155) drives the proton pump of the cytochrome oxidase (Wikström, 1977) and the Na⁺ pump coupled to the oxidation of NADH (Tokuda and Unemoto, 1984). In bacteria, the Na^+ pump is coupled to the decarboxylation of oxaloacetate (Dimroth, 1982). In plants, the hydrolysis of inorganic pyrophosphate provides the energy for the proton pump of plant vacuoles (Rea and Pook, 1985). Thus, the conformational changes required for the translocation of solutes can derive the required energy from various physical and biochemical sources. Functionally, the ionic pumps represent the primary enzyme systems capable of producing an asymmetric ionic distribution across an osmotic (solute-impermeable) membrane barrier, and may be regarded as an energy-transducing device which converts metabolic energy into the osmotic potential energy represented by the $\Delta\mu$ of a given ionic species i. Mitchells's (*cf.* 1963; 1992) chemiosmotic hypothesis originally emphasized primarily $\Delta\mu_{H^+}$ as a driving force in membrane phenomena. At present both $\Delta\mu_{H^+}$ and $\Delta\mu_{Na^+}$ appear to be the dominant driving mechanisms for performing work (chemical, osmotic or mechanical) across membranes of living matter (Fig. 2; *cf.* Skulachev, 1992); in principle, one might envisage that *any* ionic electrochemical potential may prove to be the driving force[21]. From this point of view, the coupling of downhill flow of a solute with an uphill ion flux, as in secondary active transport (see below) represents only a special case.

The study of isolated ATPases raised new, unexpected questions. It is now recognized (*cf.* Fambrough, 1988) that the sodium pump is actually a family of enzymes (first described by Sweadner in 1979) differing particularly in the properties of their α- and β-subunits and controlled by several genes. An analysis of the functional role of these isoforms awaits elucidation. More recent investigations (*cf.* Goldshleger et al., 1990) also begin to question the paradigm of a constant stoichiometry in individual ionic pumps; such idea was raised earlier (Davies and Keynes, 1961).

---

21 The Na^+/H^+ exchanger (Chapter 3, Section II.3.2) may actually serve as a coupling agent: the H^+-pump together with the Na^+/H^+ antiport is functionally a Na^+-pump.

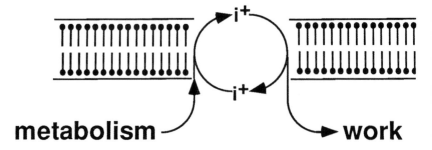

metabolism ⟍ ⟶ **work**

Fig. 2. Active ion transport and its coupling to biological energy-requiring
processes. In a cyclic process, i^+ is pumped against an electrochemical gradient of $\Delta\mu_{i+}$; by coupling to an energy-requiring process, chemical (e.g. a synthetic process such as $ADP + P_i \rightarrow ATP$, osmotic (uphill drive of solute S),
heat production, or mechanical work (e.g. flagellar motion) is performed (cf.
Chapter 6; Skulachev, 1992).

From the evolutionary point of view, the proton-pumping 'F'
ATPases may have evolved first as a device to extrude acids
generated during the anaerobic metabolism of primitive
cells[22] (cf. Serrano, 1988). ATPases of the 'P' type appear to
have developed later in response to the requirements of ionic
homeostasis in cells, offering greater versatility as to ionic
specificity (Jørgensen and Andersen, 1988).

2. TOWARDS THE MECHANISM OF THE SECONDARY
ACTIVE SOLUTE TRANSPORT

The use of membrane vesicles prepared from epithelial cells,
first introduced by Miller and Crane (1961), permitted clarification of the mechanism of transport processes in more detail
since in such preparations substrate metabolism is eliminated

---

22 Such concept would have been welcomed by Mitchell as an additional argument in
favor of the protonmotive force as the primary driving force (see Chapter 6).

due to a loss of intracellular enzymes; effects of ionic and substrate gradients can also be studied. Preparations derived from either the brush-border, or the basolateral membranes (*cf.* Kinne, 1976) permit the definition of the cellular localization of the transport.

Membrane vesicles made it possible to provide convincing evidence in favor of the gradient hypothesis as the mechanism for the secondary active transport of organic solutes, as shown at the Symposium of the New York Academy of Sciences (Semenza and Kinne, 1985). Since the brush-border vesicles of epithelial cells are impoverished in the Na^+-extruding mechanism, the imposition of a Na^+ gradient outside > inside produced an uphill flux of glucose (phlorizin-inhibitable) or of amino acids in the vesicles; the accumulation of the organic solutes was abolished when the Na^+ gradient was dissipated (Kinne, 1976). The predicted electrogenicity of the Na^+/glucose cotransport system was then established by Murer and Hopfer (1974). On the other hand, vesicles derived from the basolateral membrane did not display the Na^+/glucose cotransport system, but showed a high concentration of the $(Na^+$-$K^+)$-ATPase. Thus, the concept of secondary active organic solute transport, developed on the cellular level as a coupling between a Na^+-organic solute cotransport and an energy-dependent ion flux process was substantiated using membrane preparations. This demonstration opened the door to efforts to isolate the transport system. The electrogenicity of the cotransport system also explained the dependence of sugar transport (and other secondary active transport systems) on the membrane potential, and competition of various substrates for the electrochemical potential component of the driving force for the secondary active solute transport (Murer et al., 1975).

Conceptually, the major advance in the understanding of the process resides in the recognition that, as opposed to original ideas of a transport protein revolving in the membrane, the Na^+/glucose cotransport system may represent a channel with sites for glucose and one or two negatively charged sites

for interaction with Na⁺. New kinetic techniques, described for
the alanine-Na⁺ cotransport system (Jauch and Läuger, 1988)
also open the door for discriminating between two possible
models: The simultaneous mechanism assumes that both the
binding sites for the solute S and the cotransported cation A
are alternatively accessible at both sides of the membrane; the
conformational transitions of the carrier C which permit the
switching of the sides across the membrane then can occur
only in the empty, or fully occupied states. The alternative
transport mechanism postulates that A has to bind at the
outer side before S, and vice versa, at the cytoplasmic side S
interacts before A; such a model does not involve transitions
between two conformations of the cotransporter:

$$C_o \rightarrow AC_o \rightarrow SAC_o \rightarrow C_iAS \rightarrow C_iA \rightarrow C_i,$$

the subscripts o and i denoting the outer and inner side, respectively. Kinetics argue in favor of the simultaneous model.

Accumulated information, together with the use of phlorizin
as high affinity probe of the Na⁺/glucose cotransport system,
then served as the basis for studies directed to isolate, purify
and reconstitute the transport system. By now, the Na⁺/glu-
cose cotransporter has been cloned, and its amino acid sequence has been established (Hediger et al., 1987).

While the Na⁺-gradient appears to be the predominant driving force for the secondary active solute transport in animal
cells, it is not the only mechanism. (a) A H⁺-solute cotransport
system has been established for some solutes, at least in
subcellular organelles. (b) A role of K⁺ in organic solute
cotransport, first suggested by Riggs et al. (1958) for amino
acids, and later dismissed in favor of Na⁺, also is now being
considered (cf. Kinne and Heinz, 1987): In addition to the role
of K⁺ in the Na⁺,K⁺,Cl⁻-cotransport system (see above), a stimulatory role of K⁺ in the transport of amino acids, sugars and
organic acids has been found. In organisms with K⁺ in their
cellular environment (e.g. moths), the electrochemical gradi-

ent of this cation appears to be the primary driving force for solute transport, and a K^+-amino acid cotransport system has been documented by Giordana et al. (1982); this system is not quite specific for K^+ since gradients of Na^+ and some other alkali cations can also be used. (c) Moreover, there is a class of solutes which are accumulated against sizeable gradients, but so far the driving force could not be pinpointed. Some sugars (*e.g.* 2-deoxyhexoses) can be accumulated in various cells as free (non-phosphorylated) solutes against considerable concentration gradients by a mechanism which is Na^+- and probably H^+-independent (*cf.* Widdas et al., 1989), but cannot be dissociated from the phosphorylation of the substrate. The Na^+-independent active transport of neutral amino acids (system L of the Christensen nomenclature, (*cf.* Ohsawa et al., 1980; Heinz et al., 1981) represents a similar uncharted area.

In bacteria, the genetically-controlled sugar accumulation was first considered by Monod to be catalyzed by a metabolic enzyme (termed a permease) localized in the membrane. Subsequently, Mitchell (1963) proposed a coupling of sugar entry to a proton flux, and such mechanism was then established by West (1970). The electrogenicity of the process was recognized in the laboratories of Mitchell and of Harold. As with animal cells, the use of membrane vesicles derived from bacterial cytoplasmic membranes (Kaback, 1972) proved to be a major step forward in establishing the mechanisms of transport (*cf.* Kornberg, 1973). The proton is used as the energy coupling cation in most cotransport systems in bacteria[23], yeast, and lower and higher plants. By now, the arabinose-H^+ and xylose-

---

23 In Gram-negative bacteria, the actual active accumulation of sugars (and other nutrients) may be still more complex. (Pardee, 1968) first drew attention to the existence of high-affinity substrate binding proteins readily dissociable from the bacterial cells by osmotic shock. The relationship of such galactose binding protein to its active transport and the chemotactic phenomena was then demonstrated (*cf.* Kalckar, 1971). By now, a family of such high-affinity binding proteins has been characterized, and their role as intermediaries in the transfer of the substrate to its membrane transport site has been established (*cf.* Quiocho, 1989).

H^+ transport proteins of *E. coli* have been isolated and sequenced (Maiden et al., 1987); these proteins are homologous with each other, and unexpectedly, with the (facilitated) glucose transporter of animal cells. Thus, some basic similarities in transport systems begin to emerge. However, Na^+/sugar cotransport systems are also operative in bacteria although the rather high affinity of the transport system for this cation (with a K_m in the μM range) may make it difficult to distinguish it from other cotransport systems (Wilson and Wilson, 1987).

Available information indicates that the use of a specific cationic species as the energy-coupling mechanism in cotransport systems in various cells reflects the pertinent electrochemical gradients of the given cation as well as the structural aspects of the particular solute. In animal cells the electrochemical gradient of Na^+ is usually predominant, exceeding that of H^+. With partially dissociated solutes such as amino acids, both Na^+ and/or H^+ may serve. On the other hand, in *e.g.* bacteria or yeast, the H^+ gradient appears to be the predominant driving force.

3. ACTIVE TRANSPORT SYSTEMS AND THE MEMBRANE THEORY

The discoveries of the last decade require a reappraisal of the available evidence by critics of the membrane theory of transport processes (see above). The observations that transport proteins, isolated from membrane preparations and reconstituted in lipid bilayers, display the original transport function represent a cogent argument for the membrane theory. Furthermore, the findings that single genes express membrane transport functions are fully consistent with the membrane theory.

IV. What next?

The above outline demonstrates the significant progress achieved, from the early recognition of the phenomenon of active transport to an analysis of the underlying mechanisms. The present views postulate a better understanding on several levels.

The analytical approach raises questions about the specific properties of the transport proteins related to their function. To the extent to which the transport proteins are not directly coupled to metabolic energy (*e.g.* ATP), these points apply equally to the mobile carrier concept (see Chapter 3). Students of transport adopt the approach of protein chemists by asking questions such as: how is a solute recognized by the transport protein (*i.e.* the nature of bonds between the solute and the protein); this point includes the nature of the interaction between ATP (or other energy-yielding systems) and the transport protein; what is the tertiary structure by the transport protein which permits its anchoring in the lipid membrane and at the same time provides a channel through which solutes can pass (be exchanged); are conformational changes of the protein involved in the actual transport process; what is the structural basis for the basic asymmetry of the transport protein? The physiology of transport phenomena adds further questions at the protein level, such as modifier sites which permit a fine tuning of the function of the transport protein, and the interaction between the membrane-lipids and the transport protein.

For the transport enzymes coupled to the various energy sources metabolic energy the fundamental question still begs for an answer: how is the conformational change in the transport protein (produced in transport ATP-ases by protein phosphorylation) translated into a directional propulsion of a solute? Judging from the studies of Engelhardt et al. (1988) on bacteriorhodopsin, the mobile surface residues of aspartate may be involved in the actual propulsion of H^+ by this pump[24]; an alternative view (Henderson et al., 1990) visualizes a proton channel with a series of aspartate groups acting as sequential acceptors of the proton. Striking similarities in the amino acid sequence in several transport ATPases may eventually lead to answers to such questions.

The physiologist in the student of transport is intrigued by another set of questions, related to the cellular organization of the active transport process, its regulation, be it by the cellular environment, other transport processes, physiological or pharmacological agents, developmental and comparative aspects of the process, and its genetic expression.

Newton's postulate (Newton, 1727): 'Natura enim simplex est' (Nature is always simple) prompts yet another look at the development of the pump concept. Hitherto, the main effort was directed at the description of the various pumps, based on their specificities and other properties. Given the emerging similarities in the amino acid sequences of at least the P-class of ionic pumps (see above), the observed homologies permit enquiry into the possibility that the detailed mechanisms (including the amino acid sequence) of interaction of the pump protein with ATP, and the actual propulsion mechanism, might be very close, if not identical; basically, the pumps would then differ essentially in the recognition site for the transported solute(s).

---

24 In the bacterial H^+-substrate symport, the specific polarity and configuration of the closely located histidine and glutamic acid in the peptide chain of the transport protein appear to be essential for the lactose-coupled H^+ translocation (Kaback, 1988). This study also demonstrates the powerful tool provided by site-directed mutagenesis for the elucidation of the molecular aspects of such questions.

Acknowledgements

The author is greatly indebted to Drs. W.M. Armstrong, M.M. Civan, D.M. Fambrough, R.L. Post and D.F. Wilson for their critical comments and suggestions.

REFERENCES

Anderson, M.P., Berger, H.A., Rich, D.P., Gregory, R.J., Smith, A.E. and Welsh, M.J. (1991) Nucleotide triphosphates are required to open the CFTR chloride channel. Cell 67, 775–784.

Aronson, P.S. (1981) Identifying secondary active solute transport in epithelia. Am. J. Physiol 240, F1–F11.

Bernstein, J. (1902) Untersuchungen zur Thermodynamik der bioelektrischen Ströme. Pflüger's Arch. ges. Physiol. 192, 521–562.

Bonting, S.L. (1971) Sodium-potassium activated adenosinetriphosphatase and cation transport. In Membranes and Ion Transport, Vol. 1. (Bittar, E.E., ed.), pp. 257–363. New York, Wiley-Interscience.

Boyle, P.J. and Conway, E.J. (1941) Potassium accumulation in muscle and associated changes. J. Physiol. (Lond.) 100, 1–63.

Briggs, G.E., Hope, A.B. and Robertson, R.N. (1961) Electrolytes and Plant Cells. Oxford, Blackwell Publ.

Brodie, T.G. and Vogt, H. (1910) The gaseous metabolism of the small intestine. Part I. The gaseous exchange during absorption of water and dilute salt solutions. J. Physiol. (Lond.) 40, 135–172.

Brooks, S.C. (1938) The penetration of radioactive potassium chloride into living cells. J. Cell. Comp. Physiol. 11, 247–252.

Brooks, S.C. and Brooks, M.M. (1941) The Permeability of Living Cells. Berlin-Zehlendorf, Gebr. Borntraeger.

Chen, C.-J., Chin, S.E., Ueda, K., Clark, D.P., Pastan, I., Gottesman, M.M. and Roninson, I.B. (1986) Internal duplication and homology with bacterial transport proteins in the mdr1 (P-glycoprotein) gene from multidrug-resistant human cells. Cell 47, 381–389.

Christensen, H.N. (1975) Biological Transport. Reading, MA, W.A. Benjamin.

Christensen, H.N. (1976) Towards a sharper definition of energetic coupling through integration of membrane transport into bioenergetics. J. Theor. Biol. 57, 419–431.

Christensen, H.N. (1984) Organic ion transport during seven decades. The amino acids. Biochim. Biophys. Acta 779, 255–269.

Christensen, H.N. (1993) Amino acid nutrition: a two-step absorptive process. Nutr. Revs. 51, 1–6.

Clark, G.A. (1922) Glucose absorption in the renal tubules of the frog. J. Phyiol. (Lond.) 56, 201–205.

Clarkson, E.M. and Maizels, M. (1952) Distribution of phosphatases in human erythrocytes. J. Physiol. (Lond.) 116, 112–128.

Cohen, G.N. and Monod, J. (1957) Bacterial permeases. Bact. Rev. 21, 169–194.

Cohn, W.E. and Cohn, E.T. (1939) Permeability of red corpuscles of the dog to sodium ion. Proc. Soc. Exp. Biol. Med. 41, 445–449.

Cohnheim, O. (1898) Über Dünndarmresorption. Ztschr. f. Biol. 12, 129–153.

Cohnheim, O. (1901) Versuche über Resorption, Verdauung und Stoffwechsel von Echinodermen. Hoppe-Seyler's Z. physiol. Chem. 33, 9–54.

Collander, R. (1921) Über die Permeabilität pflanzlicher Protoplasten für Sulfosäure Farbstoffe. Jahrb. wiss. Bot. 60, 354–410.

Cori, C.F. (1925) The fate of sugar in the animal body. I. The rate of absorption of hexoses from the intestinal tract. J. Biol. Chem. 66, 691–715.

Crane, R.K., Miller, D. and Bihler, I. (1961) The restrictions on possible mechanisms of intestinal active transport of sugars. In *Membrane Transport and Metabolism*, (Kleinzeller, A. and Kotyk, A., eds.), pp. 439–449. New York, Academic Press.

Crane, R.K. (1977) The gradient hypothesis and other models of carrier-mediated active transport. Rev. Physiol. Biochem. Pharmacol. 78, 101–159.

Crane, R.K. (1986) Questions. In *Ion Gradient-Coupled Transport* (Alvarado, F. and van Oss, C.H., eds.), pp. 431–438. Elsevier, Science Publ.

Csáky, T.Z. and Thale, M. (1960) Effect of ionic environment on the intestinal sugar transport. J. Physiol. (Lond.) 151, 59–65.

Csáky, T.Z. and Wilson, J.E. (1956) The fate of $^{14}CH_3$-glucose in the rat. Biochim. Biophys. Acta 22, 185–186.

Danielli, J.F. (1954) Morphological and molecular aspects of active transport. Symp. Soc. Exp. Biol. 8, 502–516.

Danowski, T.S. (1941) The transfer of potassium across the human blood cell membrane. J. Biol. Chem. 139, 693–705.

Davies, R.E. and Keynes, R.D. (1961) A coupled sodium-potassium pump. In *Membrane Transport and Metabolism* (Kleinzeller, A. and Kotyk, A., eds.), pp. 336–340. Academic Press, New York.

Davson, H. (1940) Ionic permeability of the cat erythrocyte membrane to sodium and potassium. The comparative effects of environmental changes. J. Cell Comp. Physiol. 15, 317–330.

Dean, R.B. (1941) Theories of electrolyte equilibrium in muscle. Biol. Symp. 3, 331–348.

Dean, R.B., Noonan, T.R., Haege, L. and Fenn, W.O. (1941) Permeability of erythrocytes to radioactive potassium. J. Gen. Physiol. 24, 353–365.

DeGowin, E.L., Harris, J.E. and Plass, E.D. (1940) Studies on preserved human blood. I and II. J. Am. Med. Assoc. 114, 850–857.

De Weer, P., Gadsby, D.C. and Rakowski, R.F. (1988) Voltage dependence of the Na-K pump. Annu. Rev. Physiol. 50, 225–241.

Dimroth, P. (1982) The generation of an electrochemical potential gradient of sodium ions upon decarboxylation of oxaloacetate by a membrane-bound and Na⁺-activated oxaloacetate decarboxylase from *Klebsiella aerogenes*. Eur. J. Biochem. 121, 443–449.

Dixon, J.F. and Hokin, L.E. (1980) The reconstituted (Na,K)-ATPase is electrogenic. J. Biol. Chem. 255, 10681–10686.

Donnan, F.G. (1911) Theorie der Membrangleichgewichte und Membranpotentiale bei Vorhandensein von nicht-dialysierenden Elektrolyten. Ein Beitrag zur physikalisch-chemischen Physiologie. Z. Elektrochem. 17, 572–581.

Dowman, C.B.B., Oliver, J.O. and Young, I.M. (1940) Partition of potassium in stored blood. Brit. Med. J., pp. 559–561.

Ellory, J.C. and Tucker, E.M. (1969) Stimulation of the potassium transport system in low potassium sheep red cells by a specific antigen-antibody reaction. Nature (Lond.) 222, 477–478.

Engelhard, M., Hess, B., Emeis, D., Metz, G. and Siebert, F. (1988) Magic angle sample spinning ¹³C-nuclear magnetic resonance of isotopically labeled bacteriorhodopsin. Biochemistry 28, 3967–3985.

Erecinska, M. (1987) The neurotransmitter amino acid transport systems. A fresh outlook on an old problem. Biochem. Pharmacol. 36, 3547–3555.

Fambrough, D.M. (1988) The sodium pump becomes a family. TINS 11, 325–328.

Fenn, W.O. (1936) Electrolytes in muscle. Physiol. Rev. 16, 450–487.

Fenn, W.O. (1940) The role of potassium in physiological processes. Physiol. Rev. 20, 377–415.

Fleckenstein, A. (1948) Über den primären Energiespeicher der Muskelkontraktion. Pflüger's Arch. ges. Physiol. 250, 643–666.

Franck, J. and Mayer, J.E. (1947) An osmotic diffusion pump. Arch. Biochem. Biophys. 14, 297–313.

Gale, E.F. (1971) Don't talk to me about permeability: The Tenth Marjorie Stephenson Memorial Lecture. J. Gen. Microbiol. 68, 1–14.

Gale, E.E. and Mitchell, P.D. (1947) The assimilation of amino-acids by bacteria. 4. The action of triphenylmethane dyes on glutamic acid accumulation. J. Gen. Microbiol. 1, 299–313.

Ganser, A.L. and Forte, J.G. (1973) K⁺-stimulated ATPase in purified microsomes of bullfrog oxyntic cells. Biochim. Biophys. Acta 307, 169–180.

Gárdos, G. (1954) Akkumulation der Kaliumionen durch menschliche Blutkörperchen. Acta Physiol. Acad. Sci. Hung. 6, 191–199.

Geck, P., Pietrzyk, C., Burckhardt, B.-C., Pfeiffer, B. and Heinz, E. (1980) Electrically silent cotransport of Na⁺, K⁺, Cl⁻ in Ehrlich cells. Biochim. Biophys. Acta 600, 432–447.

Giordana, B., Sacchi, F.V. and Hanozet, G.M. (1982) Intestinal amino acid absorption in lepidopteran larvae. Biochim. Biophys. Acta 692, 81–88.

Glynn, I.M. (1956) Sodium and potassium movements in human red cells. J. Physiol. (Lond.) 134, 278–310.

Glynn, I.M. and Karlish, S.J.D. (1978) The association of biochemical events and cation movements in (Na+-K+)-dependent adenosine triphosphatase activity. Biochem. Soc. Spec. Publ. 4, 145–158.

Goldacre, R.J. (1952) The folding and unfolding of protein molecules as a basis for osmotic work. Internat. Rev. Cytol. 1, 135–164.

Goldshleger, R., Shahak, Y. and Karlish, S.J.P. (1990) Electrogenic and electroneutral transport modes of renal Na,K-ATPase reconstituted into phospholipid vesicles. J. Membrane Biol. 113, 139–154.

Hahn, L.A., Hevesy, G. and Rebbe, O.H. (1939) Do the potassium ions inside the muscle cells and blood corpuscles exchange with those present in the plasma? Biochem. J. 33, 1549–1558.

Hales, S. (1727) Vegetable Staticks, pp. 76 ff. London, W. and J. Innys and T. Woodward.

Hamburger, H.J. (1891) Über den Einfluss der Atmung auf die Permeabilität der Blutkörperchen. Z. Biol. 28, 405–416.

Hamburger, H.J. and Bubanovic, F. (1910) La permeabilité des globules rouges, specialement vis-à-vis des cations. Arch. Int. Physiol. 10, 1–36.

Harris, J.E. (1941) The influence of the metabolism of human erythrocytes on their potassium content. J. Biol. Chem. 141, 579–595.

Harvey, W.R. and Nedergaard, S. (1964) Sodium independent active transport of potassium in the isolated midgut of Cecropia silkworm. Proc. Natl. Acad. Sci. USA 51, 757–765.

Hasselbach, W. and Makinose, M. (1961) Die Calciumpumpe der 'Erschlaffungsgrana' des Muskels und ihre Abhängigkeit von der ATP-Spaltung. Biochem. Z. 333, 518–528.

Hediger, M.A., Coady, M.J., Ikeda, T.S. and Wright, E.M. (1987) Expression cloning and cDNA sequencing of the Na+/glucose co-transporter. Nature (Lond.) 330, 379–381.

Heidenhain, R. (1874) Versuche über den Vorgang der Harnabsonderung. Pflüger's Arch. ges. Physiol. 9, 1–27.

Heidenhain, R. (1894) Neue Versuche über die Aufsaugung im Dünndarm. Pflüger's Arch. ges. Physiol. 56, 579–631.

Heinz, A., Jackson, J.W., Richey, B.E., Sachs, G. and Schafer, J.A. (1981) Amino acid transport by substrates in the absence of a Na+ electrochemical potential gradient. J. Membr. Biol. 62, 149–160.

Heinz, E. (1978) Mechanics and Energetics of Biological Transport. Heidelberg, Springer Verl.

Heinz, E. (1989) The unfinished story of secondary active transport. In Membrane Transport. People and Ideas. (Tosteson, D.C., ed.), pp. 2367–250. Bethesda, MD., Am. Physiol. Soc.

Henderson, R., Baldwin, J.M., Ceska, T.A., Zemlin, F., Beckmann, E. and Denning, K.H. (1990) Model for the structure of bacteriorhodopsin based on high resolution electron cry-microscopy. J. Mol. Biol. 213, 899–929.

Hill, A.V. (1931) *Adventures in Biophysics*. Philadelphia, University of Pennsylvania Press.

Hoagland, D.R. (1937) Some aspects of the salt nutrition of higher plants. Bot. Rev. 3, 307–334.

Höber, R. (1901) Über Resorption in Darm. Pflüger's Arch. ges. Physiol. 86, 199–214.

Höber, R. (1926) *Physikalische Chemie der Zelle und der Gewebe*. 6. Aufl. Leipzig, Wilhelm Engelmann.

Hodgkin, A.L. and Keynes, R.D. (1955a) Active transport of cations in giant axons from *Sepia* and *Loligo*. J. Physiol. (Lond.) 128, 28–60.

Hodgkin, A.L. and Keynes, R.D. (1955b) The potassium permeability of a giant nerve fibre. J. Physiol. (Lond.) 128, 61–88.

Hoffman, J.F. (1962) Cation transport and structure of the red cell membrane. Circulation 26, 1201–1213.

Hoffman, J.F. and Forbush, B., III. (eds) (1983) *Structure, Mechanism, and Function of the Na/K Pump*. Curr. Top. Membr. Transp. 19.

Hokin, L.E. and Hokin, M.R. (1960) Studies on the carrier function of phosphatidic acid in sodium transport. I. The turnover of phosphatidic acid and phosphoinositide in the avian salt gland on stimulation of secretion. J. Gen. Physiol. 44, 61–85.

Horio, M., Gottesman, M.L. and Pastan, I. (1988) ATP-dependent transport of vinblastine in vesicles from human multidrug-resistant cell. Proc. Natl. Acad. Sci. USA 85, 3580–3584.

Huf, E. (1935) Über den Anteil vitaler Kräfte bei der Resorption von Flüssigkeit durch die Froschhaut. Pflüger's Arch. ges. Physiol. 236, 1–19.

Huf, E. (1936) Über aktiven Wasser- und Salztransport durch die Froschhaut. Pflüger's Arch. ges. Physiol. 237, 143–166.

Jacobs, M. (1931) The permeability of the erythrocyte. Erg. Biol. 7, 1–55.

Jauch, P. and Läuger, P. (1988) Kinetics of the Na^+/alanine cotransporter in pancreatic acinar cells. Biochim. Biophys. Acta 939, 179–188.

Jørgensen, P.L. and Andersen, J.P. (1988) Structural basis for E_1-E_2 conformational transitions in Na,K-pump and Ca-pump proteins. J. Membr. Biol. 103, 95–120.

Kaback, H.R. (1972) Transport across isolated bacterial cytoplasmic membranes. Biochim. Biophys. Acta 265, 367–416.

Kaback, H.R. (1988) Site-directed mutagenesis and ion-gradient driven active transport: On the path of the proton. Annu. Rev. Physiol. 50, 243–256.

Kalckar, H.M. (1971) The periplasmic galactose binding protein in *Escherichia coli*. Science 174, 557–565.

Kasbekar, D.H. and Durbin, R.P. (1965) An adenosine triphosphatase from frog gastric mucosa. Biochim. Biophys. Acta 105, 472–482.

Kedem, O. (1961) Criteria of active transport. In *Membrane Transport and Metabolism* (Kleinzeller, A. and Kotyk, A., eds.), pp. 87–93. New York, Academic Press.

Kinne, R. (1976) Properties of the glucose transport system in the renal brush border membrane. Curr. Top. Membr. Transp. 8, 209–267.

Kinne, R. and Heinz, E. (1987) Role of potassium in cotransport systems. Curr. Top. Membr. Transp. 28, 73–85.

Kimmich, G.A. and Randles, J. (1980) Evidence for an intestinal Na$^+$:sugar transport coupling stoichiometry of 2.0. Biochim. Acta 596, 439–444.

Kleinzeller, A. and Kotyk, A. (eds.) (1961) Membrane Transport and Metabolism. New York, Academic Press.

Koeppe, H. (1897) Der osmotische Druck als Ursache des Stoffaustausches zwischen roten Blutkörperchen und Salzlösungen. Pflüger's Arch. ges. Physiol. 67, 189–206.

Kornberg, H.L. (1973) Carbohydrate transport by microorganisms. Proc. R. Soc. Lond. B 183, 105–123.

Krogh, A. (1938) The active absorption of ions in some fresh water animals. Z. vergl. Physiol. 25, 336–350.

Krogh, A. (1946) The active and passive exchanges of inorganic ions through the surface of living cells and through living membranes generally. Proc. R. Soc. Lond. B 133, 140–200.

Läuger, P. (1991) Electrogenic Ion Pumps. Sunderland, MA., Sinauer Associates, Inc.

Liebig, J. (1847) Über die Bestandtheile der Flüssigkeiten des Fleisches. Ann. d. Chem. u. Pharmacie 62, 257–368.

Ling, G.N. (1984) In Search of the Physical Basis of Life. New York and London, Plenum Press.

Ludwig, C. (1861) Lehrbuch der Physiologie des Menschen. 2. Bd., p. 346. Leipzig u. Heidelberg, C.F. Winter'sche Verl.

Lundegårdh, H. (1937) Untersuchungen über die Anionenatmung. Biochem. Z. 290, 104–124.

Lundegårdh, H. (1945) Absorption, transport and exudation of inorganic ions by the roots. Ark. f. Bot. 32A, No 12, 2–139.

Lundsgaard, E. (1933) Die Wirkung von Phlorrhizin auf die Glucoseresorption. Biochem. Z. 264, 221–223.

Maiden, M.C.J., Davis, E.O., Baldwin, S.A., Moore, D.C.M. and Henderson, P.J.F. (1987) Mammalian and bacterial sugar transport systems are homologous. Nature (Lond.) 325, 641–643.

Miller, D. and Crane, R.K. (1961) A procedure for the isolation of the epithelial brush border membrane of hamster small intestine. Analyt. Biochem. 2, 254–286.

Mitchell, P.D. (1957) A general theory of membrane transport from studies of bacteria. Nature (Lond.) 180, 134–136.

Mitchell, P.D. (1962) Molecule, group and electron translocation through natural membranes. Biochem. Soc. Symp. 22, 142–168.

Mitchell, P.D. (1971) Reversible coupling between transport and chemical reactions. In Membranes and Ion Transport, Vol. 1, (Bittar, E.E. ed.), pp. 192–256. New York, Wiley-Interscience.

Mitchell, P.D. (1979) Compartmentation and communication in living systems. Eur. J. Biochem. 95, 1–20.

Mitchell, P.D. (1992) Foundations of vectorial metabolism and osmochemistry. Bioscience Reports 11, 297–344.

Mudge, G.H. (1951) Studies on potassium accumulation in rabbit kidney cortex slices. Effect of metabolic activity. Amer. J. Physiol. 165, 113–127.

Murer, H. and Hopfer, U. (1974) Demonstration of electrogenic Na^+-dependent D-glucose transport in intestinal brush border membrane. Proc. Natl. Acad. Sci. USA 171, 484–488.

Murer, H., Sigrist-Nelson, K. and Hopfer, U. (1975) On the mechanism of sugar and amino acid interaction in intestinal transport. J. Biol. Chem. 250, 7392–7396.

Nasse, H. (1878) Untersuchungen über den Austritt und Eintritt von Stoffen durch die Wand der Haargefässe. Pflüger's Arch. ges. Physiol. 16, 604–634.

Nellans, H.N., Frizzell, R.A. and Schultz, S.G. (1973) Coupled sodium-chloride influx across the brush border of rabbit ileum. Am. J. Physiol. 225, 467–475.

Nelson, M.T. and Blaustein, M.P. (1982) GABA efflux from synaptosomes: Effects of membrane potential, and external GABA and cations. J. Membrane Biol. 69, 213–223.

Netter, H. (1934) Die Stellung des Kaliums in Elektrolytensystem des Muskels. Pflüger's Arch. ges. Physiol. 234, 680–695.

Newton, I. (1729) The Mathematical Principles of Natural Philosophy, English translation. Vol. II. London, B. Motte, pp. 202.

Ohsawa, M., Kilberg, M.S., Kimmel, G. and Christensen, H.N. (1980) Energization of amino acid transport in energy-depleted Ehrlich cells and plasma membrane vesicles. Biochim. Biophys. Acta 599, 175–190.

Opie, E.L. (1949) The movement of water in tissues removed from the body and its relation to movement of water during life. J. Exp. Med. 89, 185–208.

Osterhout, W.J.V. (1933) Permeability in large plant cells and in models. Erg. Physiol. 35, 967–1021.

Overton, E. (1896) Über die osmotischen Eigenschaften der Zelle, in ihrer Bedeutung fur die Toxikologie und Pharmakologie. Vierteljahrschr. d. Naturf. Ges. Zürich 41, 383–406.

Overton, E. (1899) Über die allgemeinen osmotischen Eigenschaften der Zelle, ihre vermutlichen Ursachen und ihre Bedeutung für die Physiologie. Vierteljahrschr. d. Naturf. Ges. Zürich 44, 88–114.

Overton, E. (1902) Beiträge zur allgemeinen Muskel und Nervenphysiologie. II. Mitt. Über die Unentbehrlichkeit von Natrium (oder Lithium) Ionen für den Contractionsact des Muskels. Pflüger's Arch. ges. Physiol. 92, 346–386.

Pardee, A.B. (1968) Membrane transport proteins. Science 162, 632–637.

Pedersen, P.L. and Carafoli, E. (1987) Ion motive ATPases. I. Ubiquity, properties, and significance to cell function. Trends Biochem. Sci. 12, 146–150.

Pfeffer, W.F. (1877) *Osmotische Untersuchungen*, p. 160. Leipzig, W. Engelmann.

Polvani, C. and Blostein, R. (1988) Protons as substitutes for sodium and potassium in the sodium pump reaction. J. Biol. Chem. 263, 16757–16763.

Ponder, E. (1934) *The Mammalian Red Cell and the Properties of Haemolytic Systems*. Protoplasma Monographien No. 6. Berlin, Gebr. Borntraeger.

Post, R.L. (1974) A reminiscence about sodium, potassium-ATPase. Ann. N.Y. Acad. Sci. 242, 6–11.

Post, R.L. and Jolly, P.C. (1957) The linkage of sodium, potassium and ammonium active transport across human erythrocyte membrane. Biochim. Biophys. Acta 25, 118–128.

Quiocho, F.A. (1989) Atomic structures of periplasmic proteins and the high-affinity active transport systems in bacteria. Phil. Trans. R. Soc. B 326, 341–351.

Rabon, E.C. and Reuben, M.A. (1990) The mechanism and structure of the gastric H,K-ATPase. Annu. Rev. Physiol. 52, 321–344.

Rea, P.A. and Pook, R.I. (1985) Proton translocation inorganic pyrophosphatase in red beet *(Beta vulgaris L.)* tonoplast vesicles. Plant Physiol. 81, 126–129.

Reid, E.W. (1890) Osmosis experiments with living and dead membranes. J. Physiol. 11, 312–351.

Reid, E.W. (1892) Preliminary report on experiments upon intestinal absorption without osmosis. Brit. Med. J., p. 1133–1134.

Reid, E.W. (1900) On intestinal absorption, especially on the absorption of serum, peptone and glucose. Philos. Trans. R. Soc. London B. 192, 211–297.

Rickenberg, H.V., Cohen, G.N., Buttin, G. and Monod, J. (1956). La galactoside-perméase d'*Escherichia coli*. Ann. Inst. Pasteur 91, 829–857.

Rickles, E. and Quastel, J.H. (1958) Effects of cations on sugar absorption by isolated surviving guinea pig intestine. Canad. J. Biochem. 36, 347–362.

Riggs, T.R., Walker, L.M. and Christensen, H.N. (1958) Potassium migration and amino acid transport. J. Biol. Chem. 233, 1479–1484.

Ritchie, J.M. and Straub, R.W. (1957) The hyperpolarization which follows activity in mammalian non-medullated fibres. J. Physiol. (Lond). 136, 80–97.

Robinson, J.R. (1950) Osmoregulation in surviving slices of the kidneys of adult rats. Proc. Roy. Soc. (Lond.) B, 137, 378–402.

Roseman, S. (1969) The transport of carbohydrate by a bacterial phosphotransferase system. J. Gen. Physiol. 54, 138s–180s.

Rosenberg, T. (1948) On accumulation and active transport in biological systems. I. Thermodynamic considerations. Acta Chem. Scand. 2, 14–33.

Rosenberg, T. (1954) The concept and definition of active transport. Symp. Soc. exp. Biol. 8, 27–41.

Sachs, G. (1987) The gastric proton pump: the H^+,K^+-ATPase. In *Physiology of the Gastrointestinal Tract, 2nd Ed.* (Johnson, L.R., ed.), pp. 865–881. New York, Raven Press.

Sachs, J. (1865) *Handb. der Experimental-Physiologie der Pflanzen.* p. 204. Leipzig, Engelmann Verl.

Schagina, L.V., Grinfelt, A.E. and Lev, A.A. (1983) Concentration dependence of the bidirectional flux ratio as a characteristic of transmembrane ion transporting mechanism. J. Membr. Biol. 73, 203–216.

Schatzmann, H.J. (1953) Herzglykoside als Hemmstoffe für den aktiven Kalium-Natriumtransport durch die Erythrocytenmembran. Helv. Physiol. Acta 11, 346–354.

Schatzmann, H.J. (1982) The plasma membrane calcium pump of erythrocytes and other cells. In *Membrane Transport of Calcium* (Carafoli, E., ed.), pp. 41–108. London, Academic Press.

Schmidt, C. (1850) *Charakteristik der epidemischen Cholera gegenueber verwandten Transsudationsanomalien.* Leipzig u. Mittau, G.A. Reyher's Verl.

Schobert, B. and Lanyi, J.K. (1982) Halorhodopsin is a light-driven chloride pump. J. Biol. Chem. 257, 10306–10313.

Schönheimer, R. (1941) *The Dynamic State of Body Constituents.* Boston, Harvard University Press.

Schultz, S.G. (1977) Sodium-coupled solute transport by small intestine: a status report. Am. J. Physiol. 233, E249–E254.

Schultz, S.G. and Zalusky, R. (1964) Ion transport in the isolated rabbit ileum. I. Short-circuit current and Na-fluxes. J. Gen. Physiol. 47, 567–584.

Semenza, G. (1971) On the mechanism of mutual inhibition among sodium-dependent transport systems in the small intestine: A hypothesis. Biochim. Biophys. Acta 241, 637–649.

Semenza, G. and Kinne, R. (Eds) (1985) Membrane transport driven by ion gradients. Ann. N.Y. Acad. Sci. 456.

Serrano, R. (1988) Structure and function of proton translocating ATPases in plasma membranes of plants and fungi. Biochim. Biophys. Acta 947, 1–28.

Shannon, J.A. (1939) Renal tubular secretion. Physiol. Rev. 19, 63–93.

Skou, J.C. (1957) The influence of some cations on an adenosine triphosphatase from peripheral nerves. Biochim. Biophys. Acta 23, 394–401.

Skou, J.C. (1960) Further investigations on a Mg^{++} + Na^+-activated adenosinetriphosphatase, possibly related to the active, linked transport of Na^+ and K^+ across the nerve membrane. Biochim. Biophys. Acta 42, 6–23.

Skou, J.C. (1989) The identification of the sodium-pump as the membrane-bound Na^+/K^+-ATPase: a commentary. Biochim. Biophys. Acta 1000, 435–438.

Skulachev, V.P. (1992) Chemiosmotic systems in bioenergetics: H^+-cycles and Na^+-cycles. Bioscience Reports 11, 387–441.

Solomon, A.K. (1952) The permeability of the human erythrocyte to Na and K. J. Gen. Physiol. 36, 371–388.

Steinbach, H.B. (1940a) Sodium and potassium in frog muscle. J. Biol. Chem. 133, 695–701.

Steinbach, H.B. (1940b) Electrolyte balance of animal cells. In Cold Spring Harbor Symp. 8, 242–252.

Stoeckenius, W. and Bogomolni, R.A. (1982) Bacteriorhodopsin and related pigments of halobacteria. Annu. Rev. Biochem. 51, 587–617.

Straub, F.B. (1953) Über die Akkumulation der Kaliumionen durch menschliche Blutkörperchen. Acta Physiol. Acad. Sci. Hung. 4, 235–240.

Sweadner, K.J. (1979) Two molecular forms of $(Na^+ + K^+)$-stimulated ATPase in brain. J. Biol. Chem. 254, 6060–6067.

Szent-Györgyi, A. (1957) Bioenergetics, p. 57. New York, Academic Press.

Tappeiner, H. (1878) Über die Aufsaugung der gallensauren Alkalien im Dünndarm. Sitzungsber. d. k. Akademie d. Wiss. 77, 281–304.

Teorell, T. (1933) Untersuchungen über die Magensaftsekretion. Scand. Arch. Physiol. 66, 225–317.

Tokuda, H. and Unemoto, T. (1984) Na^+ is translocated at NADH: quinone oxidoreductase segment in the respiratory tract of Vibrio alginolyticus. J. Biol. Chem. 259, 7785–7790.

Troshin, A.S. (1966) Problems of Cell Permeability (Hall, M.G., transl.). London, Pergamon Press.

Turner, J. (1983) Quantitative studies of cotransport systems: Models and vesicles. J. Membr. Biol. 76, 1–15.

Ussing, H.H. (1949) The distinction by means of tracers between active transport and diffusion. Acta physiol. Scand. 19, 43–56.

Ussing, H.H. and Zerahn, K. (1951) Active transport of sodium as the source of electric current in the short circuited isolated frog skin. Acta physiol. Scand. 23, 110–127.

Van Slyke, D.D. and Meyer, G.M. (1913) The fate of protein digestion products in the body. III. The absorption of amino acids from the blood by the tissues. J. Biol. Chem. 16, 197–212.

Wearn, J.T. and Richards, A.N. (1924) Observations on the composition of glomerular urine, with particular reference to the problem of reabsorption in the renal tubules. Am. J. Physiol. 71, 299–227.

West, I.C. (1970) Lactose transport coupled to proton movements in Escherichia coli. Biochem. Biophys. Res. Commun. 41, 655–661.

Whittam, R. (1962) The asymmetrical stimulation of a membrane adenosine triphosphatase in relation to active cation transport. Biochem. J. 84, 110–118.

Widdas, W.F., Kleinzeller, A. and Thompson, K. (1989) The accumulation of free and phosphorylated sugars in adipocytes based on a dynamic diffusion barrier. Biochim. Biophys. Acta 979, 221–230.

Wikström, M.K.F. (1977) Proton pump coupled to cytochrome c oxidase in mitochondria. Nature (Lond.) 266, 271–273.

Wilbrandt, W. (1938) Die Permeabilität der Zelle. Erg. Physiol. 10, 204–291.

Wilbrandt, W. (1961) In Membrane Transport and Metabolism (Kleinzeller, A. and Kotyk, A., eds.), p. 342. New York, Academic Press.

Wilbrandt, W. and Rosenberg, T. (1961) The concept of carrier transport and its corollaries in pharmacology. Pharmacol. Rev. 13, 109–183.

Wilson, D.M. and Wilson, T.H. (1987) Cation specificity for sugar substrates of the melibiose carrier in *Escherichia coli*. Biochim. Biophys. Acta 904, 191–200.

Zadunaisky, J.A. (1979) Characteristics of chloride secretion in some intestinal epithelia. In *Mechanisms of Intestinal Secretion*. (Binder, H.J., ed.), pp. 53–64, New York, Alan Liss Publ.

A. Kleinzeller (Editor) A History of Biochemistry: Exploring the Cell Membrane.
(Comprehensive Biochemistry Vol. 39) © 1995 Elsevier Science B.V.

Chapter 5

Membrane Receptors

M.D. HOLLENBERG[1] and A. KLEINZELLER[2]

[1]Department of Pharmacology and Therapeutics, University of Calgary Faculty of Medicine, 3330 Hospital Drive, Calgary, Alberta, Canada T2N4N1
[2]Department of Physiology, University of Pennsylvania, Philadelphia, PA 19104–6085, U.S.A.

I. The receptor concept

This chapter will outline (1) the development of the receptor concept, as it relates to the actions of toxins, drugs, neurotransmitters and hormones, and (2) the description of the structures and mechanisms of action of a number of membrane-localized receptors. These membrane-localized pharmacologic receptors play a unique role in the communication network that conveys information from one cell to another.

1. TOXINS, DRUGS AND THE ROLE OF RECEPTIVE SUBSTANCES

The study of the actions of a number of potent toxins (nicotine, atropine, curare) was central to the development of the receptor concept. For instance, in his studies of the antagonism by atropine of the action of pilocarpine at sites of nerve/gland cell junctions, Langley (1879) suggested that

'... there is some substance in the nerve endings of gland cells with which both atropine and pilocarpine are capable of forming compounds.'

In his parallel studies of the action of nicotine and curare on muscle, Langley (1906a) enlarged upon his concept of reversible drug antagonism and went on to postulate the existence of a 'receptive substance' with which either nicotine or curare must combine. This concept was clearly summarized in Langley's (1906b) Croonian lecture. In parallel studies, Ehrlich (1900, 1908) developed a comparable receptor concept to explain the interaction between toxins and their target cells. This interaction was envisioned as a chemical reaction between the 'haptophoric moiety' of the toxin and a selective chemical structure at the cell surface (side chain), with an implied specificity (the lock-and-key concept of E. Fischer) and stoichiometry. Ehrlich (1908, 1909) described the cellular constituents that interacted with the toxin as 'poison receptors', or simply as 'receptors'. In the course of his study of the antitrypanosomal activities of triphenylmethane dyes and arsenicals, in which drug resistance was seen to emerge, so as to include families of agents to which the organism was insensitive, Ehrlich expanded his concept of receptors to include the rudiments of a structure-activity requirement for receptor function. He clearly realized the importance of the field 'concerned with the connection between constitution and action, which forms the basis for a rational development of Therapeutics' (Ehrlich, 1909). Ehrlich's conceptual framework extended to include not only antibodies and antitrypanosomal agents but also the action of various cell nutrients ('nutriceptors') and a variety of drugs, including arsenicals. From his perspective, Ehrlich (1900, 1908, 1909) saw that

> 'we may regard the cell quite apart from its familiar morphological aspects, and contemplate its constitution from a purely *chemical* standpoint. We are obliged to adopt the view that the protoplasm is equipped with certain atom groups, whose function especially consists in fixing to themselves certain food-stuffs, of importance in cell life. Adopting the nomenclature of organic chemistry, these groups may be designated *side chains*.'

In essence, Ehrlich's 'side chains' or 'toxin receptors' represented in their way the receptive substances proposed by Langley. Thus, at the turn of the century, the concept of a pharmacologic receptor was well established as a chemically specific cellular entity that could combine with active agents according to the laws of mass action, so as to cause cellular activation. It is this combined recognition-activation function that distinguishes the membrane-localized pharmacologic receptors to be described in this chapter from other targets of drug action, such as nutrient transporters or cytoplasmic enzymes.

2. EARLY RECEPTOR MODELS AND DRUG ACTION

In concert with the work of Langley on the actions of nicotine and curare on striated muscle, Hill (1909) developed a quantitative approach to the mode of drug action, employing the law of mass action, whereby an agonist, A, was allowed to interact reversibly with a receptive substance, R, according to the equation

$$A + R \rightleftharpoons AR \tag{1}$$

Using this approach, Clark (1926, 1927, 1937) and Gaddum (1926, 1937) developed an 'occupancy' model of drug action, wherein the magnitude of a response (Q) is seen to be proportional to the concentration of ligand-receptor complex formed (AR) in equation (1):

$$Q \propto (AR), \text{ or } Q = \beta(AR) \tag{2}$$

where β represents a proportionality constant. Clark (1937) emphasized that his model of a mass-action relationship between the agonist and the receptor was conceptually related to emerging views on enzyme-substrate interactions.

Although it was possible to develop a number of models of agonist action based on equations (1) and (2) (*e.g.* see Clark,

1937), the lack of knowledge of the amplification factors relating receptor occupancy to the exact magnitude of a biological response made it impossible, for agonists, to determine the precise receptor parameters (*e.g.* rate constants of binding, affinity constants or numbers of binding sites for receptor-ligand interactions), based solely on the shapes of dose-response curves. Nonetheless, relative agonist affinities could be estimated from the models, based on the relative median effective dose (ED_{50}) of a series of congeners in a specific bioassay system. Surprisingly, models of antagonist action, along with dose-response curves could be used more effectively to determine the affinity of interaction of antagonists with the receptor (Arunlakshana and Schild, 1959). The affinity of an antagonist for its receptor could be determined either from 'occupancy' models of drug action (Arunlakshana and Schild, 1959) or from a rate model of drug action, wherein the intensity of response was postulated to be proportional to the rate of formation of the agonist-receptor complex (Paton, 1961). Thus, by the early 1960s, the models of drug action along with bioassay measurements had led to estimates of the affinities with which antagonists interact with receptors and to approximations of the relative affinities of certain families of agonists for specific receptor systems. However, the mechanism of agonist action was still something of an enigma, since it was realized (1) that only a fraction of the available receptors needed to be occupied to generate a maximum response, *i.e.* there were 'spare receptors' in most systems (Nickerson, 1956) and (2) that for the same degree of receptor occupancy, receptors could be activated with variable efficiency by different agents, ranging from zero intrinsic activity for pure antagonists to very high degrees of efficacy for certain agonists (Ariëns, 1954; Stephenson, 1956). Nonetheless, from the bioassay measurements, the essential properties a ligand-receptor interaction were clearly understood in terms of (1) strict structural and steric specificity, (2) saturability, (3) tissue specificity, (4) high affinity and (5) reversibility. These criteria for a ligand-receptor interaction, similar to those put forward for membrane carriers (*cf.* Chapter 3), proved of considerable value in the molecular characterization of membrane receptor molecules. In view of the above properties, membrane receptors were thought of, and later shown to be proteins (see below). In most, if not all instances, the receptor molecules were found to contain polysaccharidic moieties (*cf.* Sharon and Lis, 1989; also Chapter 2, Section II.1.1.).

3. MEMBRANE-LOCALIZED RECEPTORS AND THE 'MOBILE' OR 'FLOATING' MODEL OF DRUG ACTION

It was suspected early on that the cell membrane was a target for drug action because of: the staining of the cell surface by pharmacologically active agents such as methylene blue (Ehrlich, 1886); the distribution and pharmacological effects of methylene blue (Cook, 1926; *cf.* Chapter 2, Section I); the rapid hemolytic effects of toxic agents such as snake venoms (Flexner and Noguchi, 1902; Keyes, 1902) or saponin (Ransom, 1901); and the marked difference in narcotic action found upon external cell applications, as opposed to micro-injection of opiates (Marsland, 1934). This site of drug action and the very small numbers of membrane receptors required for drug action were highlighted by the early quantitative considerations of Clark (1927), who in essence performed the first ligand binding measurements for agonist action in intact tissue, using a bioassay for acetylcholine. Clark (1927, 1937) estimated that only tens of thousands of receptors needed to be occupied on the cell surface to generate a cellular response. The stage was thus set for studies that began in the early 1960s to identify membrane receptors by a combined pharmacologic/biochemical approach.

Two key developments contributed enormously to the biochemical understanding of receptor function. The first development comprised the appropriate use of radioactively la-

belled ligand probes to measure directly the binding of agonists and antagonists to their receptors. The description of
the binding of radiolabelled atropine to the muscarinic receptor in smooth muscle (Paton and Rang, 1965) represented a
landmark study in this regard. This approach was dramatically amplified shortly thereafter, when it was realized that
[125]I-labelled peptide agonists with much higher specific radio-
activities, such as ^{125}I-insulin could be used to measure agonist-receptor interactions (Freychet et al., 1971; Cuatrecasas, 1971; Cuatrecasas and Hollenberg, 1976). An essential
component of these early ligand binding studies was the correlation of the measured ligand affinity constants with the relative biological activities of the agonists or antagonists that
were being studied. For the antagonists, the affinity measured
by binding experiments could be compared directly with the
K_D measured by bioassay, thereby validating the interpretation of binding measurements (*e.g.* see Paton and Rang, 1965).
This receptor-specific binding could be identified in any system of choice, and an analysis of the binding data (*e.g.* according to Scatchard, 1949) was able to yield the numbers and
affinities of receptors present in a membrane preparation. The
plasma membrane was rapidly identified biochemically as the
primary site of action of many (but by no means all) agonists,
fully vindicating the concepts developed by Ehrlich and Langley. Parenthetically it should be added that comparable studies of ligand binding were able to identify the cytoplasmic and
nuclear location of receptors for agonists such as corticoster-
oids, thyroid hormone and retinoids.

The kinetic postulates expressed by eqs. (1) and (2), above,
directed the focus on underlying molecular events. Crucial
contributions in this direction were made well before more
sophisticated kinetic models were presented in the early sev-
enties (see below). Hokin and Hokin (1955) found that acetylcholine and its analogs increased the turnover of some
membrane phospholipids, especially phosphoinositides, thus
providing the first instance of a hormone producing its efffect

on membrane-localized reactions. This discovery, later extended by Michell (1975) and Berridge (1983, 1988, 1993) established that phosphoinositides represent precursors for one of the second messenger systems by which agonist action is translated into a cellular signal. A parallel breakthrough was the discovery of the cyclic AMP (adenylate cyclase) system by Sutherland and collaborators (*cf.* Sutherland, 1962). The chain of postulated events occurring between the agonist-receptor interaction at the membrane and the physiological/pharmacological effects could now be explored more directly. The actual cellular response to external signals was found to be either direct (as for the operation of ligand-gated ionic channels, see below), or more indirect, involving processes such as phosphorylation and dephosphorylation of membrane proteins (Krebs, 1973; Greengard, 1976; Cohen, 1988), hydrolytic processes such as the action of phospholipases C and D (*cf.* Billah and Anthes, 1990), or methylation and demethylation of membrane proteins (*cf.* Springer et al., 1979).

A second development that had a large impact on understanding receptor function came from studies (*cf.* Chapter 2) of the mobilities of membrane-localized proteins (Frye and Edidin, 1970; Edidin, 1974; Edidin et al., 1975) and from the evolution of membrane structure theories (Singer and Nicolson, 1972). Once it was realized that receptors, like other membrane proteins, were mobile constituents in the plasma membrane it was possible to envision a highly dynamic mechanism of function of receptors, that had been originally thought of (either explicitly or implicitly) as comparatively static entities localized at specific anatomic sites, such as the neuromuscular junction (*e.g.* nicotinic cholinergic receptors). The idea that receptors could diffuse freely in the plane of the membrane allowed for an explanation of a hitherto puzzling set of observations, indicating that in a single cell type (*e.g.* adipocyte) a variety of hormones (*e.g.* glucagon, adrenocorticotropic hormone (ACTH), epinephrine) could independently activate membrane associated adenylate cyclase. To account

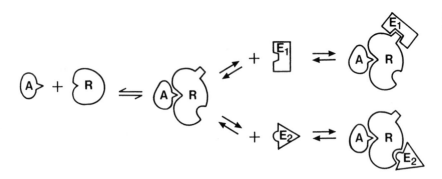

Fig. 1. Interaction of mobile hormone receptors with multiple membrane effectors. In the mobile or floating model of hormone action, a hormone receptor (R) (e.g. for insulin), upon binding its ligand, A, is depicted as changing receptor conformation. The hormone-receptor complex is shown to be capable of interacting in the plane of the membrane with two separate membrane effector molecules (E_1 and E_2). Neither the ligand nor the receptor membrane effector molecules are drawn to scale.

for these data, since receptors were now thought to be mobile in the plane of the membrane, it was hypothesized that each receptor could 'float' independently to interact with the same adenylate cyclase. The 'floating' or 'mobile' model of agonist action, conceived independently by a number of investigators (Boeynaems and Dumont, 1977; 1980; De Haen, 1976; Cuatrecasas, 1974a; Bennet et al., 1975; Jacobs and Cuatrecasas, 1976), and illustrated in Figure 1 provided a novel way of modelling the biochemical steps of receptor-mediated cell activation, as will be elaborated upon in the following sections.

II. Receptor-mediated transmembrane signalling

Although the principal function of the cell membrane is to maintain an effective barrier between the extracellular and intracellular milieu, there are highly specialized membrane-localized constituents (*e.g.* ion channels, nutrient transporters

and pharmacologic receptors) that can be singled out as play-
ing particularly pivotal roles in terms of selectively transmitting information from the external to the internal cellular environment (and in some cases, vice versa). The pharmacologic
receptors situated in the plasma membrane possess not only
the ability to recognize extracellular hormonal signalling molecules with high affinity and specificity, but also the capacity,
once combined with the specific ligand, to transmit information across the cell membrane, so as activate intracellular signalling pathways. The possibility of transmitting information
from surface receptors to intracellular sites was visualized
early on by Peters (1937) who was impressed by Clark's observations (1937). It is this dual recognition-transmembrane signalling property, by which the receptor per se acts as a mes-
sage generating system, that distinguishes receptors from
other cell surface recognition/transport constituents. The sections below will, with selected examples, focus on the general
mechanisms whereby cell surface receptors generate a transmembrane signal. Receptors for agonists that act via intracellular receptors (e.g. steroid hormones) will not be discussed.

1. GENERAL MECHANISMS OF TRANSMEMBRANE
SIGNALLING AND THE MOBILE RECEPTOR PARADIGM

To fulfil their recognition/action function, membrane receptors must be able to generate an intracellular signal that can
be greatly amplified. In general, it can be said that the receptors capable of generating a transmembrane signal actually
have transmembrane domains that can either communicate
with (e.g. via an ion channel) or interact directly with the in-
tracellular/submembraneous environment. Thus a physical
contiguity appears to be a prerequisite for a transmembrane
signalling molecule. Any constituent that may exist solely on
the external aspect of the membrane would thus, of necessity,
not be able to generate an intracellular signal except via inter-

action with a second membrane/localized component that possesses a transmembrane domain. Thus, one can think either in terms of receptors that communicate directly with cell interior (*e.g.* receptor-ion channels or receptor-enzymes, to be discussed below) or terms of receptors that communicate via 'signal adapter molecules' that couple ligand-mediated receptor activation to intracellular signalling processes. This adapter/coupling mechanism is a key feature of the mobile receptor paradigm to be outlined below.

1.1. Receptor mobility and transmembrane signalling: The mobile receptor model of agonist action

As alluded to above, it is now realized that in perhaps the majority of cases, receptors for agents as diverse as insulin, epinephrine and epidermal growth factor-urogastrone (EGF-URO) are dynamic cell surface constituents that are mobile in the plane of the membrane. In terms of ligand-triggered transmembrane signalling, receptor mobility is viewed as a key property that enables the receptor to interact with a variety of other membrane-localized constituents during the course of cell activation. It is a basic tenet of the 'mobile' or 'floating' receptor paradigm of agonist action (Boeynaems and Dumont, 1977; Cuatrecasas, 1974a; De Haen, 1976; Cuatrecasas and Hollenberg, 1976; Hollenberg, 1979; Jacobs and Cuatrecasas, 1976) that the binding of a ligand to its receptor dramatically alters the ability of a receptor (1) to migrate in the plane of the membrane and (2) to interact with other membrane components. Thus, the entity [AR], resulting from a combination of an agonist, A with its receptor, R: $[A] + [R] \rightleftharpoons [AR]$, becomes an 'activated' receptor complex ($[AR] \rightleftharpoons [AR^*]$) that can, as illustrated in Figure 1, go on to form ternary complexes of the kind, $[AR^*E]$:

$$[AR^*] + [E] \rightleftharpoons [AR^*E^*] \qquad (3)$$

wherein E can be viewed as an effector or 'signal adapter' molecule involved in the process of cell activation. In certain instances E may comprise a second receptor molecule forming an enzymatically activated dimer that leads to cell activation. An alternative possibility is a 'dissociation' model in which an effector (E), that is held as an inactive complex with the receptor (RE), is dissociated when the agonist binds to yield a free active effector, E^*:

$$[AR] + [RE] \rightleftharpoons [AR] + [E^*] \tag{4}$$

In principle, the mobile receptor model (Figure 1) does not restrict the number of distinct effector moieties with which the agonist-receptor complex may interact. In this manner, by interacting with multiple effectors, a single agonist-receptor complex could liberate simultaneously a variety of intracellular mediators. These types of multiple interactions are particularly relevant to the action of G-protein-coupled receptors (see below), which can couple via a number of distinct G-protein signal adapter proteins that can coexist in the same cell and that can couple simultaneously to distinct signal pathways. The mobile receptor paradigm provides enormous flexibility in modelling processes of cell activation, ranging from the modulation of an ion channel *via* ligand-regulated subunit interactions to the activation of G-protein-coupled processes.

1.2. Receptor microclustering, receptor internalization and agonist action

Apart from allowing ligand-occupied receptors to interact with effector moieties, receptor mobility allows for the formation of receptor microclusters (dimers, up to decamers) and for receptor internalization. Both of these aspects of receptor mobility are believed to be relevant to the process of transmembrane signalling. The importance of receptor microclustering for receptor activation was realized when it was observed that divalent (but *not* monovalent) antibodies directed either against the receptor (Kahn et al., 1981) or against a receptor-bound antagonist (Hopkins et al., 1981; Gregory et al., 1982; Conn et al., 1982; Hazum and Keinan, 1985) could mimic ligand-triggered cell activation. Whilst ligand-induced receptor dimerization/transphosphorylation is seen to be a key event in

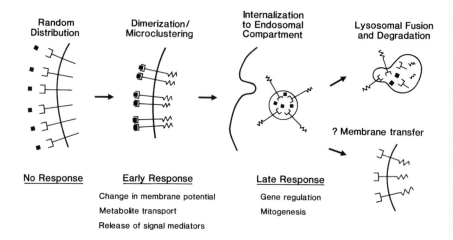

Fig. 2. Receptor dynamics and cell activation. A general scheme for cell activation is depicted, pointing out the receptor dynamics thought to be involved in the process of cell activation. Very likely, cell responses that are rapidly regulated by neurotransmitters or hormones involve the initial dimerization/ microclustering event. Delayed effects of the ligand may be caused by receptor that is internalized in the endosomal organelle. The topography of the endosome would permit the intracellular portion of the receptor (designated as 'ʌʌʌ') to interact with a variety of intracellular constituents located at considerable distances from the plasma membrane; the endosomal form of the receptor represents an ideal vehicle to carry the receptor rapidly to selected sites of intracellular action. Possibly the intracellular portion of some neurotransmitter receptors will be found to contain a kinase domain such as the ones present in receptors for EGF-URO and insulin. The kinase may regulate a variety of enzymatic processes via phosphorylation reactions. In the course of its intracellular migration, the receptor-bearing endosome could ultimately fuse with either lysosomes or other membrane structures (e.g. Golgi elements, nuclear membranes), resulting in further relocation of the receptor.

cell activation by the receptors for growth factors such as insulin or EGF-URO (Schlessinger and Ullrich, 1992; Ullrich and Schlessinger, 1991), the role of receptor dimerization/microclustering in the activation of cells by G-protein-coupled receptors, such as the one for luteinizing hormone-releasing hormone (LHRH), is still an open question.

Once activated by an agonist, the agonist-receptor complex may not only participate in cell surface interactions but may also undergo internalization (*i.e.* receptor-mediated endocytosis, involving clathrin-coated membrane pits) and translocation to other cellular compartments such as the lysosome and the nucleus. Such a process has been documented for some agonist peptides and hormones (Pastan and Willingham, 1981). The removal of the receptor from the cell surface by this process may play a role in cellular desensitization (so-called receptor down-regulation), whereas the internalized receptor, either in a pre-lysosomal compartment (Posner et al., 1985) or in a nuclear compartment (*e.g.* for nerve growth factor: Yankner and Shooter, 1982; Bothwell, 1991), may play a role in cell activation that is complementary to the cell-surface-generated signal. In all of these events, both at the cell surface and in the cytoplasmic milieu, the constituents of the plasma membrane (both receptor and non-receptor) can be seen to play important functions. In general terms, the process of cell activation can be visualized as occurring in two phases: (A) receptor occupancy, followed by microclustering and primary signal generation (steps 1 and 2, Fig. 2) and (B) receptor internalization, followed by intracellular translocation and possibly, intracellular signal generation (steps 3 and 4, Fig. 2). Thus, receptor mobility could provide for two anatomic tiers of signalling (one at the cell surface, and the second, intracellularly) and two associated time frames, one immediate, and the other, delayed.

In the above sections, general mechanisms of receptor-mediated signalling were discussed. In the sections that follow, specific examples of three general classes of receptors will be given: (1) ligand-modulated ion channels, (2) ligand-regulated transmembrane enzymes and (3) ligand-regulated, adapter-coupled receptors, such as those that act via G-proteins.

2. LIGAND-GATED ION CHANNELS

2.1. The nicotinic cholinergic receptor

The study of this class of receptors can be said to begin in the early 1850s, with an analysis of the action of curare. Independently, Bernard (1856) and Kölliker (1856) arrived at the conclusion that the poison acts on the junction between nerve and muscle. As pointed out above, Langley (1906a,b) deduced that curare and nicotine acted competitively with the same 'receptive substance' in muscle. Structure-function studies showed the 'onium' moiety of the agonist molecule to be responsible for the action of these poisons (Crum Brown and Fraser, 1868). This 'nicotine' receptor at the neuromuscular junction, which was distinguishable from the 'muscarinic' receptor (present at parasympathetic nerve endings (Dale, 1914) was seen to be the target for the neurohumoral factor (termed 'Vagusstoff' by Loewi, 1921) subsequently identified as acetylcholine (Dale and Gaddum, 1930; Feldberg, 1933).

An involvement of ions in the action of an acetylcholine receptor was first indicated by Howell and Duke (1908) by demonstrating an increased release of K^+ into the blood circulation on stimulation of the vagus nerve with consequent inhibition of heart action. When the localization of the acetylcholine (nicotinic) receptors at the neuromuscular junction (Nastuk, 1953; del Castillo and Katz, 1955) and the cell membrane of the electroplax (Cartaud et al., 1973) was better understood, several laboratories reported that acetylcholine produced a transitory depolarization[1] of these cell membranes (Fatt and Katz, 1951; Nastuk, 1953). These observations led to the hypothesis of an agonist-induced non-specific increase in membrane permeability. Such view was then modified by the demonstration that the hormone increased the membrane permeability for Na^+, K^+ and to some extent also Ca^{2+}, but not for Cl^- (*cf.* Fatt and Katz, 1951; Nastuk, 1953; Takeuchi, 1960, 1963). Such specificity argued in favor of a ligand-induced permeabilization of a cationic channel.

The nicotinic receptor for acetylcholine became the first ligand-gated ion channel receptor to be characterized in molecular terms. The biochemical identification of the nicotinic receptor was greatly facilitated by two observations: (1) that the electric organs of eels (*Electrophorus electricus*) and rays (*Torpedo marmorata*) served as extraordinarily rich sources of receptor protein (Nachmanson, 1959) and (2) that radiolabelled snake venom toxins (α-bungarotoxin) could almost irreversibly and specifically bind to the nicotinic receptor (Changeux et al., 1970; Weber and Changeux, 1974). The isolation (Meunier et al., 1974), microsequencing (Raftery et al., 1980) and cloning (Numa, 1989) of the oligomeric chains of the pentameric ($\alpha_2\beta\gamma\delta$) nicotinic receptor (summarized by Marx, 1983) illustrated in Figure 3, represents a remarkable achievement that no doubt would have intrigued both Langley and Ehrlich. The nicotinic receptor found at the neuromuscular junction is a transmembrane oligomeric structure composed of five glycosylated proteins (two of which are identical) that interact to form a cation channel ($\alpha_2\beta\gamma\delta$) with about 50% of its 10 nm length projecting above the membrane surface (Brisson and Unwin, 1985). The binding sites for acetylcholine reside on the α-subunits. The comparable receptor found in the central nervous system appears to have fewer distinct oligomers forming the ligand-regulated channel (Lindstrom et al., 1987). Although individual chains of the oligomeric structure (*e.g.* the α-chain), when expressed alone in frog oocytes can self-assem-

---

1 Such a view was for a long period of time vigorously contested by Nachmanson (1959) who maintained that acetylcholine and its turnover is directly responsible for membrane excitation.

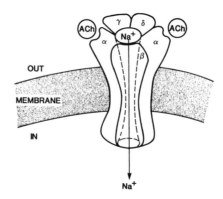

Fig. 3. Model of the nicotinic acetylcholine receptor. The pentameric structure of the receptor, comprising two α-subunits, along with the β, γ and δ subunits, is shown schematically, in keeping with the data of Brisson and Unwin (1985). Acetylcholine (Ach) is depicted as binding to the two α-subunits to facilitate the entry of sodium through the ion channel pore.

ble to form an ion channel, the complete complement of oligomers is required to form a channel regulated by acetylcholine in a manner akin to that observed *in vivo* (Changeux, 1980; Tank et al., 1983; Conti-Troconti and Raftery, 1982; Taylor et al., 1983; Stevens, 1985; Karlin, 1991). The reconstitution of the receptor in liposomes (*cf.* Tank et al., 1983) represented a milestone in clarifying the mechanism of action of this receptor. Structure-activity studies of the several receptor subunits are unravelling the roles of a number of the subunit functional domains. In a sense, the activation of the nicotinic receptor can be envisioned as a special case of the mobile receptor model discussed above, wherein the binding of acetylcholine to the α-subunits induces these subunits to interact cooperatively with the other receptor oligomers to regulate channel function. In essence, the entire channel then becomes the 'effector' that conveys the transmembrane signal via increased cation flux.

2.2. Other ligand-gated channels

The nicotinic acetylcholine receptor represents a superfamily of ligand-gated ion channels (Stroud et al., 1990), as indicated by the isolation and cloning of receptors for γ-aminobutyric acid (the so-called GABA-A/benzodiazepine receptor: Sieghart (1992); Schofield et al. (1987); Pritchett et al. (1989), glycine (Grenningloh et al. 1987; Karlin, 1991), glutamic acid (Gregor et al., 1989; Hollmann et al., 1989) and serotonin (the 5HT$_3$ receptor: Maricq et al., 1991; Peters et al., 1992). Interestingly, there is surprising structural homology between the protein subunits of these quite distinct receptors that not only bind different ligands, but regulate the transport of quite distinct ions (Na$^+$ for the nicotinic receptor, *cf.* Numa, 1989); Cl$^-$ for the GABA-A/benzodiazepine receptor, (*cf.* Pritchett et al., 1989). As will be seen below, ion channels can also be regulated by receptors indirectly, via the effects of G-protein oligomers (section 4.4).

3. LIGAND-MODULATED TRANSMEMBRANE ENZYMES

3.1. Receptor tyrosine kinases

The observation that activation of the EGF-URO receptor by its ligand led to the phosphorylation on tyrosine of a variety of membrane proteins (Carpenter and Cohen, 1979) provided the first evidence that hormone receptors might be tightly coupled to protein kinase activity and led subsequently to the discovery that the EGF-URO receptor itself possesses intrinsic ligand-regulated tyrosine kinase activity (Carpenter and Cohen, 1990; Cohen et al., 1980). Concurrently, it was discovered that the insulin receptor as well possessed intrinsic tyrosine kinase activity (Rosen et al., 1983; Roth and Cassell, 1983; Goldfine, 1987; Rosen, 1987). It is now realized that an enlarging number of receptors for peptide agonists including the ones for platelet-derived growth factor, fibroblast growth factor, insulin-like growth factor-I and colony stimulating factor-1 are transmembrane tyrosine kinase receptors, belonging to a large receptor superfamily (Ullrich and Schlessinger, 1990; Schlessinger and Ullrich, 1992). In keeping with the mobile receptor paradigm, the binding of EGF-URO or insulin to its receptor not only induces receptor dimerization and autophosphorylation, but also confers on the receptor the ability to interact with and phosphorylate a wide variety of cellular constituents. A key element in this process, as envisioned by the mobile receptor model, is the ability of the phosphorylated tyrosine-phosphate domains of the tyrosine kinase receptors to interact specifically and with high affinity with so-called SH-2 peptide domains on other target proteins, such as phospholipase C-γ and GTPase-activating protein (or GAP) (Koch et al., 1992). The 'SH-2 domains' (from src-homology-2) are protein sequences that share homology with the phosphotyrosyl-interacting domain-2 present in the viral or cellular counterpart of the sarcoma virus tyrosine kinase. In essence, the SH-2 domains on receptor target proteins can form non-covalent bridges between phosphorylated tyrosines on the receptor and other cellular constituents at or near the intracellular domain of the ligand-occupied tyrosine kinase.

3.2. Other receptors with intrinsic enzymatic activity

To date, two other transmembrane receptor families with distinct enzymatic activity have been discovered: (1) The receptor for atrial natriuretic factor (ANF) (and presumably its related family of peptides) has been discovered to possess intrinsic guanylate cyclase activity (Chinkers et al., 1989; Garbers, 1992). (2) The receptor for Activin and presumably for the family of transforming growth factor-β (TGF-β) polypeptides has been found to possess intrinsic protein kinase activity targeted to serine and threonine, rather than tyrosine residues

(Mathews and Vale, 1991). In principle there is no reason to restrict the type of intrinsic transmembrane enzymatic activity that a receptor may possess. In this regard, one may point to the discovery of a transmembrane phosphotyrosine phosphatase moiety, which may represent receptors for which a ligand has yet to be found (Weaver et al., 1992; Fischer et al., 1991).

4. G-PROTEIN-COUPLED RECEPTORS

The discovery of cyclic adenosine monophosphate as a 'second messenger' for the action of a variety of hormones, including epinephrine, glucagon and adrenocorticotrophic hormone (Sutherland, 1962) clearly points to the central role of adenylate cyclase for receptor-mediated transmembrane signalling. As outlined above, it was the multiplicity of hormones that could independently regulate adenylate cyclase in a single target cell that necessitated the development of the mobile receptor model of hormone action. Yet, for some time, the link between the receptors and the cyclase remained an enigma. Early on, Rodbell and colleagues noted that guanine nucleotides (GTP) were essential for linking receptor-occupation to the stimulation of adenylate cyclase in membrane preparations (Rodbell et al., 1971; Rodbell, 1980). These seminal observations, followed by the elegant biochemical/genetic approach of Gilman and colleagues, using S49 lymphoma cell mutants (Ross et al., 1978; Gilman, 1987) led to the discovery that a non-receptor guanine nucleotide binding protein oligomer (so-called G_s) was an essential adapter for the coupling of receptor occupation to the activation of adenylate cyclase. Thus, in keeping with mobile receptor model, the ligand-occupied receptor, via the G-protein adapter complex, is able to regulate the cellular 'effector', adenylate cyclase. This G-protein-coupled mechanism (Gilman, 1987; Levitzki and Bar-Sinai, 1991; Spiegel et al., 1992), outlined in Figure 4, repre-

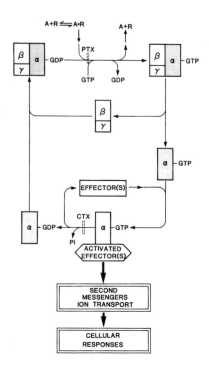

Fig. 4. General scheme for transmembrane signalling mediated by heterotrimeric GTP-binding proteins (G-proteins). G-proteins consist of unique α-subunits with a guanine nucleotide binding site and intrinsic GTPase activity, that are combined with βγ subunits. In the basal state, the oligomeric G-protein has GDP bound tightly to the α subunit. An agonist-receptor complex (A·R) facilitates nucleotide exchange (dissociation of GDP and association of GTP), producing a conformational change that then results in dissociation of the agonist-receptor complex which can then interact with another G-protein. The activated G-protein oligomer also dissociates to give βγ subunits and the GTP-bound α subunit which then activates an effector (enzymes or ion channels) to generate intracellular second messengers. Effector activation is terminated by the intrinsic GTPase activity of the α subunit; the α-GDP subunit then recombines with the βγ subunits to regenerate the inactive heterotrimer. Covalent modification (ADP ribosylation) of specific α subunits by pertussis toxin (PTX) uncouples the G-protein heterotrimer from interacting with receptors, thus inhibiting G-protein-mediated transmembrane signalling. In contrast, cholera toxin (CTX)-catalyzed ADP ribosylation of certain α subunits inhibits their intrinsic GTPase activity; as a result, the effector system is constitutively activated.

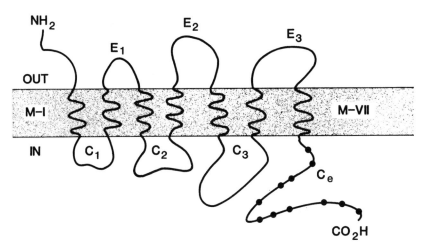

Fig. 5. Schematic representation of the β-adrenergic receptor. In keeping with the synopsis by Lefkowitz and Caron (1988), the receptor modeled on the structure of rhodopsin is shown as having seven transmembrane domains (M-I to M-VII), predicted from hydropathy analysis. The amino acid sequences of the β₂-adrenergic receptor thought to comprise the three extracellular (E₁ to E₃) and three intracellular (C₁ to C₃) loops are shown. Potential sites of threonine or serine phosphorylation (●) in the COOH-terminal cytoplasmic domain, Cₑ, of the receptor are also depicted.

sents one of the most versatile methods for transmembrane signalling by a myriad of receptors, as will be elaborated upon below.

4.1. G-protein-coupled receptors and the regulation of adenylate cyclase

As indicated above, a considerable number of agonists can activate adenylate cyclase via G-protein-coupled receptors, such as the one for β-adrenergic agonists (Dohlman et al., 1987). The β-adrenergic receptor, modeled in Figure 5 as having seven transmembrane domains, in keeping with studies of bacterial rhodopsin, is representative of a very large superfamily of G-protein-linked receptors that not only regulate adenylate cyclase but also modulate a variety of other membrane processes (see below). Accumulated data indicate that these 'serpentine' receptors, once occupied by their specific agonists can interact with the membrane-associated G-proteins via the cytoplasmic receptor domains. Some receptors, via interactions with the G-protein, denoted G_s, stimulate adenylate cyclase, whereas other receptors, such as the α_2-adrenergic receptors, by interacting with a distinct G_i oligomer can cause an inhibition of the membrane-bound cyclase. The basic mechanism whereby the ligand-occupied receptor activates the heterotrimeric G-protein ($\alpha_s\beta\gamma$ or $\alpha_i\beta\gamma$) comprises a receptor-mediated exchange of GTP for G-protein-bound GDP, and a dissociation of the heterotrimeric G-protein into its α- and $\beta\gamma$ subunits (Fig. 4). The α_s-GTP subunit in its GTP-liganded state is then able to interact with and activate adenylate cyclase. The inactivation of the α_s-GTP complex comprises the hydrolysis of GTP (the α-subunit possesses intrinsic GTP-ase activity) and a complexing of α-subunits by free $\beta\gamma$ substituents. The negative regulation of the cyclase by G_i oligomers is more complex, in that in the majority of cases, the active α_i-GTP complex triggered by the receptor does not directly inhibit adenylate cyclase, but targets other membrane constituents; rather, the $\beta\gamma$-subunits released from α_i are able to complex with whatever active α_s-GTP may be present, so as to reduce cyclase activity. It is known that there are multiple forms of adenylate cyclase, and it is believed that in certain cell systems, the $\beta\gamma$-subunits, upon receptor activation can inhibit the type I enzyme directly (Tang and Gilman, 1991; Taussig et al., 1993). This direct inhibition of the cyclase involving $\beta\gamma$ subunits represents another indirect mechanism for cyclase inhibition.

The paradigm worked out for the G-protein-mediated regulation of adenylate cyclase is now known to apply to the regulation of a much wider spectrum of membrane processes than was first suspected. Not only does a specific G-protein (transducin, G_t) play a key role in the control of the visual system by rhodopsin, but other G-proteins are known to be involved in the processes of ion channel regulation and membrane phospholipase activation (*e.g.* G_g, G_k, etc.). At least 18 distinct G-protein α-subunits have now been cloned (Hepler and Gilman, 1992; Simon et al., 1991), all of which can form complexes with a number of distinct β- and γ-subunits that have been identified (Gautam et al., 1990). Thus, there is enormous flexibility possible for the receptor-mediated signals that are generated by G-protein-coupled receptors.

4.2. G-protein-coupled phospholipase activation

Because of the prominence of cyclic AMP as a second messenger of hormone action, it is understandable that the G-protein-mediated control of adenylate cyclase has received intense scrutiny. Nonetheless, early on it was appreciated that many agonists could stimulate phosphoinositide turnover in their target tissues (Hokin and Hokin, 1955). The relationship between the action of specific agonists and phosphoinositide hydrolysis, summarized by Michell and colleagues (Michell, 1975) led to the identification of inositol-trisphosphate (IP_3) and diacylglycerol (DAG) as important mediators of hormone action, subsequent to their release via receptor-mediated phospholipase activation (Berridge, 1983, 1988, 1993). It was known at the time that the G_s and G_i oligomers were targets for the action, respectively, of cholera toxin (G_s) and pertussis toxin (G_i). Yet, since neither of these toxins affected the activity of agonists, known by other criteria to work via G-proteins (like other G-protein-coupled agonists, their receptor binding was modulated by GTP), it was realized that a third class of G-proteins (designated G_p or G_q) was required for receptor-mediated phospholipase activation. Such proteins have now been identified (Simon et al., 1991) and studied in reconstitution systems (Smrcka et al., 1991), wherein a mechanism akin to the one worked out for the regulation of adenylate cyclase is believed to apply (Simon et al., 1991).

The cloning and sequencing of members of the mammalian receptor-related G-protein superfamily has revealed remarkable diversity, which along with rhodopsin-coupled transducin, comprise at least 18 distinct α-subunit species, comprising five subfamilies (Simon et al., 1991). To add to this complexity, there are at least four β-subunit species and six species of γ-subunits (Gautam et al., 1990). Since the βγ subunits are tightly associated and since this dimer is thought to be able to form complexes with a variety of α-subunits, the combinatorial possibilities for distinct G-protein heterotrimers are thus truly overwhelming. Each unique G-protein heterotrimer may be able to regulate a specific membrane effector.

The family of G-proteins extends to a wide range of organisms ranging from yeast to *Drosophila* and also includes the superfamily of so-called 'small' GTP-binding proteins related to the oncogene, *ras*. These small G-proteins have functions quite distinct from the receptor-coupled G-proteins (Satoh et al., 1992; Bourne et al., 1991).

4.3. The visual system

The phototransduction mechanism of the visual system represents one of the best characterized G-protein-coupled processes (Stryer, 1991). In this instance, capture of a photon of light by rhodopsin causes a conformational change in the protein that promotes an interaction between the 'light' receptor and the heterotrimeric G-protein, transducin. The receptor G-protein interaction, promoting a GTP/GTP exchange, thereby activating α_t, results in a transducin (G_T)-mediated stimulation of cyclic GMP phosphodiesterase. The resulting fall in cellular cGMP leads to a closing of a membrane cation channel, with a consequent hyperpolarization-induced nerve impulse that is processed by the retinal neural network for transmission to the visual cortex. In many ways the phototransduction system typifies the floating receptor model of cell signalling.

4.4. G-protein-regulated ion channels

Membrane enzymes (adenylyl cyclase, phospholipase, phosphodiesterase) are not the only membrane effectors regulated by G-proteins. Both the α (Hamilton et al., 1991) and $\beta\gamma$ (Logothetis et al., 1987) constituents can directly modulate ion channels (Strosberg, 1991). For instance, G_s can directly stimulate voltage-gated Ca^{2+} channels in skeletal muscle (Hamilton et al., 1991) and G_i can open the inwardly rectifying K^+ channel in atrial tissue as a result of muscarinic M_2 receptor stimulation. This mechanism of receptor-mediated ion channel regulation is complementary to the mechanism described above for the nicotinic cholinergic receptor.

5. OTHER SIGNAL ADAPTER-COUPLED RECEPTORS

There are a growing number of receptors that have been characterized relatively recently, for which none of the mechanisms described above applies (Miyajima et al., 1992). That is to say, these receptors, such as the ones for the interleukins or granulocyte-macrophage colony stimulating factor (GM-CSF), do not form ion channels, do not possess intrinsic enzyme activity and do not regulate G-protein activity. Nonetheless, these receptor systems can be seen to fit the model of the mobile receptor paradigm, especially since it appears that the ligand binding subunits (e.g. GM-CSF receptor α-subunit) require a second subunit (the GM-CSF receptor β-subunit) for conferring a biological response on the ligand-receptor interaction. A common event in the stimulation of receptor systems of this class is the activation of cellular tyrosine kinase. This recruitment of enzyme activity via receptors that do not of themselves possess intrinsic enzyme activity is akin to the regulation of membrane enzymes via G-protein-coupled receptors. In a sense, both G-protein-coupled receptors and some of the cytokine receptors can be considered to be 'signal adapter-coupled' receptors. In the case of receptors such as the β-adrenergic receptor, the G-protein 'adapter' is well characterized. However, the precise adapter constituents that link the cytokine receptors to the various effectors has yet to be identified with precision. The search for these putative adapter proteins involved in the action of GM-CSF, growth hormone and other regulatory polypeptides represents a most fruitful area for future research.

6. MEMBRANE RECEPTORS, METHYLATION AND SIGNALLING

Evidence for a role of cell membrane methylation (and demethylation) in signal transduction comes principally from studies done with prokaryotes. Studies with gram-positive bacteria pointed the way in terms of identifying periplasmic high-affinity binding proteins for a variety of nutrients including amino acids and galactose (Pardee, 1958; Piperno and Oxender, 1966; Anraku, 1968). The list of high affinity proteins from Gram-negative bacteria has grown considerably since the mid-1960s (see Furlong, 1987). A relationship between substrate recognition (*i.e.* carbohydrate, amino acids, etc.) and bacterial chemotaxis[2] was pointed out by Adler as early as 1966 (see Adler, 1975); by the 1970s, the interaction of such nutrients or chemicals with binding (chemoreceptor) proteins was recognized as a first step either in their transport or in their effects (positive or negative) or chemotypes (Hazelbauer

---

2 Movement of flagellar bacteria towards (positive chemotaxis) or away from (negative chemotaxis) nutrients or other chemicals.

and Adler, 1971; Kalckar, 1971). The role of specific 'chemotaxis' receptors was then established and the reversible methylation of a membrane protein was determined to be the first step in a signalling process leading to the activation of directional flagellar rotation in bacteria. Glutamic acid residues at the inner face of the 'receptor' were identified as the target for the carboxymethylation reaction leading to chemotaxis (see Springer et al., 1979). The level of methylation of the membrane proteins, determined by the respective rates of methylation and demethylation, is believed to regulate membrane protein conformation, thereby acting as a signal for the set of events leading to positive or negative chemotaxis (Weis and Koshland, 1988). Moreover, the chemotactic process is also thought to have a 'memory component' comprising the state of membrane protein methylation that in turn regulates the overall state of cellular adaptation (Weis and Koshland, 1988). Membrane protein methylation has also been observed in eukaryotic cells and is believed to play a role in leukocyte chemotaxis (O'Dea et al., 1978) and in signal transduction in retinal rods (Pérez-Sala et al., 1991). The receptor-mediated process that regulates membrane methylation in mammalian cells has yet to be determined.

III. Regulation of receptor function

In keeping with the well-recognized feedback regulatory mechanisms for cytoplasmic enzyme pathways, it is now realized that receptor-driven processes too are subject to feedback/feed-forward regulation. As outlined in more detail elsewhere (Hollenberg, 1985), there are a number of levels at which receptor function can be modulated: (1) at the transcriptional level, whereby receptor activation can lead to the induction of receptor biosynthesis, (2) at the post-translational level, wherein receptor function may be altered by processes such as glycosylation, phosphorylation or proteolysis and (3) at the level of the membrane environment wherein receptor properties can be altered by changes in membrane polarization, altered lipid composition or by allosteric interactions with other membrane constituents such as G-proteins.

At the level of the membrane, receptor regulation can be either (1) 'homospecific', where a ligand can regulate the number and/or function of its own receptors or (2) 'heterospecific' (so-called transmodulation), wherein the activation of one receptor modulates the numbers or functions of a second distinct receptor. Such regulatory reactions can cause either an acute receptor desensitization, due to a covalent modification of the receptor itself, or a more prolonged down-regulation of the receptor, due to its internalization and disappearance from the cell surface.

Receptor function is thus no longer viewed as a static process localized solely at the plasma membrane, but rather as a dynamic event, involving receptor activation, feedback regulation, internalization and, either via *de novo* synthesis or recycling, re-insertion of the receptor into the membrane. The manner in which receptor-mediated cell activation (*e.g.* triggering of a phosphorylation cascade) can feed back on the receptor itself to turn off a receptor-driven process can be illustrated by work that has been done with the β-adrenergic reception (Hausdorff et al., 1990; Kobilka, 1992).

Activation of the β-adrenergic receptor results in a rapid increase in intracellular cyclic AMP, because of the G_s-dependent stimulation of adenylyl cyclase. However, in spite of the continued extracellular presence of a β-adrenergic agonist, cyclic AMP levels decline over a period of minutes, and the tissue rapidly becomes refractory to a second stimulus by the β-receptor agonist. This rapid phase of receptor desensitization is believed to be mediated by at least two distinct phosphorylation reactions, one of which involves cyclic AMP-dependent protein kinase (protein kinase A) and a second of which involves a second receptor-targeted kinase (so-called, βARK) that preferentially phosphorylates the ligand-occupied receptor. At low concentrations of agonist, phosphorylation of the β-receptor by protein kinase A appears to uncouple the receptor from interacting with the cyclase-stimulatory G_s protein. At higher agonist concentration, phosphorylation of the receptor by βARK kinase facilitates the interaction of the receptor with a protein termed β-arrestin, which further prevents the receptor from coupling to the G-protein. These rapid receptor-desensitization events can be reversed by phosphatase-catalyzed dephosphorylation, so as to provide for a dynamic equilibrium between 'desensitized' and 'resensitized' receptor. A more prolonged exposure of a tissue to β-adrenergic stimulation can lead to redistribution and sequestration of the receptor away from the plasma membrane and ultimately to a lysosomal site of degradation. This decline in cell surface receptor numbers due to prolonged agonist action can also lead to a reduction in receptor mRNA, thereby reducing further the cellular content of receptor. Overall, this loss of receptor due to prolonged agonist exposure may require appreciable time (*e.g.* tens of hours) for recovery. The process of receptor internalization and desensitization may also be due to the activation of other protein kinases, such as kinase C and to enzyme-catalyzed reactions at cystine sulfhydryl groups (oxidation, fatty acylation) present in the cytoplasmic portions of the receptor.

The above discussion represents but one example of the processes whereby a receptor can be regulated via the same set of reactions that are activated by the ligand-occupied receptor itself, when it stimulates its target tissue. The 'regulation of the regulator' represents an elegant example of the complex way in which overall membrane function can be integrated with a membrane-triggered cellular response.

IV. Summary and view to the future

Membrane receptors for hormones, neurotransmitters and other signalling molecules provide an essential interface be-

tween the extracellular and intracellular environment. Although there are a considerable number of distinct membrane
receptor species now known, the recognition/signalling function of these cellular constituents can be seen mechanistically
to fall into three broad categories: (1) The receptor is a ligand-
regulated ion channel, (2) The receptor is a transmembrane
ligand-regulated enzyme or (3) The receptor is a recognition
protein closely linked to a signal-adapter protein, such as a
G-protein, that couples receptor occupation to the activation of
membrane-localized enzymes or ion channels. The regulation
of receptor function at the biochemical level, ranging from receptor synthesis and turnover to receptor modulation by phos-
phorylation-dephosphorylation reactions is every bit as complex as the regulation of conventional enzyme pathways (e.g.
see Hollenberg, 1985; 1987). The knowledge of receptor function at the biochemical and cellular level that has been
achieved over the past 30 years now provides an excellent
foundation for future work aimed at understanding the overall
integrated function of multicellular organisms. Surely, this
picture of receptor function that has been developed would
have delighted Clark, Ehrlich, Langley and their colleagues.

In a sense, the information that is presently being published
about receptor structure and function now represents an em-
barassment of riches. The rate of appearance of new information has begun to exceed the rate of assimilating the data,
even for a single superfamily of receptors, such as the G-protein-linked 'serpentine' receptor class. Thus, in the near future, it is hoped that within a receptor superfamily, it will be
possible to identify discrete domains in common that may be
associated with functions that are in common between related
receptors. For instance, certain sequences within the G-protein-coupled receptor family are highly conserved (see Savarese and Fraser, 1992). It is expected that the precise function of many of these conserved residues and sequences will
soon been clarified. Alternately, it is anticipated that the
unique receptor domains that result in the ligand specificity of

a given receptor will be readily identified by a variety of molecular approaches, as outlined elsewhere (Hollenberg, 1991).

As rapidly as information is appearing about receptor structure, so is information about receptor-triggered signal transduction pathways expanding almost exponentially. What is becoming clear is that there may be multiple crossover points in the signal pathways that are initiated by the three distinct mechanisms of membrane receptor-mediated cell activation outlined above (*e.g.*, see Lange-Carter et al., 1993). One challenge for the future will be to trace the common and distinct signal pathways activated by different receptor systems, such as the tyrosine kinase class on the one hand, as opposed to the G-protein-coupled family on the other.

Why, one may ask, has there been such an explosion of information and interest in the field of membrane receptor structure and function over the past decade? Two prime reasons come to mind, one of which relates to the advances in molecular biology technology and a second of which relates to an intense focus on the problem of cancer and the puzzle of the cell cycle and cellular differentiation. In essence it would appear that the rapid acquisition of molecular information made possible by recent advances in molecular biology technology has worked efficiently in concert with a heightened focus on the biochemical properties of cancer cells and on the genetics of the cell cycle in organisms such as *S. cerevisiae* and *S. pombe.* Surprisingly, many of the so-called oncogenes identified either in retroviruses or in tumor-derived tissues (*e.g.* H-*ras*, *erb*-B) have turned out to represent modified versions of normal cellular counterparts that play key roles in the signal transduction processes for a variety of agonists. It can thus be said that the perceptive focus of Langley, Ehrlich and Clarke on events that are initiated via 'receptors' that are triggered by agonists at the cell membrane has now led to an avalanche of information about cellular signalling processes. This information is leading to a deeper understanding of the way cells receive and generate signals, to a greater appreciation of the 'communication' networks both between and within cells and ultimately to greater insights related to the mysteries of cellular differentiation and the problem of cancer.

Acknowledgements

The advice and help of Dr. S. Erulkar on a portion of this essay is greatly appreciated by A.K.; M.D.H. is grateful for the input of Dr. D.L. Severson. Work in one of the author's laboratories (MDH), supported largely by funds from the Medical Research Council of Canada, the Heart and Stroke Foundation of Alberta and the Canadian Diabetes Association, has led to many of the insights presented in the text.

REFERENCES

Adams, P.R., Brown, D.A.and Constanti, A. (1982) Pharmacological inhibition of the M-current. J. Physiol. (Lond.) 332, 223–262.

Adler, J. (1975) Chemotaxis in bacteria. Annu. Rev. Biochem. 44, 341–356.

Ariëns, E.J. (1954) Affinity and intrinsic activity in the theory of competitive inhibition. Arch. Pharmacodyn. 99, 32–49.

Ariëns, E.J. and Simonis, A.M. (1976) Receptors and receptor mechanisms. In *Beta-adrenoreceptor blocking agents* (Saxena, P.R. and Forsyth, R.P., eds.) pp. 3–27. Amsterdam, North-Holland Publ. Comp.

Anraku, Y. (1968) Transport of sugars and amino acids in bacteria. I. Purification and specificity of the galactose- and leucine-binding proteins. J. Biol. Chem. 243, 3116–3122.

Arunlakshana, O. and Schild, H.O. (1959) Some quantitative uses of drug antagonists. Br. J. Pharmacol. 14, 48–58.

Barbacid, M. (1987) *ras* Genes. Annu. Rev. Biochem. 56, 779–827.

Benham, C.P. and Tsien, R.W. (1987) A novel receptor-operated Ca^{2+}-permeable channel activated by ATP in smooth muscle. Nature (Lond.) 328, 275–278.

Bennett, V., O'Keefe, E. and Cuatrecasas, P. (1975) Mechanisms of action of cholera toxin and the mobile receptor theory of hormone receptor adenylate cyclase interactions. Proc. Natl. Acad. Sci. USA 72, 33–37.

Berg, D.K. and Hall, Z.W. (1974) Fate of α-bungarotoxin bound to acetylcholine receptors of normal and denervated muscle. Science 184, 473–475.

Bernard, C. (1856) Analyse physiologique des propriétés des systèmes musculaire et nerveux au moyen du curare. C.R. l'Acad. Sci. Paris 43, 825–829.

Berridge, M.J. (1983) Rapid accumulation of inositol triphosphate reveals that agonists hydrolyse polyphosphoinositides instead of phosphatidylinositol. Biochem. J. 212, 849–858.

Berridge, M.J. (1988) Inositol lipids and calcium signalling. Proc. Roy. Soc. B 234, 359–378.

Berridge, M.J. (1993) Inositol triphosphate and calcium signalling. Nature (Lond.) 361, 315–325.

Billah, M.M. and Anthes, J.C. (1990) The regulation and cellular functions of phosphatidylcholine hydrolysis. Biochem. J. 269, 281–291.

Birnbaumer, L. (1973) Hormone-sensitive adenylyl cyclases. Useful models for studying hormone receptor functions in cell-free systems. Biochim. Biophys. Acta 300, 129–158.

Birnbaumer, L., Abramowitz, J. and Brown, A.M. (1990) Receptor-effector coupling by G-proteins. Biochim. Biophys. Acta 1031, 163–224.

Birnbaumer, L., Pohl, S.L., Krans, M.H.J. and Rodbell, M. (1970) The action of hormones on the adenyl cyclase system. Adv. Biochem. Psychopharmacol. 3, 185–208.

Black, J.W. and Leff, P. (1983) Operational models of pharmacological agonism. Proc. Roy. Soc. B. 220, 141–162.

Boeynaems, J.M. and Dumont, J.E. (1977) The two-step model of ligand-receptor interaction. Mol. Cell. Endocrinol. 7, 33–47.

Boeynaems, J.M. and Dumont, J.E. (1980) *Outlines of Receptor Theory.* Elsevier/North-Holland, Amsterdam.

Bothwell, M. (1991) Keeping track of neurotrophin receptors. Cell 65, 915–918.

Bourne, H.R., Sanders, D.A. and McCormick, F. (1990) The GTPase superfamily: A conserved switch for diverse cell functions. Nature (Lond.) 348, 125–132.

Bourne, H.R., Sanders, D.A. and McCormick, F. (1991) The GTPase superfamily: conserved structure and molecular mechanism. Nature (Lond.) 349, 117–127.

Brisson, A. and Unwin, P.N.T. (1985) Quaternary structure of the acetylcholine receptor. Nature (Lond.) 315, 474–477.

Carpenter, G. (1992) Receptor tyrosine kinase substrates; src homology domains and signal transduction. FASEB J. 6, 3283–3289.

Carpenter, G. and Cohen, S. (1979) Epidermal growth factor. Ann. Rev. Biochem. 48, 193–216.

Carpenter, G. and Cohen, S. (1990) Epidermal growth factor. J. Biol. Chem. 265, 7709–7712.

Cassell, D. and Selinger, Z. (1977) Mechanism of adenylate cyclase activation by cholera toxin: inhibition by GTP hydrolysis at the regulatory site. Proc. Natl. Acad. Sci. USA 74, 3307–3311.

Changeux, J.-P. (1980) The acetylcholine receptor: an 'allosteric' membrane protein. Harvey Lect. Ser. 75, 85–254.

220

REFERENCES

Changeux, J.-P., Kasai, M. and Lee, C.-M. (1970) Use of snake venom to characterize the cholinergic receptor protein. Proc. Natl. Acad. Sci. USA 67, 1241–1247.

Chinkers, M., Garbers, D.L., Chang, M.-S., Lowe, D.G., Chin, H., Goeddel, D.V. and Schultz, S. (1989) A membrane form of guanylate cyclase is an atrial natriuretic peptide receptor. Nature (Lond.) 338, 78–83.

Clapham, D.E. (1990) Arachidonic acid and its metabolites in the regulation of G-protein-gated K⁺ channels in atrial myocytes. Biochem. Pharmacol. 39, 813–815.

Clark, A.J. (1926) The antagonism of acetylcholine by atropine. J. Physiol. (Lond.) 61, 547–566.

Clark, A.J. (1927) The reaction between acetylcholine and muscle cells. Part II. J. Physiol. (Lond.) 64, 123–143.

Clark, A.J. (1937) General Pharmacology. Handb. d. exp. Pharmakol., Ergänzungswerk, Band IV. (Heubner, W. and Schüller, J., eds.) Berlin, Springer.

Cohen, P., Carpenter, G. and King, L. Jr. (1980) Epidermal growth factor-receptor-protein kinase interactions: copurification of receptor and epidermal growth factor-enhanced phosphorylation activity. J. Biol. Chem. 255, 4834–4842.

Cohen, P. (1988) Protein phosphorylation and hormone action. Proc. Roy. Soc. B. 234, 115–144.

Conn, P.M., Rogers, D.C., Stewart, J.M., Neidel, J. and Sheffield, T. (1982) Conversion of a gonadotropin-releasing hormone antagonist to an agonist. Nature (Lond.) 296, 653–655.

Conti-Tronconi, B.M. and Raftery, M.A. (1982) The nicotinic cholinergic receptor: correlation of molecular structure with functional properties. Ann. Rev. Biochem. 51, 491–530.

Cook, R.P. (1926) The antagonism between acetylcholine by methylene blue. J. Physiol. (Lond.) 62, 160–165.

Crum-Brown, A. and Fraser, T.R. (1868) On the connection between chemical constitution and physiological action; with special reference to the physiological action of the salts of the ammonium bases derived from strychnia, brucia, thebaia, codeia, morphia and nicotia. J. Anat. Physiol. 1, 224–245.

Cryer, P.E., Jarrett, L. and Kipnis, D.M. (1968) Nucleotide inhibition of adenyl cyclase activity in fat cell membranes. Biochim. Biophys. Acta 177, 586–590.

Cuatrecasas, P. (1971) Insulin-receptor interactions in adipose tissue cells: direct measurement and properties. Proc. Natl. Acad. Sci. USA 68, 1264–1268.

Cuatrecasas, P. (1974a) Membrane receptors. Annu. Rev. Biochem. 43, 169–214.

Cuatrecasas, P. (1974b) Commentary. Insulin receptors, cell membranes, and hormones. Biochem. Pharmacol. 23, 2353–2361.

Cuatrecasas, P. (1985) Internalization and processing of peptide hormone receptors. In *Mechanism of Receptor Regulation* (Poste, G. and Crooke, S.T. eds.), pp. 207–223. New York, Plenum Press.

Cuatrecasas, P. and Hollenberg, M.D. (1976) Membrane receptors and hormone action. Adv. Protein Chem. 30, 251–451.

Dale, H.H. (1914) The actions of certain esters and ethers of choline, and their relation to muscarine. J. Pharmacol. Exp. Ther. 6, 147–190.

Dale, H.H. and Gaddum, J.H. (1930) Reactions of denervated voluntary muscle and their bearing on the mode of action of parasympathetic and related nerves. J. Physiol. (Lond.) 70, 109–144.

De Haen, C. (1976) The non-stoichiometric floating receptor model for hormone-sensitive adenylate cyclase. J. Theor. Biol. 58, 383–400.

De Lean, A., Stadel, J.M. and Lefkowitz, R.J. (1980) A ternary complex model explains the agonist-specific binding properties of the adenylate cyclase-coupled β-adrenergic receptor. J. Biol. Chem. 255, 7108–7117.

Del Castillo, J. and Katz, B. (1954) The membrane change produced by the neuromuscular transmitter. J. Physiol. (Lond.) 125, 546–565.

Dionne, V.E., Steinbach, J.H. and Stevens, C.F. (1978) An analysis of the dose-response relationship at voltage-clamped frog neuromuscular junctions. J. Physiol. (Lond). 281, 421–444.

Dohlman, H.G., Caron, M.G. and Lefkowitz, R.J. (1987) A family of receptors coupled to guanine nucleotide regulatory proteins. Biochem. 26, 2657–2664.

Edidin, M. (1974) Rotational and translational diffusion in membranes. Annu. Rev. Biophys. Bioeng. 3, 179–291.

Edidin, M., Zaganasky, Y. and Lardner, T.M. (1976) Measurement of membrane protein lateral diffusion in single cells. Science 19, 466–468.

Ehrlich, P. (1886) Über die Methylenblaureaction der lebenden Nervensubstanz. Dtsch. med. Wschr. 12, 49–52.

Ehrlich, P. (1990) On immunity with special reference to cell life. Proc. Roy. Soc. 66, 424–448.

Ehrlich, P. (1902) Über die Beziehungen von chemischer Constitution, Vertheilung und pharmacologischer Wirkung. In *The Collected Papers of Paul Ehrlich* (Himmelweit, F., Marquardt, M. and Dale, H.H., eds.), Vol. II, pp. 570–595. London 1960, Pergamon Press.

Ehrlich, P. (1908) Nobel lecture on partial functions of the cell. In *The Collected Papers of P. Ehrlich, Vol. III.* (Himmelweit, F., Marquardt, M., Dale, H.H., eds.), Oxford. Pergamon Press pp. 183–194.

Ehrlich, P. (1909) Über Partialfunktionen der Zelle. Nobel Vortrag. Münch. Med. Wschr. 56, 217–222.

Fatt, P. and Katz, B. (1951) An analysis of the end-plate potential recorded with an intra-cellular electrode. J. Physiol. (Lond.) 115, 320–370.

Feldberg, W. (1933) Der Nachweis eines acetylcholinähnlichen Stoffes im Zungenvenenblut des Hundes bei der Reizung des Nervus lingualis. Pflüger's Arch. Ges. Physiol. 232, 88–104.

Fischer, E.H., Charbonneau, H. and Tonks, N.K. (1991) Protein tyrosine phosphatases: a diverse family of intracellular and transmembrane enzymes. Science 253, 401–406.

Flexner, S. and Noguchi, H. (1902) The constitution of snake venom and snake sera. Univ. Pennsylvania Med. Bull. 16, 345–362.

Freychet, P., Roth, J. and Neville, D.M. Jr. (1971) Insulin receptors in the liver: specific binding of [^{125}I]insulin to the plasma membrane and its relation to insulin bioactivity. Proc. Natl. Acad. Sci.USA 68, 1833–1837.

Frye, L.D. and Edidin, M. (1970) The rapid intermixing of cell surface antigens after formation of mouse-human heterokaryons. J. Cell Sci. 7, 319–335.

Furchgott, R.F. (1955) The pharmacology of vascular smooth muscle. Pharmacol. Revs. 7, 183–265.

Furlong, C.E. (1987) Osmotic shock-sensitive transport systems. In *Escherichia coli and Salmonella typhimurium. Cellular and Molecular Biology* (Neidhardt, F.C., ed.), pp. 760–767. Washington D.C., Am. Soc. Microbiol.).

Gaddum, J.H. (1926) The action of adrenalin and ergotamine on the uterus of the rabbit. J. Physiol. (Lond.) 61, 141–150.

Gaddum, J.H. (1937) The quantitative effects of antagonistic drugs. J. Physiol. (Lond.) 89, 7P–9P.

Garbers, A.G. (1992) Guanylyl cyclase receptors and their endocrine, paracrine, and autocrine glands. Cell 71, 1–4.

Gautam, N., Northup, J., Tamir, H. and Simon, M.I. (1990) G-protein diversity is increased by associations with a variety of γ subunits. Proc. Natl. Acad. Sci. USA 87, 7973–7977.

Gilman, A.G. (1987) G-proteins: transducers of receptor-generated signals. Ann. Rev. Biochem. 56, 615–649.

Gilman, A.G., Rall, T.W., Nies, A.S. and Taylor, P. (1990) *Goodman and Gilman's The Pharmacological Basis of Therapeutics*. New York, Pergamon Press.

Goldfine, I.D. (1987) The insulin receptor: molecular biology and transmembrane signalling. Endocr. Rev. 8, 235–255.

Greengard, P. (1976) Possible role for cyclic nucleotides and phosphorylated membrane proteins in postsynaptic actions of neurotransmitter. Nature (Lond.) 260, 101–108.

Gregor, P., Mano, I., Maoz, I., McKeown, M. and Teichberg, V.I. (1989) Molecular structure of the chick cerebellar kainate-binding subunit of a putative glutamate receptor. Nature (Lond.) 342, 689–692.

Gregory, H., Taylor, C.L. and Hopkins, C.R. (1982) Luteinizing hormone release from dissociated pituitary cells by dimerization of occupied LHRH receptors. Nature (Lond.) 300, 269–271.

Grenningloh, G., Rienitz, A., Schmitt, B., Methfessel, C., Zensen, M., Beyreuther, K., Gundelfinger, E.D. and Betz, H. (1987) The strychnine-binding subunit of the glycine receptor shows homology with nicotinic acetylcholine receptors. Nature (Lond.) 328, 215–220.

Hamilton, S.L., Codina, J., Hawkes, M.J., Yatani, A., Sawada, T., Strickland, F.M., Froehner, S.C., Spiegel, A.M., Toro, L., Stefani, E., Birnbaumer, L. and Brown, A.M. (1991) Evidence for direct interaction of $G_s\alpha$ with the Ca^{2+} channel of skeletal muscle. J. Biol. Chem. 266, 19528–19535.

Hartzell, H.C., Kuffler, S.W., Stickgold, R. and Yoshikami, D. (1977) Synaptic excitation and inhibition resulting from direct action of acetylcholine on two types of chemoreceptors in individual amphibian parasympatic neurones. J. Physiol. (Lond.) 271, 817–846.

Hausdorff, W.P., Caron, M.G. and Lefkowitz, R.J. (1990) Turning off the signal: desensitization of β-adrenergic receptor function. FASEB J. 4, 2881–2889.

Hazelbauer, G.L. and Adler, J. (1971) Role of the galactose binding proteins in chemotaxis of Escherichia coli toward galactose. Nature New Biology 230, 101–104.

Hazum, E. and Keinan, D. (1985) Gonadotropin releasing hormone activation is mediated by dimerization of occupied receptors. Biochem. Biophys. Res. Comm. 133, 449.

Hepler, J.R. and Gilman, A.G. (1982) G-proteins. Trends Biochem. Sci. 17, 383–387.

Hill, A.V. (1909) The mode of action of nicotine and curari, determined by the form of the contraction curve and the method of temperature coefficients. J. Physiol. (Lond.) 39, 361–373.

Hokin, L.E. and Hokin, M.B. (1955) Effects of acetylcholine on the turnover of phosphoryl units in individual phospholipids of pancreas slices and brain cortex slices. Biochim. Biophys. Acta 18, 102–110.

Hollenberg, M.D. (1979) Membrane receptors and hormone action. Pharmacol. Rev. 30, 393–410.

Hollenberg, M.D. (1985) Control of Receptor Function by Homologous and Heterologous Ligands. Mechanisms of Receptor Regulation. G. Poste and S.T. Crooke (Eds) New York, Plenum Publishing Corp., pp. 295–322.

Hollenberg, M.D. (1987) Receptor regulation and receptor-receptor communication. In: Wenner-Gren Center International Symposium Series, Stockholm, Sweden, October 9–11, 1986. MacMillan Press, U.K. Vol. 48, pp. 546–554.

Hollenberg, M.D. (1991) Structure-activity relationships for transmembrane signalling: the receptor's turn. FASEB J. 5, 178–186.

Hollmann, M., O'Shea-Greenfield, A., Rogers, S.W. and Heinemann, S. (1989) Cloning by functional expression of a number of the glutamate receptor family. Nature (Lond.) 342, 643–648.

Hopkins, C.R., Šemoff, S. and Gregory, H. (1981) Regulation of gonadotropin secretion of the anterior pituitary. Philos. Trans. R. Soc. Lond. B296, 73–81.

Houslay, M.D., Hesketh, T.R., Smith, G.A., Warren, G.B. and Metcalfe, J.C. (1976) The lipid environment of the glucagon receptor regulates adenylate cyclase activity. Biochim. Biophys. Acta 436, 495–504.

Howell, W.H. and Duke, W.W. (1908) The effect of vagus inhibitors on the output of potassium from the heart. Am. J. Physiol. 21, 51–63.

Huganir, R.L., Miles, K. and Greengard, P. (1984) Phosphorylation of the nicotinic acetylcholine receptor by an endogenous tyrosine-specific protein kinase. Proc. Natl. Acad. Sci. USA 81, 6968–6972.

Hunt, R. and Taveau, R. de M. (1906) On the physiological action of certain cholin derivatives and new methods for detecting cholin. Brit. Med. J. 1906, 2, 1788–1791.

Jacobs, M.H. (1950) Surface properties of the erythrocyte. Ann. New York Acad. Sci. 50, 824–834.

Jacobs, S. and Cuatrecasas, P. (1976) The mobile receptor hypothesis and 'cooperativity' of hormone binding. Application to insulin. Biochim. Biophys. Acta 433, 482–495.

Jensen, E.V. and Jacobson, H.I. (1962) Basic guides to the mechanism of estrogen action from binding to liver proteins. Recent Progr. Horm. Res. 18, 387–414.

Kahn, C.R., Baird, K.L., Flier, J.S., Grunfeld, C., Harmon, J.T., Harrison, L.C., Karlsson, F.A., Kasuga, M., King, G.L., Lang, U.C., Podskalny, J.M. and Obberghen, E. (1981) Insulin receptor, receptor antibodies and the mechanism of insulin action. Recent Prog. Horm. Res. 37, 477–538.

Kalckar, H. (1971) The periplasmic galactose binding protein of Escherichia coli. Science 174, 557–565.

Karlin, A. (1991) Explorations of the nicotinic acetylcholine receptor. Harvey Lect. Ser. 85, 71–107.

Kasai, M. and Changeux, J.-P. (1971) In vitro excitation of purified membrane fragments by cholinergic agonists. I. The permeability change caused by cholinergic agonists. J. Membr. Biol. 6, 24–57.

Keller, E.B. and Zamecnik, P.C. (1956) The effect of guanosine diphosphate and triphosphate in the incorporation of labeled amino acids into proteins. J. Biol. Chem. 221, 45–59.

Kenakin, T.P., Bond, R.A. and Bonner, T.I. (1992) Definition of pharmacological receptors. Pharmacol. Revs. 44, 351–362.

Kobilka, B. (1992) Adrenergic receptors as models for G-protein-coupled receptors. Annu. Rev. Neurosci. 15, 87–114.

Kölliker, M.A. (1856) Note sur l'action du curare sur le système nerveux. C.R. l'Acad. Sci. Paris 43, 791–792.

Krebs, E.G. (1973) The Mechanism of Hormonal Regulation by Cyclic AMP. Excerpta Medica Int. Congr. Ser. 273, 17–29.

Keyes, P. (1902) Über die Wirkungsweise des Cobragiftes. Berl. Klin. Wschr. 39, 918–922.

Koch, C.A., Anderson, D., Moran, M.F., Ellis, C. and Pawson, T. (1991) SH2 and SH3 domains: elements that control interactions of cytoplasmic signalling proteins. Science 252, 668–674.

Lange-Carter, C.A., Pleiman, C.M., Gardner, A.M. and Johnson, G.L. (1993) A divergence in the MAP kinase regulatory network defined by MEK kinase and Raf. Science 260, 315–319.

Langley, J.N. (1879) On the physiology of the salivary secretion. Part II. On the mutual antagonism of atropin and pilocarpin, having especial reference to their relations in the sub-maxillary gland of the cat. J. Physiol. (Lond.) 1, 339–369.

Langley, J.N. (1906a) On the reaction of cells and of nerve-endings to certain poisons, chiefly as regards the reaction of striated muscle to nicotine and to curari. J. Physiol. (Lond.) 33, 374–413.

Langley, J.N. (1906b) On nerve endings and on special excitable substances in cells. Proc. Roy. Soc. B 70, 170–194.

Lefkowitz, R.J. (1992) The subunit story thickens. Nature (Lond.) 358, 372–373.

Lefkowitz, R.J., Mullikin, D. and Caron, M.G. (1976) Regulation by beta-adrenergic receptors by guanyl-5′-yl imidophosphate and other purine nucleotides. J. Biol. Chem. 251, 4686–4692.

Lefkowitz, R.J. and Caron, M.G. (1988) Adrenergic regulator. Models for the study of receptors coupled to guanine nucleotide regulatory proteins. J. Biol. Chem. 263, 4993–4996.

Levitzki, A. and Bar-Sinai, A. (1991) The regulation of adenylyl cyclase by receptor-operated G-proteins. Pharmacol. Ther. 50, 271–283.

Lindstrom, J., Schoepfer, R. and Whiting, P. (1987) Molecular studies of the neuronal nicotinic acetylcholine receptor family. Mol. Neurobiol. 1, 281–336.

Litosch, I., Wallis, C. and Fain, J.N., (1985) 5-Hydroxytryptamine stimulates inositol phosphate production in a cell-free system from blowfly salivary glands. J. Biol. Chem. 260, 5464–5470.

Loewi, O. (1921) Über humorale Übertragbarkeit der Herznervenwirkung. Pflüger's Arch. Ges. Physiol. 189, 239–242.

Loewi, O. and Navratil, E. (1926) Über humorale Übertragbarkeit der Herznervenwirkung. X. Über das Schicksal des Vagusstoff. Pflüger's Arch. Ges. Physiol. 214, 678–688.

Logothetis, D.E., Kurachi, Y., Galper, J., Neer, E.J. and Clapham, D.E. (1987) The $\beta\gamma$ subunits of GTP-binding proteins activate the muscarinic K^+ channel in heart. Nature (Lond.) 325, 321–326.

McLaughlin, S.K., McKinnon, P.J. and Margolskee, R.F. (1992) Gustducin is a taste-cell-specific G-protein closely related to the transducins. Nature (Lond.) 357, 563–569.

Maricq, A.V., Peterson, A.S., Brake, A.J., Myers, R.M. and Julius, D. (1991) Primary structure and functional expression of the $5HT_3$ receptor, a serotonin-gated ion channel. Science 254, 432–437.

Marsland, D. (1934) The site of narcosis in a cell; the action of a series of paraffin oils in Amoeba dubia. J. Cell. Comp. Physiol. 4, 9–29.

Marx, J.L. (1983) Cloning the acetylcholine receptor genes. Science 219, 1055–1056.

Mathews, L.S. and Vale, W.W. (1991) Expression cloning of an activin receptor, a predicted transmembrane serine kinase. Cell 65, 973–982.

Meldolesi, J. and Pozzan, T. (1987) Pathways of Ca^{2+} influx at the plasma membrane: Voltage-, receptor- and second messenger-operated channels. Exp. Cell. Res. 171, 271–283.

Meunier, J.-C., Sealock, R., Olsen, R. and Changeux, J.-P. (1974) Purification and properties of the cholinergic receptor protein from *Electrophorus electricus* electric tissue. Eur. J. Biochem. 45, 371–394.

Michell, R.H. (1975) Inositol phospholipids and cell surface receptors. Biochim. Biophys. Acta 415, 81–147.

Miki, N., Keirns, J.J., Markus, F.R., Freeman, J. and Bitensky, M.W. (1973) Regulation of cyclic nucleotide concentrations in photoreceptors: An ATP-dependent stimulation of cyclic nucleotide phosphodiesterase by light. Proc. Natl. Acad. Sci. USA 70, 3820–3824.

Miyajima, A., Kitamura, T., Harada, N., Yokota, T. and Arai, K.-I. (1992) Cytokine receptors and signal transduction. Ann. Rev. Immunol. 10, 295–331.

Monod, J., Wyman, J. and Changeux, J.-P. (1965) On the nature of allosteric transitions: A plausible model. J. Mol. Biol. 12, 88–118.

Murayama, T. and Ui, M. (1985) Receptor-mediated inhibition of adenylate cyclase and stimulation of arachidonic acid release in 3T3 fibroblasts. J. Biol. Chem. 260, 7226–7233.

Nachmanson, D. (1959) *Chemical and Molecular Basis of Nerve Activity.* New York, Academic Press.

Nastuk, W.L. (1953) Membrane potential changes at a single muscle endplate produced by transitory application of acetylcholine with an electrically controlled micropipet. Feder. Proc. 12, 102.

Nickerson, M. (1956) Receptor occupancy and tissue response. Nature (Lond.) 178, 697–698.

Northup, J.K., Smigel, M.D., Sternweis, P.C. and Gilman, A.G. (1983) The subunits of the stimulating regulatory component of adenylate cyclase. J. Biol. Chem. 258, 11369–11376.

Numa, S. (1989) A molecular view of neurotransmitter receptors and ionic channels. Harvey Lect. 83, 121–165.

O'Dea, R.F., Viveros, O.H., Axelrod, J., Aswanikumar, S., Schiffmann, E. and Corcoran, B.A. (1978) Rapid stimulation of protein carboxymethylation in leukocytes by a chemotactic peptide. Nature (Lond.) 272, 462–464.

Pace, U., Hanski, E., Salomon, Y. and Lancet, D. (1985) Odorant-sensitive adenylate cyclase may mediate olfactory receptors. Nature (Lond.) 316, 255.

Pardee, A.B. (1958) Membrane transport proteins. Science 162, 632–637.

Pastan, I.H. and Willingham, M.C. (1981) Journey to the center of the cell: role of the receptosome. Science 214, 504–509.

Paton, W.D.M. (1961) A theory of drug action based on the rate of drug-receptor combination. Proc. R. Soc. Lond. Ser. B 154, 21–69.

Paton, W.D.M. and Rang, H.P. (1965) The uptake of atropine and related drugs by intestinal smooth muscle of the guinea pig in relation to acetylcholine receptors. Proc. R. Soc. Lond. Ser. B 163, 1–44.

Perez-Sala, O., Tan, E.W., Canada, F.J. and Rando, R.R. (1991) Methylation and demethylation reactions of guanidine nucleotide-binding proteins of retinal rod outer segment. Proc. Natl. Acad. Sci. USA 88, 3043–3046.

Peters, J.A., Malone, H.M. and Lambert, J.J. (1992) Recent advances in the electrophysiological characterisation of 5-HT$_3$ receptors. Trends Pharmacol. Sci. 13, 391.

Peters, R.A. (1937) Proteins and cell-organization. In *Perspectives in Biochemistry* (Needham, J. and Green, D.E., eds.) Cambridge, University Press, pp. 36–44.

Pfeuffer, T. (1977) GTP-binding proteins in membranes and the control of adenylate cyclase activity. J. Biol. Chem. 252, 7224–7234.

Piperno, J.R. and Oxender, D.I. (1966) Amino acid binding protein released from *Escherichia coli* by osmotic shock. J. Biol. Chem. 241, 5732–5734.

Posner, B.I., Khan, M.N. and Bergeron, J.J.M. (1985) Receptor-mediated uptake of peptide hormones and other ligands. In: *Polypeptide Hormone Receptors*. B.I. Posner, ed., Marcel Dekker Inc., New York, pp. 61–90.

Pritchett, D.B., Sontheimer, H., Shivers, B.D., Ymer, S., Kettenmann, H., Schofield, P.R. and Seeberg, P.H. (1989) Importance of a novel GABA$_A$ receptor subunit for benzodiazepine pharmacology. Nature (Lond.) 338, 583–585.

Quastel, J.H. (1930) The mechanism of bacterial action. Trans. Faraday Soc. 26, 853–861.

Raftery, M.A., Vandlen, R., Reed, R. and Lee, T. (1976) Structure and functional characteristics of an acetylcholine receptor. Cold Spring Harb. Symp. Quant. Biol. 40, 193–202.

Raftery, M.A., Hunkapiller, M.W., Strader, C.D. and Hood, L.E. (1980) Acetylcholine receptor: complex of homologous subunits. Science 208, 1454–1456.

Ransom, F. (1901) Saponin und sein Gegengift. Dtsch. Med. Wschr. 194–196.

Reuter, H. (1983) Calcium channel modulation by neurotransmitter, enzymes and drugs. Nature (Lond.) 301, 569–574.

Rodbell, M. (1980) The role of hormone receptors and GTP-regulatory proteins in membrane transduction. Nature 284, 17–22.

Rodbell, M., Birnbaumer, L., Pohl, S.L. and Krans, H.M.J. (1971) The glucagonsensitive adenyl cyclase system in plasma membranes of rat liver. J. Biol. Chem. 246, 1877–1882.

Rosen, O.M. (1987) After insulin binds. Science 237, 1452–1458.

Rosen, O.M., Herrera, R., Olowe, Y., Petruzzelli, L.M. and Cobb, M.H. (1983) Phosphorylation activates the insulin receptor tyrosine protein kinase. Proc. Natl. Acad. Sci. USA 80, 3237–3240.

Ross, E.M., Howlett, A.C., Ferguson, K.M. and Gilman, A.G. (1978) Reconstitution of hormone-sensitive adenylate cyclase activity with resolved components of the enzyme. J. Biol. Chem. 253, 6401–6412.

Ross, E.M. and Gilman, A.G. (1980) Biochemical properties of hormone-sensitive adenylate cyclase. Annu. Rev. Biochem. 49, 533–564.

Roth, R.A. and Cassell, D.J. Evidence that the insulin receptor is a protein kinase. Science 219, 299–301.

Sansom, M.S.P. and Usherwood, P.N.R. (1990) Single-channel studies of glutamate receptors. Int. Rev. Neurobiol. 32, 51–106.

Satoh, T., Nakafuku, M. and Kaziro, Y. (1992) Function of ras as a molecular switch in signal transduction. J. Biol. Chem. 267, 24149–24152.

Savarese, T.M. and Fraser, C.M. (1992) In vitro mutagenesis and the search for structure-function relationships among G-protein-coupled receptors. Biochem. J. 283, 1–19.

Scatchard, G. (1949) The attraction of proteins for small molecules and ions. Ann. N.Y. Acad. Sci. 51, 660–672.

Schlessinger, J. and Ullrich, A. (1992) Growth factor signalling by receptor tyrosine kinases. Neuron 9, 383–391.

Schofield, P.R., Darlison, M.G., Fujita, N., Burt, D.R., Stephenson, F.A., Rodriguez, H., Rhee, L.M., Ramchandran, J., Reale, V., Glencorse, T.A., Seeberg, P.H. and Barnard, E.A. (1987) Sequence and functional expression of the $GABA_A$ receptor show a ligand-gated receptor superfamily. Nature (Lond.) 328, 221–227.

Schuetze, S.M. and Role, L.W. (1987) Developmental regulation of nicotinic acetylcholine receptors. Annu. Rev. Neurosci. 10, 407–457.

Sharon, N. and Lis, H. (1989) Lectins as cell recognition molecules. Science 246, 227–234.

Sieghart, W. (1992) $GABA_A$ receptors: ligand-gated CL^- ion channels modulated by multiple drug-binding sites. Trends Biochem. Sci. 13, 446–450.

Singer, S.J. and Nicolson, G.L. (1972) The fluid mosaic model of the structure of cell membranes. Science 175, 720–731.

Simon, M.I., Strathmann, M.P. and Gautam, N. (1991) Diversity of G-proteins in signal transduction. Science 252, 802–808.

Smrcka, A.V., Hepler, J.R., Brown, K.O. and Sternweis, P.C. (1991) Regulation of polyphosphoinositide-specific phospholipase C activity by purified G_q. Science 251, 804–807.

Somlyo, A.V. and Somlyo, A.P. (1968) Electromechanical and pharmacomechanical coupling in vascular smooth muscle. J. Pharmacol. exp. Ther. 159, 129–145.

Spiegel, A.M., Shenker, A. and Weinstein, L.S. (1992) Receptor-effector coupling by G-proteins: implications for normal and abnormal signal transduction. Endocrine Reviews 13, 536–565.

Springer, M.S., Goy, M.F. and Adler, J. (1979) Involvement of reversible methylation of membrane-bound receptors. Nature (Lond.) 280, 279–284.

Stephenson, R.P. (1956) A modification of the receptor theory. Brit. J. Pharmacol. 11, 379–393.

Stevens. C.F. (1985) Molecular tinkerings that tailor the acetylcholine receptor. Nature (Lond.) 313, 350–351.

Strosberg, A.D. (1991) Structure/function relationship of proteins belonging to the family of receptors coupled to GTP-binding proteins. Eur. J. Biochem. 196, 1–10.

Stroud, R.M., McCarthy, M.P. and Shuster, M. (1990) Nicotinic acetylcholine receptor superfamily of ligand-gated ion channels. Biochem. 29, 11009–11023.

Stryer, L. (1991) Visual excitation and recovery. J. Biol. Chem. 266, 10711–10714.

Sturgill, T.W. and Wu, J. (1991) Recent progress in characterization of protein kinase cascades for phosphorylation of ribosomal protein S_6. Biochim. Biophys. Acta 1092, 350–357.

Sutherland, E.W. (1962) The biological role of adenosine-3′,5′-phosphate. Harvey Lect. Ser. 51, 17–33.

Sutherland, E.W. (1972) Studies on the mechanism of hormone action. Science 177, 401–409.

Takeuchi, A.T. and Takeuchi, N. (1960) On the permeability of end-plate membrane during the action of transmitter. J. Physiol. (Lond.) 154, 52–67.

Takeuchi, N. (1963) Effects of calcium on the conductance change of the endplate membrane during the action of transmitter, J. Physiol. (Lond.) 167, 141–155.

Tang, W.-J. and Gilman, A.G. (1991) Type-specific regulation of adenylyl cyclase by G-protein $\beta\gamma$ subunits. Science 254, 1500–1503.

Tank, D.W., Huganir, R.L., Greengard, P. and Webb, W.W. (1983) Patch-recorded single-channel currents of the purified and reconstituted Torpedo acetylcholine receptor. Proc. Natl. Acad. Sci. USA 80, 5129–5133.

Taussig, R., Quarmby, L.M. and Gilman, A.G. (1993) Regulation of purified type I and type II adenylylcyclases by G-protein $\beta\gamma$ subunits. J. Biol. Chem. 268, 9–12.

Taylor, P., Brown, R.D. and Johnson, D.A. (1983) The linkage between ligand occupation and responses of the nicotinic acetylcholine receptor. In: Current Topics in Membranes and Transport, 18, 407–444.

Trautwein, W., Kuffler, S.W. and Edwards, C. (1956) Changes in membrane characteristics of heart muscle during inhibition. J. Gen. Physiol. 40, 135–145.

Ullrich, A. and Schlessinger, J. (1990) Signal transduction by receptors with tyrosine kinase activity. Cell 61, 203–212.

Weaver, C.T., Pingel, J.T., Nelson, J.O. and Thomas, M.L. (1992) CD45, a transmembrane protein tyrosine phosphatase involved in the transduction of antigenic signals. Biochem. Soc. Trans. 20, 169–174.

Weber, M. and Changeux, J.-P. (1974) Binding of Naja nigricollis [^{3}H]α-toxin to membrane fragments from Electrophorus and Torpedo electric organs. Mol. Pharmacol. 10, 1–14.

Weis, R.M. and Koshland, D.E.Jr. (1988) Reversible receptor methylation is essential for normal chemotaxis of Escherichia coli in gradients of aspartic acid. Proc. Natl. Acad. Sci.USA. 85, 83–87.

Yankner, B.A. and Shooter, E.M. (1982) The biology and mechanism of action of nerve growth factor. Ann. Rev. Biochem. 51, 845–868.

Yatani, A., Codina, J., Brown, A.M. and Birnbaumer, L. (1987a) Direct activation of mammalian muscarinic potassium channels by GTP regulatory protein G_k. Science 235, 207–211.

Yatani, A., Codina, J., Imoto, Y., Reeves, J.P., Birnbaumer, L. and Brown, A.M. (1987b) A G-protein directly regulates mammalian cardiac calcium channels. Science 238, 1288–1292.

A. Kleinzeller (Editor) A History of Biochemistry: Exploring the Cell Membrane.
(Comprehensive Biochemistry Vol. 39) © 1995 Elsevier Science B.V.

Chapter 6

Energy Metabolism in Cellular Membranes

DAVID F. WILSON

Department of Biochemistry and Biophysics, University of Pennsylvania,
Philadelphia, PA 19104-6059, U.S.A.

I. Introduction

Energy metabolism is central to the survival of all organisms, providing the chemical energy for biosynthesis, maintaining ionic balance and doing mechanical work. In higher organisms, energy metabolism is dominated by oxidative phosphorylation, a controlled combustion of foodstuffs by oxygen. The enzymes of oxidative phosphorylation are not only an integral part of the structure of cellular membranes but also require an intact membrane for function.

Gaining an understanding of the chemical and physical events involved in oxidative phosphorylation and the mechanisms by which it is regulated has been a study in membrane structure and of the physiological role of membranes. It should be kept in mind that before about 1960 it was difficult to prove that a preparation isolated from cells consisted of membranes. Even when the investigators were convinced that they were working with membranes, they typically referred to their preparations as particulate or insoluble.

The study of respiration and its role in providing energy for survival is one of the active research problems in modern biochemistry with a direct lineage to questions about life as addressed by early philosophers. Galen (A.D. 131–210) considered conversion of 'natural spirits' to 'vital spirits' the essence of life and this conversion involved controlled combustion (heat production). During the intervening years, much has been learned, but it is clear that there remain unanswered questions of fundamental importance. The published papers documenting the progression of scientific thought concerning animal respiration contain a record of the internal working of the discipline we call science. Keilin (1966) eloquently presented his personal view of this progress:

> 'In looking back at the development of our knowledge of respiration we can see that it did not follow a straight or logical course. On the contrary, it can be compared to the flow of several convergent streams winding in their course, here moving separately, then sending off branches to join each other; being crystal clear in some places, turbid in others; sometimes hidden in the mist arising from themselves; now moving slowly, then rushing suddenly past many obstacles to slow down again before finally joining together to follow a straight course.'

Equally importantly, an examination of this record can provide a perspective on the character of the individual scientists, the individuality of their approach to science, and of the contribution of the scientific community (environment) in the advancement of knowledge. The picture is of a distinctly human endeavor. Individuals of greatly differing capabilities and backgrounds wrestled with questions without clear answers. Scientists were generally aided by, but sometimes impeded by, their colleagues and the scientific community. The great strength of science, however, is that there is a steady, if not continuous, progression toward increased understanding. We can look forward with confidence to having the answers to our questions, albeit not always as soon as we would like.

II. Discovery and characterization of the respiratory chain

1. EARLY BACKGROUND

By the end of the 19th century what may be called the modern phase of the study of tissue oxidations was well underway. The role of the blood in transporting oxygen had been established. The chemical reactions carried out by living organisms (fermentations) were known to be associated with the content of cells but studies on the nature of biological catalysis was in its infancy. Two major hypotheses had been put forward for the oxygen requiring ferment. Traube (1882) believed that these were oxidative reactions in which the initial chemical event was activation of molecular oxygen. Once the oxygen was activated it reacted with readily oxidizable substances (M) in the cell to form oxidized products and hydrogen peroxides.

$$M + 2H_2O \rightarrow M(OH)_2 + 2H + O_2 \rightarrow H_2O_2 \tag{1}$$

Hoppe-Seyler (1879) on the other hand, proposed that the initial chemical event was activation of hydrogen (formation of 'nascent' hydrogen). The nascent hydrogen then reacted with oxygen, splitting the molecule to form water and an active oxygen atom. This active oxygen then reacted with easily oxidizable substrates and otherwise behaved in a manner similar to that proposed by Traube.

$$MH \rightarrow H + M + O_2 + H_2O \rightarrow MO + H_2O_2 \tag{2}$$

These proposed mechanisms were important primarily because they emphasized the possible role of hydrogen peroxide in metabolism. The first two enzymes identified catalyzed reactions involving hydrogen peroxide, i.e. peroxidase (Linossier, 1898) and catalase (Loew, 1901).

Two rather speculative hypotheses had also been put forward in which oxygen was less directly involved. Ehrlich (1885) studied the state of reduction and oxidation of dyes such as indophenol blue injected subcutaneously into animals *in vivo*. He observed that the dyes became mostly oxidized in some tissues and mostly reduced in others. Since the state of reduction or oxidation was tissue specific, he proposed that the difference was due to different degrees of 'oxygen saturation'. He believed dye reduction correlated with low oxygen saturation and acidic conditions whereas dye oxidation correlated with high oxygen saturation and alkaline conditions. MacMunn (1884, 1887), on the other hand, carried out a detailed characterization of the intrinsic pigments of muscle and observed a four banded absorption spectrum (613–593 nm, 567.5–561.5 nm, 554.5–546 nm, and about 532–511.5 nm). This spectrum was attributed to the reduced form of a pigment (myohaematin) which was mostly in the oxidized form in the aerobic sample. MacMunn reported that once the pigment was reduced it could be oxidized by strong oxidizing agents such as H_2O_2 but he was unable to observe reoxygenation by molecular oxygen. He concluded from his observations that myohaematin was capable of oxidation and reduction and was therefore respiratory, *i.e.*, involved in the process of respiration. These hypotheses are indicative of the variety of ideas being considered as explanations of the phenomenon of respiration. The interpretation of MacMunn was in many ways the most correct, but remained unaccepted and was mostly forgotten. It was not a factor in later developments (for discussion of the reasons see Keilin, 1966).

Each of the above hypotheses was partially true, experimentally testable, and consistent with a selected set of available data. None, however, was clearly advanced over the others. The available data of the time, including that provided by MacMunn, were insufficient to establish the correctness of any one of the proposals. Most critically, MacMunn was unable to generate new and more convincing evidence of redox cycling of myohaematin. In this case, there was a rather extreme result, *i.e.*, myohaematin and its proposed role in respiration was forgotten.

2. TISSUE RESPIRATION IN THE PERIOD FROM 1900 TO DISCOVERY OF THE RESPIRATORY CHAIN

As noted earlier, Ehrlich (1885) had observed that when redox-active dyes were taken up by cells and tissue some were accumulated in the reduced form and some in the oxidized form. Ehrlich was not a chemist by training and his hypothesis of respiration based on oxygen saturation, alkaline and acid pH did not contribute substantially to understanding the process of respiration. The dye systems he described, however, were the basis of much further experimentation. Thunberg (for review, see Thunberg, 1930), while studying dye reactions in tissue homogenates, extended the technology of the time by designing a colorimeter tube which could be conveniently sealed or connected to a vacuum pump and evacuated.

The oxygen could be effectively removed from the sample by repeated evacuations, each time replacing the gas with oxygen free nitrogen. Once the oxygen had been removed, a reagent could be added from an internal compartment to initiate the reaction. This 'Thunberg tube' allowed easy, precise colorimetric measurements of the reduction and oxidation of the dyes by cells and tissue homogenates in the presence and absence of oxygen, a technical advance of great value. Wieland (see for example (1912a,b)) proposed that tissue contained dehydrogenases, catalysts which activated hydrogen atoms on substrates such that they can be readily transferred to hydrogen acceptors. These hydrogen acceptors could be either dyes which were reduced to their leuco form (most of the dyes were colored when oxidized but not when reduced) or oxygen which was reduced to hydrogen peroxide. This hypothesis developed rapidly from 1900 to 1924, when Fleisch (1924) suggested the

dehydrogenase itself accepted the hydrogen atoms and transferred them to the hydrogen acceptor:

$$\text{Succinate} + \text{dehydrogenase} \rightarrow \text{fumarate} + \text{dehydrogenase-H}_2 \tag{3}$$

$$\text{Dehydrogenase-H}_2 + \text{MB} \rightarrow \text{dehydrogenase} + \text{MBH}_2 \tag{4}$$

where MB is the oxidized form of methylene blue and MBH_2 is
reduced methylene blue.

Warburg, in the period from 1908 to 1930, developed a theory of cellular respiration based on the catalytic powers of
iron. It was known that solutions of iron salts could catalyze
the oxidation of various organic compounds by molecular oxygen and could rapidly decompose hydrogen peroxide. Warburg
realized that biological systems generally contained substantial amounts of iron and proposed catalysis of respiration by
iron salts in tissue. He prepared charcoal from tissue and
showed it was active in catalyzing the oxidation of many organic compounds by oxygen. The catalytic activity was cyanide sensitive, a property in common with the respiration of
viable animal tissue. Although this hypothesis was far from
being correct, it generated a series of experiments of great
significance. Haldane and Smith (1896) had reported that the
CO-hemoglobin compound could be dissociated by light, and
Warburg demonstrated that inhibition of respiration by CO
was similarly reversed by illumination of the samples with
strong light. Warburg and Negelein (1928, 1929) placed living
cells in an atmosphere containing a ratio of CO and O_2 for
which, in the dark, the respiratory rate was severely inhibited
by the CO. The rate of oxygen consumption was measured for
the cell suspensions in the dark and then while they were exposed to light. The light-induced reversal of the inhibition of
respiration by CO was dependent on both the wavelength and
intensity of the light. Absorption of light resulted in dissociation of the CO and while the CO was dissociated, respiration

could occur. This was a dynamic (steady state) phenomenon in which the CO was continuously being dissociated by light and then recombining with a rate characteristic of CO reaction with the hemoprotein. Plotted against the wavelength of the irradiating light, the data result in a 'photochemical action spectrum' which directly reproduced the absorption spectrum of the CO complex of the pigment responsible for tissue respiration, which Warburg called Atmungsferment. The photochemical action spectrum had a strong maximum at near 428 nm (Soret band) with weaker maxima at 560 nm (β band) and 590 nm (α band). This spectrum was clearly that of an iron porphyrin complex with CO, and established the crucial role of iron porphyrin in tissue respiration.

The single most important advance in understanding cellular respiration was the discovery of the respiratory chain by Keilin, elaborated in a series of papers beginning in 1925. These papers changed, in a very fundamental way, not only understanding of respiration but also of enzymes in general. Instead of enzymes catalyzing single reactions, respiration utilized a multi-reaction process which occurred in a highly organized, specially designed, insoluble enzyme complex. The first paper (Keilin, 1925) described the presence, in many different organisms and their tissues, of a group of pigments which he named cytochromes. The absorption of the cytochromes in the visible part of the spectrum was observed primarily when the cytochromes were reduced. They gave four distinct, narrow bands at approximately 605 nm, 562 nm, 550 nm and 524 nm which were shown to belong to 3 different cytochromes. Keilin chose to name the cytochromes sequentially by the position of their absorption bands. Thus, the cytochromes with absorption bands at 605 nm, 562 nm, and 550 nm, respectively, were named cytochromes a, b, and c. The absorption at 524 nm was shown to arise partially from the b cytochrome and partly from the c cytochrome. The bands of the reduced cytochromes were observed under anaerobic conditions and rapidly disappeared when the tissue or cells were oxygenated. The cycle of reduction and oxidation was readily repeatable and many cycles could be carried out without altering the spectra. Observations in living organisms, such as suspensions of yeast or insect flight muscle, under aerobic conditions showed a partial reduction of the cytochromes, consistent with their participation in respiration by active redox cycling.

The role the cytochromes in respiration was further demonstrated by their response to various respiratory inhibitors. Cyanide, hydrogen sulfide, sodium azide and carbon monoxide, all inhibitors of tissue respiration, caused reduction of all three of the cytochromes. Urethane, a narcotic and respiratory inhibitor, induced oxidation of cytochromes c and a and reduction of cytochrome b. The sequence of electron transfer through the 'respiratory chain' and the sites of inhibitor interaction (inhibitors in italics) was established as:

$$
\begin{array}{ccc}
& & H_2S \\
\text{dehydrogenase} \rightarrow \text{cytochrome } b \rightarrow \text{cytochrome } c \rightarrow \text{cytochrome } a \rightarrow O_2 \\
& Urethane & HCN \\
& & HN_3, \\
& & CO
\end{array}
$$

While searching the literature for other references relevant to the cytochromes for his 1925 paper, Keilin became aware of earlier reports by MacMunn (1884, 1887) of a pigment which the author had called myohaematin and identified as a respiratory pigment. Myohaematin was reported to have a four banded spectrum like that of the cytochromes and to be widely distributed in tissue. Despite these similarities to cytochrome, in the intervening 35 years myohaematin had been largely ignored and was, for practical purposes, forgotten. This was in part due to criticism of the work of MacMunn by Levy, a student in the laboratory of Hoppe-Seyler (Levy, 1889; Hoppe-Seyler, 1890). A telling criticism was that MacMunn reported myohaematin was the only pigment in washed pigeon breast muscle, and Levy was able to demonstrate the muscle contained large amounts of myoglobin. Unfortunately, Levy used a high resolution spectrograph which was suitable only for measurements in clear solutions. While he could observe the spectra of pigments after extraction into clear solutions he could not make measurements in tissue samples. As a result, he demonstrated the presence of myoglobin, which appeared in aqueous extracts of pigeon breast muscle, but could not observe the myohaematin which remained in the tissue residue. Although Levy did not properly address the myohaematin question, he showed one part of the experimental data was in error and thereby undermined confidence in the rest of the work. Hoppe-Seyler, who should certainly have known better, focused only on MacMunn's error and dismissed all of the other data, placing his considerable influence against MacMunn's work.

In contrast to the case of myohaematin, the evidence presented for cytochromes and a respiratory chain was so extensive and comprehensive that their existence and importance were quickly accepted. The experimental results and discussion presented by Keilin were more convincing than those of MacMunn for several reasons: (1) the absorption bands were shown to arise from combination of the absorption of three different pigments (cytochrome) and not the absorption of a single pigment (myohaematin). When the spectra of the three pigments were separated, each had absorption characteristics similar to known heme compounds; (2) the appearance of the absorption bands upon hypoxia and disappearance upon reoxygenation were reversible and rapid for cytochrome whereas myohaematin could be reoxidized only by adding strong oxidants such as hydrogen peroxide. Therefore the cytochromes, but not myohaematin, truly 'cycled' in a manner compatible with their function in respiration; and (3) inhibitors of tissue respiration were shown to affect the cytochrome redox state in a manner consistent with their participation in respiration. In addition, Keilin emphasized: (A) That it was important to use the Hartridge reversion microscope to make observations in cells and tissue. (This was the only spectroscopic system of that time capable of making absorption measurements in turbid samples and tissue because of its low dispersion and excellent light-gathering properties.) (B) That the cytochromes were present in a wide range of different species of insects, birds and mammals, including species without hemoglobin or myoglobin, the pigments which Levy (1889) had suggested gave rise to myohaematin as a decomposition product. (Note that MacMunn also made this point, although not so comprehensively, but this argument was erroneously dismissed by Hoppe-Seyler.)

3. IDENTIFICATION OF ADDITIONAL REDOX COMPONENTS OF THE RESPIRATORY CHAIN

Keilin and Hartree (1939) discovered an additional cytochrome when they extended the spectral observations on the respiratory chain to include the near ultra-violet region of the spectrum. The reduced cytochromes a, b and c had absorption bands (Soret or γ bands) at 445–450 nm, 430 nm, and 424 nm respectively. The presence of an additional a type cytochrome (cytochrome a_3) was identified by the effect of respiratory inhibitors on the absorption band at 445–450 nm. In the presence of cyanide, azide and CO the absorbance at 445–450 nm was much less than for the fully reduced (anaerobic) preparations. (The subscripts 1, 2, 3 were used to distinguish cytochromes by the sequence in which they were discovered. Cytochrome a_2, a cytochrome with an absorption band near 620 nm, was observed in certain bacterial cells between discoveries of cytochromes a and a_3). Approximately half of the 445 nm absorbance of the reduced cytochromes was due to an additional cytochrome which reacted directly with cyanide and CO to form distinct inhibitor complexes. This pigment (cytochrome a_3) was considered responsible for reduction of oxygen.

Thus cytochrome a_3 was the same as Warburg's 'Atmungsferment'. The respiratory chain sequence was extended to:

Subs. $\rightarrow$ dehydrogenase $\rightarrow$ cytochrome b $\rightarrow$ cytochrome
c $\rightarrow$ cytochrome a $\rightarrow$ cytochrome a_3 $\rightarrow$ O_2 (5)

It was recognized very early (Elvehjem, 1931) that copper in the medium of microorganisms or the diet of animals was required for the development of 'indophenol oxidase' and cytochrome a in yeast and mice. Keilin seriously considered the possibility that cytochrome oxidase was a copper enzyme, noting that the properties of the reaction were consistent with the catalytic activity of copper and that copper was always present in the preparations having oxidase activity. The presence of copper as an integral part of the enzyme was not fully established until the work of Wainio and coworkers (Person et al., 1953; Sands and Beinert, 1959). Convincing evidence was obtained that copper was present in amounts equal to that the total heme a (cytochrome a + cytochrome a_3) and that one of the two copper atoms (later called the low potential copper or copper a) was oxidized and reduced in a manner consistent with its being an active electron transfer component of the enzyme. The presence of another redox component in cytochrome oxidase was shown by measuring the number of reducing equivalents accepted or given up per heme a (van Gelder and Beinert, 1969) bringing the total to 4 electrons per cytochrome a or 2 per heme a. This new redox component was identified as a second copper atom, and evidence was obtained for an electronic interaction between this copper and cytochrome a_3. The half-reduction potential was measured to be 350 mV at pH 7.2 (Lindsay and Wilson, 1974) and it is often referred to as the high potential copper or as copper a_3. The speculations of Keilin concerning the role of copper in cytochrome c oxidase were to be realized, although in a more complex manner than he had considered. The high potential copper and cytochrome a_3 were shown to cooperate in the binding of CO (Lindsay and Wilson, 1974; Lindsay et al., 1975) *i.e.* in the presence of CO the two metal atoms titrated as a single two-electron acceptor instead of two independent one-electron acceptors. This was a rather simple observation, made during an otherwise routine examination of the binding of CO and other ligands to cytochrome oxidase using potentiometric methods. Its importance was great, however, and this was immediately recognized. Since carbon monoxide is a competitive inhibitor of the oxygen reaction and binds to the oxygen reaction site, the active site for oxygen reduction involved both the iron and copper atoms. Carbon monoxide, which normally ligands to only one metal atom at a time, preferentially binds the heme iron when both metal ions are reduced. Oxygen, however, can readily form a bridged, doubly liganded species and the bridged compound can be reduced in a two electron process to form an intermediate bridged peroxide compound. This bypasses the thermodynamically unfavorable one electron reduction of oxygen allowing the first step of reduction to be the thermodynamically favorable two electron reduction. Further reduction to water, a total of 4 electrons and 4 protons added per O_2, involves reduction of the bound peroxide which is tightly bound and does not leave the enzyme active site. The cytochrome oxidase active site thus combines the catalytic activities of both copper and iron.

The number of respiratory enzymes and knowledge of the sequence of reactions expanded steadily during the period from 1938 to 1970. In addition to the copper components of cytochrome oxidase discussed above, several other redox components have been discovered. Cytochrome c_1 was discovered by Yakushiji and Okunuki (1940) in cell-free ox heart muscle preparations after extraction of cytochrome c. They observed that only part of the cytochrome c was readily extracted from the Keilin-Hartree heart muscle preparation (a preparation of submitochondrial particles). The remaining c cytochrome had an absorption spectrum similar to that of cytochrome c, but the maxima were at slightly different wavelengths. Cytochrome c_1, unlike cytochrome c, is an integral membrane protein. Detergents were required to extract the cytochrome and it could then be partially purified by precipitation with ammonium sulfate. Independently, Keilin (1949) identified a new cytochrome (cytochrome e) by the presence of an additional absorption band seen in samples of tissue from various organisms observed at liquid air temperature. At the temperature of liquid air, the absorption bands were much narrower than at room temperature and the absorption maximum of the new cytochrome was at a wavelength sufficiently different from that of cytochrome c for both to be observed in the same sample. General recognition of cytochromes c_1 and e was delayed by claims of Slater (1949) that the former was cytochrome c modified by the treatment of the muscle preparation and of Chance and Pappenheimer (1954) who believed the latter was actually cytochrome b_5 (a cytochrome found in the endoplasmic reticulum). It remained for Keilin and Hartree (1955) to establish that cytochromes c_1 and e were one and the same and to accept the name given by the Japanese workers. The last cytochrome of the respiratory chain to be identified was discovered in 1970. Potentiometric titrations of mitochondrial cytochrome b showed it to actually consisted of two different b cytochromes (Wilson and Dutton, 1970). Initially called cytochromes b_K and b_T, these have become more commonly referred to as cytochromes b_{561} and b_{565} respectively (subscripts indicating position of the alpha maxima of the reduced form) with occasional designation as cytochromes b_{30} and b_{-30} respectively (subscripts indicating the half-reduction potential at pH 7.2).

A quinone was isolated from mitochondria by Crane et al. (1957) and its structure determined by Fahmy et al. (1958). Several years were required to establish the participation of this quinone in the respiratory chain and its position in the electron transfer sequence. This *ubiquinone* proved to be the most abundant member of the respiratory chain and is present in amounts about 10 times greater than any other component.

Warburg had focused on the role of iron in respiration, but with the discovery of the cytochromes, attention became centered on heme proteins rather than iron *per se* as the oxidation-reduction transfer agents in respiration. Green (1956), however, called attention to the fact there were substantial amounts of non-heme iron in mitochondria and proposed that this iron was active in electron transfer. In the same year, Kearney and Singer (1956) isolated a form of succinate dehydrogenase having 2 moles of iron per mole of flavin and suggested this non-heme iron was also involved in electron transfer by the enzyme. There was, however, no direct evidence that oxidation and reduction of the iron played an essential role in the catalytic cycle. Initial attempts to demonstrate redox cycling of the non-heme iron in the respiratory chain utilized experiments in which the mitochondria or submitochondrial particles were rapidly 'quenched', the iron extracted and its redox state measured. These measurements produced ambiguous results, due in part to the insensitivity of the method of analysis and in part to the fact that only a fraction of the iron actually changed redox state. The function of the non-heme iron in the respiratory chain was not established until electron spin resonance measurements were made on samples at low temperatures (Beinert and Sands, 1959). These revealed signals centered at g = 1.94 which were later shown to arise from a new class of iron atoms in a distinctive environment. This environment included 'active sulfide', sulfur bound in a labile complex such that upon acidification of the protein it was released as H_2S. The generic name 'iron-sulfur protein' was adopted and several chemically different species have been identified as being active in electron transfer in the respiratory chain (for review, see for example Ohnishi and Salerno, 1982). Although most are associated with dehydrogenases (succinate dehydrogenase, NADH dehydrogenase), they are also associated with other regions such as the cytochrome c, c_1 region (Rieske iron-sulfur protein; Rieske et al., 1964).

4. REDOX COMPONENTS OF THE RESPIRATORY CHAIN AS STRUCTURAL ELEMENTS OF THE MITOCHONDRIAL MEMBRANE

Respiratory activity of cells was shown to be associated with an insoluble fraction of tissue homogenates by Battelli and Stern (1911, 1914). These authors measured the oxidation of p-phenylenediamine and succinate by different tissue homogenates and extracts of tissue. Both of these oxidase activities were associated with material which could be precipitated by mildly acidic conditions and resuspended without substantial loss of activity. This insoluble oxidase was named *oxydone*, and although the oxidase activities were not shown to be properties of the same enzyme, the two activities closely paralleled one another. The activities of oxydone paralleled the total respiratory activity of tissue homogenates and both were inhibited by cyanide. Warburg (1912, 1915) confirmed the association of respiration with a particulate fraction of cells. He further reported this activity was destroyed if the particles were exposed to pure water, and concluded it required an organized particle structure. After the discovery of the respiratory chain, Keilin and Hartree (see for example 1938) reported a method for preparing 'respiratory particles' from heart muscle. This 'heart muscle preparation' had substantially increased cytochrome content and increased activities of succinate oxidase and cytochrome c oxidase per mg protein relative to the initial homogenates of heart tissue. Cytochrome c, because it was soluble and easily extracted into the phosphate buffer, was isolated from yeast and substantially purified by Keilin by 1930 (Keilin, 1930). The other known components of the respiratory chain, flavoprotein dehydrogenases and cytochrome, were strongly associated with the insoluble material in the preparation. Although this insoluble material consisted primarily of cellular membranes, it should be remembered that until about 1960 all insoluble preparations from cells and tissue were referred to as particulate or insoluble, not membranous. This was true even when the investigators were convinced that the preparations consisted of membranes. Electron microscopy was required to prove the presence of membranes and this technique was not available except to a few specialists until the 1960s.

Increases in the understanding of the relationship of membrane structure to oxidative phosphorylation has been closely correlated with understanding of the role of membrane structure in other biological problems. Before about 1960, primary emphasis was on the physical positions of the proteins in relation to the lipid bilayer and to each other and on the effect of protein-lipid interactions in determining enzyme activity. More recently, emphasis has been on understanding the organization within the membrane and the role of the membrane in determining the activities of the system, *i.e.*, do reactants cross the membrane permeability barrier and/or couple to the transmembrane ion and electrical potentials?

Before the component parts of the respiratory chain could be isolated, it was necessary to extract the proteins from the insoluble lipid-protein membrane preparations into a form soluble in aqueous buffers. The bile salts, sodium deoxycholate and sodium cholate, were first used for this purpose. Deoxycholate was used by Hopkins and coworkers (1939) to prepare a succinate-methylene blue oxidoreductase and by Wainio et al. (1947) to prepare soluble cytochrome *c* oxidase. Cholate was used by Okunuki and coworkers (see Yakushiji and Okunuki, 1940) to prepare soluble cytochrome *c* oxidase. These compounds were the prototypes for a wide range of detergents which can be used to 'solubilize' the respiratory chain components. Treatment with detergents is able to free the redox components from the membrane but unless very strong, usually denaturing, conditions are used the solubilized proteins are still associated with detergent and residual lipids.

Association of the respiratory chain with membranes raised several questions concerning the mechanisms of the involved reactions. Each of the energy transduction sites of the respiratory chain, NADH dehydrogenase, ubiquinone-cytochrome c reductase and cytochrome c oxidase, isolates as a large multiprotein complex with molecular weights from 180,000 to 600,000 Daltons. Respiration, however, requires that reducing equivalents are transferred from one to the other of these complexes at greater than 1000 sec^{-1}, when expressed as a first order rate constant. The dehydrogenases and cytochromes were considered relatively large and immobile membrane components and therefore such high rates of electron transfer suggested fixed multicomponent complexes in the membrane. Green (1962) suggested ubiquinone served as a 'mobile' redox carrier. In this hypothesis, ubiquinone was dissolved in the fluid hydrophobic core of the membrane and provided a small, rapidly diffusing redox carrier. Its role was therefore to move (diffuse) among the cytochromes and dehydrogenases, attaining the high electron transfer rates observed for the respiratory chain. This 'mobile carrier' picture became more complex when King (see for example Yu et al., 1978) proposed there were ubiquinone-binding proteins, for which the ubiquinone would play a role similar to that of the pyridine nucleotide cofactors for soluble dehydrogenases. Initial demonstration of preferential binding sites for ubiquinone came with discovery of a stabilized, partially reduced form of ubiquinone (ubisemiquinone) in preparations of the cytochrome bc_1 complex. In solution, the ubisemiquinone, which is a free radical, is maximal when the ubiquinone is half reduced but at physiological pH values it constitutes only a tiny fraction of the total ubiquinone. The amount of semiquinone observed was much larger than the expected equilibrium concentration. This indicated some of the ubiquinone was in an environment (protein binding site) in which the semiquinone form was stabilized $i.e.$ bound with an affinity substantially greater than that of the fully oxidized or reduced forms. From this beginning, there has been a steady proliferation of reported quinone binding proteins and/or quinone binding sites on proteins. At the present time, it is generally accepted that ubiquinone is a multifunctional cofactor interacting with specific binding sites on the NADH and succinate dehydrogenases as well as on reduced ubiquinone-cytochrome c reductase (cytochrome bc_1 complex).

The extent to which large protein and/or lipid components move about in the plane of the membrane is clearly of great interest, but until the technique of FRAP (fluorescence recovery after photo bleaching) was developed it could not be effectively measured. In this method, one or more components of the membrane are labeled with a fluorescent dye and a region of the membrane observed through a microscope. A high intensity pulsed laser with a wavelength absorbed by the fluorescent dye is focused on a small area of membrane in the field of the microscope. When the laser is discharged, the fluorophore in the region illuminated by the laser is bleached (photochemically altered to a nonfluorescent species), greatly decreasing the fluorescence in the illuminated area. Fluorescence recovery occurs through diffusion of fluorophores from surrounding membrane into the area which has been photobleached. The rate at which the fluorescence recovers is a measure of lateral diffusion of the fluorescent labeled species in the membrane. Fluorescent labels were attached to cytochrome c and other components of the respiratory chain (see for examples Hackenbrock et al., 1987; Vanderkooi et al., 1985) and their lateral diffusion rates calculated. Hackenbrock and coworkers proposed a model in which the electron transport complexes were dissolved in the membrane lipids and diffused freely in the plane of the membrane (see 1987 ref. for summary). In this model, electron transfer was dependent on the frequency at which the complexes collided in an orientation appropriate to electron transfer. Although such a model seems greatly oversimplified, it is generally believed that there is a high degree of lateral mobility of membrane components, including those of the respiratory chain.

Fernandez-Moran (1962) initiated a period of intense effort to determine the contribution of the enzymes of oxidative phosphorylation to the structure of the inner mitochondrial membrane. Electron micrograph pictures of negatively stained beef heart mitochondria showed that projecting from the inner (matrix) side of the membrane at regular spacing were spheres 80–100 Å in diameter attached to the membrane by narrow 'stalks' 50 Å long and 35 Å in diameter. These structural elements became generally accepted as diagnostic of the inner mitochondrial membrane and the spheres were shown to be the mitochondrial ATPase (see for example Kagawa and Racker, 1966). Among the components of the respiratory chain, only cytochrome c could be readily isolated and crystallized, a prerequisite to determination of structure by X-ray diffraction. Those requiring detergents for solubilization resisted forming crystals, and the crystals formed generally contained large amounts of detergent. Quite fortuitously, Sun et al. (1968) observed that some preparations of cytochrome c oxidase in detergent formed membranous sheets in which the protein was arranged in close packed, ordered arrays, $i.e.$ as two dimensional crystals. Henderson and coworkers (for summary see Unwin and Henderson, 1984) were able to obtain X-ray diffraction patterns from stacks of these membranes from which low resolution structures could be calculated. The oxidase in the membranes was in the form of dimers with peptide alpha helices extending through the membrane. The protein extended substantially further into the aqueous medium on the cytoplasmic side of the membrane (55 Å) than on the matrix side (20 Å). Measurements of the absorption of polarized light and the electron spin resonance (EPR) signals of the respiratory components in oriented multilayers of mitochondrial membrane showed the chromophores were highly ordered with respect to the plane of the membrane (see Blasie et al., 1978; Erecinska et al., 1978). The hemes of cytochromes a, a_3, c_1 and the b cytochromes were all oriented with an angle of 90° between the heme normal and the plane of the membrane, $i.e.$ the edge of the porphyrin ring was toward the plane of the membrane. A useful mental image of the mitochondrial inner membrane is of a dense field of protein icebergs, surrounded by a layer of associated phospholipids, floating in thin phospholipid sea. The protein icebergs are highly structured groups of peptides and associated bound lipids which protrude both above and below the sea.

The complexity of the electron transport and energy coupling mechanisms is such that the strong treatments required to solubilize the individual redox components often gave rise to doubts about how well isolated components retained their *in vivo* characteristics. Keilin and King (1958) used reconstitution as a test of the quality of their soluble succinate dehydrogenase preparation. The soluble dehydrogenase, in contrast to that associated with the particulate starting material, reacted only with artificial electron acceptors such as phenazine methosulfate and even this activity was lost within a few hours. In order to determine if these alterations in enzyme properties were due to solubilization or to irreversible structural changes induced by the solubilization procedures, they devised a method for reconstituting the original activity. Respiratory particles were treated to selectively destroy the succinate dehydrogenase activity. Addition of their soluble succinate dehydrogenase to particles devoid of dehydrogenase activity resulted in the soluble enzyme associating with the particles, in the process recovering its stability and regenerating the succinate oxidase activity characteristic of the untreated respiratory particles. This ability to reconstitute the succinate oxidase activity was found only when the enzyme was solubilized in the presence of succinate (King, 1963), although enzyme isolated in the absence of succinate also had activity with artificial acceptors. King became a strong advocate for using reconstitution to determining whether a solubilized part of oxidative phosphorylation had been irreversibly modified during the process. Only by reconstituting the original activity could the isolated parts be shown to be still capable of carrying out their original function and thereby be considered to representative of the 'native' enzyme and/or structure.

III. Coupling of respiration to metabolic work

1. EARLY OBSERVATIONS

It was clear from early times that animals breathed more strongly and used more oxygen when working than at rest. By 1900 measurements of heat production and oxygen consumption by isolated muscle had confirmed that both were increased when the muscle was induced to do work. Although there was little doubt that respiration was directly coupled to metabolic work, nothing was known about the mechanisms involved. Evidence that the energy available in respiration was conserved in the synthesis of organic phosphates and these in turn provided the energy needed for muscle contraction and other metabolic work was provided by Eggleton and Eggleton (1927). They discovered a new organic phosphate compound in muscle that they called phosphagen (identified by Fiske and Subbarow (1927) as creatine phosphate). Phosphagen was decomposed to generate inorganic phosphate during anaerobic work and resynthesized when the muscle was supplied with oxygen. Lundsgaard (1930, 1931) carried out a series of experiments which were extraordinarily well designed and brilliantly interpreted. He showed that after muscle was poisoned with iodoacetate, an agent which under carefully controlled conditions could block glycolysis without significant effect of respiration, the energy for performing work was provided by the breakdown of phosphagen, generating inorganic phosphate. In unpoisoned anaerobic muscle less phosphagen was broken down than for control muscles, and the decreased heat production was quantitatively accounted for by lactate formation. When iodoacetate poisoned muscle was stimulated in oxygenated media, it was able to do more

252

sustained work than if it were stimulated in media bubbled with nitrogen. He concluded that

'a normal muscle, in which oxidation is sufficiently rapid relative to the rate of work, is able to perform work without any intermediary lactic acid formation.'

His schematic presentation correctly summarized the relative contributions of oxidative phosphorylation and anaerobic glycolysis to muscle energy metabolism. Since heats of reaction (H) were used and no consideration was given to the thermodynamic efficiency of ATP synthesis, the ratios only approximate the ratio of ATP production by the two pathways. It is interesting to note that Lundsgaard reported that in muscle more than one mole of creatine phosphate was formed per mole of lactate produced by anaerobic glycolysis. Later studies were to establish that only one mole of ATP was synthesized per lactate produced from free glucose, and for many years the data of Lundsgaard were considered by many workers in the field to be in error. His measurements were, however, near the value of 1.5 ATP/lactate for lactate originating from glycogen, as was the case for intact muscles *i.e.* the conditions in Lundsgaard's experiments.

Adenosine triphosphate was first identified as a cellular component by Lohmann (1934) and this compound quickly became recognized as the probable primary source of energy for metabolic work including muscle contraction, but proof of the latter was not to be obtained until much later (see Cain and Davies, 1962). Although the existence of a coupling between oxidation and phosphorylation was established well before by 1930, how this coupling occurred was not known. Kalckar (1939) and Belitzer and Tsybakova (1939) demonstrated oxidative phosphorylation in tissue homogenates, a necessary starting point for the study of the mechanism of oxidative phosphorylation. Kalckar reported that extracts of kidney and liver carried out very active synthesis of organic phosphates from inorganic phosphate and formation of organic phosphates was proportional to oxygen consumed. Belitzer and Tsybakova used minces of pigeon breast muscle and of rabbit muscle to demonstrate respiration-dependent synthesis of organic phosphates, including creatine phosphate. The data of the latter workers indicated at least four moles of creatine phosphate were synthesized per mole of O_2 consumed. In unwashed frog muscle preparations, but not in those from mammalian muscle, as the rate of oxygen consumption increased there was a direct correlation between the rate of respiration and the rate of creatine phosphate synthesis. This established that the rate of respiration *in vitro* was metabolically determined, *i.e.* respiration was accelerated when the cells were provided with phosphate and phosphate acceptor.

A very interesting and significant observation was made by Bumm et al. (1934) while studying the intestinal mucosa of guinea pig. Incubation of the mucosa at below normal oxygen pressures which were still high enough that the rate of respiration was not significantly decreased, resulted in an increase in anaerobic glycolysis (lactate production). Pasteur (1861), had reported that the rate of fermentation increased when oxygen was excluded from yeast suspensions, but it was difficult to explain how increased glycolysis could occur without a decrease in respiration. The increase in lactate production at low oxygen pressures was confirmed by Laser (1937a) and extended to other tissues, including retina, chorion, liver, and mouse sarcoma. These data demonstrated that when oxygen was decreased, but not so far as to limit significantly the respiratory rate, anaerobic glycolysis was considerably increased. Laser (1937b) followed these observations with a study of the effect of carbon monoxide on respiration and lactate production. Carbon monoxide, again at CO/O_2 ratios and concentration which did not significantly affect the respiratory rate, stimulated glycolysis. The stimulatory effect on glycolysis was reversed by light. The photochemical action spectrum, determined by Stern and Melnick (1941) was similar to that of 'Atmungsferment' or cytochrome a_3. The interpretation given was that a heme compound was involved in glycolysis, possibly in the role of maintaining a glycolytic enzyme in an oxidized, inactive state. Efforts to verify the existence of a heme protein, other than cytochrome oxidase, with the characteristics required for the 'Pasteur enzyme' of glycolysis were unsuccessful. The effects of moderate decrease in oxygen pressure, CO and light on glycolysis remained unexplained. As a result, the data were largely ignored or were considered artifactual. I was introduced to this work and the 'Pasteur enzyme' in a discussion with an eminent scientist, who used it as an example of 'poor' science. The negative comments all related to the failure of the data to fit a currently popular hypothesis, however, and my reaction was one of great interest and excitement. Although the concept of a heme-containing enzyme which regulated glycolysis was very unlikely (this was about 1965), the data clearly demonstrated coupling of respiration and glycolysis which occurred at oxygen pressures of up to several mm Hg. Such coupling could reasonably occur through common products of the pathways, with the cellular ATP (energy) levels as the logical 'coupling agent'. The failure to observe a decrease in oxygen consumption corresponding to the increase in lactate production, an important point of criticism at the time, could readily be explained. Lactate production from glucose can provide only 1 ATP per lactate (2 per glucose) while respiration provides about 6 ATP per oxygen (O_2) and under aerobic conditions ATP production from lactate production in these tissues is only a tiny fraction of that from respiration. Thus, the work of Laser and others had provided evidence that the oxygen concentration dependence of mitochondrial oxidative phosphorylation *in vivo* extended to values much higher than those which caused a limit in the oxygen consumption rate. Moreover, they suggested there was a very sensitive and dynamic coupling between respiration and glycolysis over a wide range of oxygen pressures and metabolic rates. If this were true (as was to be shown) there were many important physiological consequences. The idea that lactate production in tissue *in vivo* was an indication of regions of tissue anaerobiosis, a view particularly widespread among investigators studying muscle, for example, would be untenable. An important lesson is never to dismiss experimental data and/or the work of individual investigators because they don't appear to 'fit in'. Consensus evaluations of data are determined more by how well they fit the currently 'accepted' hypothesis than by merit, and this often leads to missed opportunities for better understanding.

2. IDENTIFICATION OF MITOCHONDRIA AS THE 'POWER PLANT' OF CELLULAR METABOLISM

Mitochondria were implicated in cellular respiration for many years, initially from their unique staining by vital dyes (dyes applied to living tissue). Thus mitochondria (micro bodies) were observed as morphologically and chemically distinct entities by histological stains such as the vital dye Janus green. Janus green is a dye which stains the tissue in the region where it is oxidized to give a characteristic red color, and in tissue it is the mitochondria which are stained. Keilin and Hartree (1939) prepared subcellular particles, which were later to be shown to arise from mitochondria, in which the components of the respiratory chain were significantly enriched. Isolation of large particles (mitochondria) from tissue by differential centrifugation was attempted very early by Claude (1938) who also did much to establish the mitochondria as the 'intracellular power plants' (see Claude, 1946). Particles with the morphological appearance of mitochondria were obtained first, but demonstration of the catalytic activity was more difficult. Success awaited the technical ability to carry out differential centrifugation at temperatures near 4°C. Hogeboom and coworkers (1948) published a definitive isolation method, one in which both the morphological and catalytic functions were preserved. Claude (1946) had recognized the sensitivity of mitochondria to the osmolarity of the suspending medium. This indicated the mitochondria were surrounded by a semipermeable membrane (note that the concept of a membrane was not well developed at this time) and there was controlled access of cytoplasmic compounds to the intramitochondrial compartment.

With the development of reliable methods for isolation of mitochondria there was rapid progress in determining their enzymatic functions. The first activities reported were succinate dehydrogenase and cytochrome oxidase but Lehninger (1946) and Kennedy and Lehninger (1948, 1949), in a series of papers, established that oxidation of the citric acid cycle intermediates and of fatty acids was associated with the mitochondria. Shortly thereafter, Lehninger (1951) reported that oxidation of reduced pyridine nucleotide (NADH) by mitochondria was accompanied by phosphorylation of more than 1.5 ATP for each NADH oxidized. This filled in a 'missing link'. It had been realized that more than one phosphorylation occurred per oxygen atom reduced to water but it was not clear how much of this was associated with the molecular rearrangements of the metabolites and how much with the respiratory chain. The production of a greater than stoichiometric amount of ATP in the oxidation of NADH indicated the mitochondrial respiratory chain was coupled to ATP synthesis at more than one step along the electron transport pathway.

3. ESTABLISHING THE STOICHIOMETRIC RELATIONSHIP OF ELECTRON TRANSPORT AND ATP SYNTHESIS

The stoichiometric relationship between reduction of oxygen and formation of organic phosphate required very precise methods for measuring the reactants. The intense interest in cellular and tissue respiration had resulted in development of a very effective method for measuring oxygen consumption and accumulation of metabolic products, the 'Warburg apparatus'. This apparatus incorporated developments in manometric techniques by a number of workers and deserves special attention because it played a key role in almost all developments in respiration and oxidative phosphorylation from 1920 to about 1960. The apparatus consisted of specially designed glass flasks which were attached to high precision manometers for measuring changes in gas volume and then suspended, with shaking, in constant temperature water bath. Cellular respiration results in production of CO_2 which had to be removed to keep the increase in volume of CO_2 gas from interfering with accurate measurement of oxygen consumed. This CO_2 was quantitatively trapped in an internal chamber of glass flasks containing a solution of potassium hydroxide, where it was converted to potassium carbonate. Mitochondria could be incubated for defined periods of time with different oxidizable substrates (intermediates of the citric acid cycle, fatty acids etc.) and the amount of oxygen consumption precisely measured. Glucose and hexokinase were added to convert any ATP synthesized to glucose-6-phosphate and regenerate ADP. The 'trapping' of ATP as the stable reaction product glucose-6-phosphate minimized possible hydrolysis of ATP by ATPase and the amount of glucose-6-phosphate formed could be accurately measured using glucose-6-phosphate dehydrogenase. The result was expressed as the ratio of ATP produced to the oxygen atoms consumed (ATP/O) or the inorganic phosphate disappearing to the oxygen atoms consumed (P/O). The combination of high accuracy manometers, precise temperature control and the glucose-hexokinase trap for ATP was ideally suited for quantitating the stoichiometry of the relationship between ATP synthesized and oxygen consumed (see, for example, Coopenhaver and Lardy, 1952). The respiratory rate of mitochondrial suspensions in the presence of the glucose-hexokinase trap was substantially higher than that in the presence of ATP (Lardy and Wellman, 1952). The ratio of these two respiratory rates was called the respiratory control and became an important measure of the quality of the mitochondrial preparation. There was good agreement that the ATP/O ratio when succinate was oxidized to fumarate and for oxidation of substrates with NAD linked dehydrogenases were approximately 2 and 3, respectively. In general, the 'better' the mitochondrial preparations, the closer the measured ATP/O approached integer values. By about 1960, oxygen electrodes became the method of choice for measuring oxygen. Oxygen electrodes are inherently less accurate than manometric techniques but the electrodes are less expensive and easier to use than the manometric apparatus.

4. THERMODYNAMICS AND MITOCHONDRIAL OXIDATIVE PHOSPHORYLATION

Thermodynamics played an important role in the study of respiration. The initial studies were very dependent on measurements of heat, a good indicator of metabolic work, because precise calorimetry was a well-established art before 1900. A combination of data on heat production and chemical analysis was considered by Meyerhoff (1925), Lundsgaard (1930a,b; 1931) and others in evaluation of the chemistry of muscle contraction. The respiratory chain, on the other hand, is composed of components which undergo cycles of oxidation and reduction, an electrochemical process. The energetics of redox reactions are best described as electron transfer processes, *i.e.* in terms of their half-reduction potentials and the degree of reduction:

$$E_h = E_m + nRT \ln [Ox]/[Red] \qquad (6)$$

where [Ox] and [Red] are the concentrations of the oxidized and reduced forms respectively, n is the number of reducing equivalents transferred per molecule, R is the gas constant and T is the absolute temperature. E_h is the oxidation-reduction potential relative to a standard hydrogen electrode and

E_m is the characteristic half-reduction potential of the redox component. The energetics of electron transfer between two redox components, A and B, can be described:

$$E = E_{ma} - E_{mb} + nRT \ln([A_{Ox}][B_{Red}]/[A_{Red}][B_{Ox}]) \qquad (7)$$

where E is the voltage developed by the electrochemical cell, E_{ma} and E_{mb} are the characteristic half-reduction potentials of components A and B respectively. The free energy change (also called ΔG; maximal work function at constant temperature; and Gibbs free energy) is directly related to the voltage of the electrochemical cell and the number of electrons transferred per mole of redox compound oxidized or reduced:

$$\Delta G = -nF \, E \qquad (8)$$

where F is the Faraday constant, equal to 23.06 kcal/electron equivalent volt (96.4 kJoule/electron equivalent volt). A knowledge of the free energy available in the electron transfer reactions makes it possible to determine the feasibility of coupling of the electron transfer to ATP synthesis. It was Ball (1938) who first attempted a systematic evaluation of the half-reduction potentials of the redox components of the respiratory chain. He measured the half-reduction potentials of cytochromes c, b, and a for selected conditions and established the first, albeit rudimentary, view of the energetics of electron transfer within the respiratory chain. The half-reduction potentials for solutions of soluble redox cofactors, such as NAD, FAD, ubiquinone were measured over the next few years, but the next comprehensive program for measuring the thermodynamics of the respiratory chain was undertaken beginning about 1970.

Several modifications were made in the methods for measuring the redox properties of molecules as developed by earlier workers (see, for example, Clark, 1960). The changes were designed to allow rapid measurements in systems which contained small amounts of endogenous reducing agents and in which the redox centers are largely buried in protein (for more experimental detail see Wilson et al., 1974a and Dutton, 1978). These methods allowed measurements not only of the components which could be measured in samples at room temperature (such as the cytochromes) but also those which needed to be 'freeze trapped' and the measurements made on samples at low temperatures (such as electron spin resonance measurements of the iron-sulfur proteins and copper a). The half-reduction potentials of the redox components of the respiratory chain were systematically measured and then the degree to which each was reduced in the steady state of oxidative phosphorylation was determined. The data allow calculation of the oxidation-reduction potential of each component (E_h) under conditions of the physiological steady state of respiration. This thermodynamic profile represented a major advance in understanding of both the thermodynamics of oxidative phosphorylation and regulation of the rate of mitochondrial respiration. First it defined, in a thermodynamic sense, the coupling of respiration to synthesis of ATP. Second, it allowed identification of the rate controlling step of the respiratory chain and thereby made possible a quantitative understanding of the mechanism(s) of respiratory control. The redox profile of the respiratory chain is that of three equi-potential batteries in series coupled by an efficient energy conversion system to a common pool of adenine nucleotides and inorganic phosphate.

It was observed in 1959–1961 (see for examples, Klingenberg et al., 1959; Chance and Hollunger, 1960) that oxidation of succinate to fumarate or of 1-glycerol phosphate to dihydroxyacetone phosphate caused rapid reduction of intramitochondrial NAD^+ when an appropriate source of metabolic energy was provided. Both substrate couples donate reducing equivalents at the level of cytochrome b and ubiquinone, and reduction of NAD^+ effectively shows reversing the first site of oxidative phosphorylation. These observations were extended to show that upon addition of reducing equivalents at the level of cytochrome c (such as by addition of ascorbate plus, N,N,N',N' tetramethyl-p-phenylenediamine), there was a respiration dependent reduction of NAD^+ (Chance and Fugmann, 1961). This reduction of NAD^+ was inhibited by Antimycin A, an inhibitor of electron transfer between cytochrome b and cytochrome c, demonstrating it occurred through reversal of the first two sites. These and later studies established that reducing equivalents from the cytochrome c could be transferred to the NADH at the expense of the free energy derived from respiration or from hydrolysis of ATP (reversal of the first two steps of oxidative phosphorylation):

$$NADH + 2 \text{ cyt } c + 2 \text{ ADP} + 2 \text{ Pi} = NAD^+ + 2 \text{ cyt } c^{2+}$$
$$+ 2 \text{ ATP} \qquad (9)$$

Quantitation of reaction 9 under conditions of both phosphorylation (forward reaction) and ATP hydrolysis (back reaction) indicated that the reaction approached equilibrium from both directions (see, for example, Wilson et al., 1974b). The observation that the first two sites of oxidative phosphorylation were near equilibrium under physiological conditions was consistent with the measurements thermodynamic profile of the respiratory chain described above. The difference in oxidation-reduction potential between the 'isopotential groups' of redox components was energetically equivalent to the energy required for ATP synthesis. The partial reversal of the third phosphorylation site was observed using CO to bind cytochrome a_3. Addition of ATP to mitochondria in which cytochrome a_3 was reduced and bound to CO resulted in oxidation of cytochrome a_3 with transfer of the reducing equivalents to oxidized cytochrome c. Thus even the third energy conservation site of oxidative phosphorylation could be at least partially reversed if appropriate experimental conditions were used. Under aerobic conditions with net ATP synthesis, however, only the first two approach equilibrium and the third site, combined with oxygen reduction, is irreversible. In a metabolic pathway the metabolite flux through the reactions which are far displaced from equilibrium (irreversible) must be, in the steady state, equal to the flux through the pathway. Thus, any change in the rate of mitochondrial respiration (rate of ATP synthesis) must reflect a change in the rate of the cytochrome oxidase reaction:

$$2 \text{ cyt } c^{2+} + 1/2 \text{ O}_2 + 2 \text{ H}^+ + \text{ADP} + \text{Pi} \rightarrow$$
$$2 \text{ cyt } c^{3+} + \text{ATP} + \text{H}_2\text{O} \tag{10}$$

The kinetic behavior of reaction 10 has been examined in intact mitochondria by using artificial electron donors to directly reduce cytochrome c. These studies established that, at physiological levels of reduction of cytochrome c, the reaction rate increased more than 50-fold as the [ATP]/[ADP][Pi] was decreased (Wilson et al., 1977). This hypothesis for the physiological regulation of mitochondrial oxidative phosphorylation was consistent with the measured behavior of mitochondria *in vivo* and *in vitro*.

IV. Compartmentation of the matrix enzymes by the inner mitochondrial membrane

The first preparation of morphologically intact mitochondria was accomplished by workers utilizing differential centrifugation techniques to fractionate cells. This approach followed that of Bensley and Hoerr (1934) and Claude (1938, 1946) and culminated in isolation of mitochondria which retained biological activity by Hogeboom and coworkers (1946, 1948). Rapid progress was then made in establishing the enzymatic functions of the mitochondria, led by the aforementioned workers as well as Lehninger (1946); Kennedy and Lehninger (1949); Schneider (1946); and others. Claude (1946) observed that mitochondria were sensitive to the osmolarity of the isolation and suspending media. Raaflaub (1955) reported that EDTA and other synthetic chelating agents for multivalent metal ions inhibited non-osmotic swelling of isolated mitochondria and stabilized their capacity to oxidize substrates. He defined non-osmotic swelling as swelling which occurred shortly after diluting a concentrated stock suspension of mitochondria into an aerobic, isotonic medium. ATP also inhibited this swelling (Slater and Cleland, 1953) and Raaflaub realized that ATP and other chelators acted through their ability to bind divalent metal ions, particularly Ca^{2+}. Inhibition of the swelling correlated with suppression of lipase activity and stabilization of the ability of the mitochondria to oxidize substrates, particularly intermediates of the citric acid cycle.

Lehninger (1951) provided evidence for the importance of the membrane permeability by showing that the rate of oxidation of externally added NADH was greatly enhanced when the mitochondria were suspended in a hypotonic medium. Thus the membrane provided a barrier to access of extramitochondrial NADH to the dehydrogenase of the respiratory chain. Isolated mitochondria were shown to maintain internal K^+ concentrations higher than the external concentrations, a thermodynamically unfavorable concentration difference, using energy from oxidative phosphorylation. When deprived of this energy source, K^+ leaked out of the mitochondria. After K^+ was depleted from mitochondria, incubation in media containing either ATP and Mg^+ or glutamate plus inorganic phosphate, resulted in uptake of K^+ (slowly) against the concentration gradient. Addition of the antibiotic valinomycin to a mitochondrial suspension induced rapid uptake of potassium ions (as well as Rb^+, Cs^+) with the energy provided by either respiration or ATP hydrolysis (Moore and Pressman, 1964). Compounds like valinomycin, which induce cation permeability of membranes, were shown to act by forming lipid soluble complexes with cations (thus the name ionophores). These compounds can diffuse across the lipid 'core' of the membrane (see Pressman et al., 1967) both as the free ionophore and as an ionophore-cation complex. The ionophores therefore provide a pathway for cations (such as K^+ in the case of valinomycin) to diffuse down their electrochemical gradients. If transport of these ions is an energy linked process in which the ion can be concentrated against its electrochemical gradient, the ion 'leak' induced by the ionophore dissipates the energy which would otherwise be available for other energy requiring processes such as ATP synthesis.

V. Mechanism in the coupling of respiration and phosphorylation

1. CHEMICAL COUPLING MECHANISMS AND MITOCHONDRIAL OXIDATIVE PHOSPHORYLATION

The energy available in the oxidation of reduced coenzymes (NADH and $FADH_2$) must be 'transduced' from electrical energy to chemical energy in a process described as energy coupling. Little progress in understanding the processes involved could be made until after isolation of functional mitochondria. This occurred well after the role of organic phosphates in the catabolism of glucose, including 'substrate level phosphorylation', the synthesis ATP from ADP and inorganic phosphate in the conversion of glyceraldehyde-3-phosphate to 3-phosphoglycerate, had been elucidated. Thus, coupling of an exergonic redox reaction (oxidation of an aldehyde to an acid) to an endergonic reaction (synthesis of ATP) was known to be an effective method for biological synthesis of ATP. Oxidative phosphorylation was generally assumed, therefore, to utilize a multistep chemical reaction having similarities to substrate level phosphorylation. Since transport of ions against their concentration gradients required coupling of the ion movement across the mitochondrial membrane to an exergonic reaction, this transport was assumed to be coupled to breakdown of one or more of the intermediate reactions of oxidative phosphorylation. Slater (1953) first pointed out that the coupling of respiration to some energy requiring reactions of mitochondria did not require inorganic phosphate. He proposed that there were several steps in the energy coupling process and those closest to electron transport were not phosphorylated, *i.e.*, there were nonphosphorylated 'high energy' chemical intermediates. These nonphosphorylated 'high energy' intermediates were considered to provide energy to many of the energy dependent reactions of mitochondria, including ion transport and reversed electron transfer.

Once the involvement of non-phosphorylated intermediates in oxidative phosphorylation was proposed, great effort was devoted to attempts to experimentally determine their chemical properties. These experiments were largely unsuccessful and many hypotheses were generated as to their chemical nature. One, the 'conformational hypothesis', proposed that respiration provided the energy to give altered 'high energy' protein conformations (Boyer, 1965; King et al., 1965). Return of these proteins to their resting conformation was then coupled to ATP synthesis (in analogy to actomyosin in muscle). This hypothesis, in modified form, provides the basis for current mechanisms of ATP synthase. Evidence indicates that the primary energy requiring step in ATP synthesis is a protein conformation change which induces a large decrease in the affinity of the active site for ATP (see Boyer, 1974). ATP can be readily formed from ADP and Pi bound to the active sites of the ATP synthase because ATP is preferentially and strongly bound. A substantial fraction of the energy required for synthesis of unbound ATP (free in solution) from unbound ADP and Pi must be invested after the formation of bound ATP. This energy is needed to alter the conformation of the protein to a structure (conformation) in which the ATP is only weakly bound and therefore leaves the enzyme surface.

2. CHEMIOSMOTIC MECHANISM AND MITOCHONDRIAL OXIDATIVE PHOSPHORYLATION

The chemiosmotic mechanism for mitochrondrial oxidative phosphorylation developed from research at the interface between the study of transport of substrates into and out of cells and the study of oxidative phosphorylation. Lundegård (1945) had recognized that if two redox components operating by hydrogen atom transfer were placed on opposite sides of a membrane, electron transfer between them could extract hydrogen ions from one side of the membrane and release them on the other side, generating a difference in pH across the membrane. This concept was applied only to systems with a respiration dependent generation of acid, such as in the stomach. Between 1945 and about 1960, however, transport of ions, sugars and other metabolites into and out of cells became widely recognized as often occurring against the prevailing concentration gradient and active transport became a widely accepted concept (see Chapter 4). A particularly important paper is that of Davies and Ogston (1950) in which the thermodynamics and mechanistic requirements were described for transport coupled to either redox reactions or hydrolysis of high energy phosphates. The specific example used was the excretion of H^+ ions across the gastric mucosa, with acidification occurring on the mucosal side and alkalization on the other side of the membrane. Another important advance was recognition that movement of glucose, and uncharged molecule, across the membrane could be coupled to movement of Na^+ down its electrochemical gradient (see Crane et al., 1961) resulting in glucose being actively accumulated against its concentration gradient. Mitchell, who was actively studying both metabolite transport and oxidative phosphorylation, applied the principles of coupled transport in proposing a chemiosmotic mechanism for coupling the energy in the redox reactions of the respiratory chain to synthesis of ATP (see, for example, Mitchell, 1961). His proposal was particularly innovative in that it suggested that transport of hydrogen ions across the inner mitochondrial membrane was coupled to the synthesis of ATP, a reversal of the role of ion transport. This mechanism was based on three primary assumptions: (1). The mitochondrial inner membrane is impermeable to ions, in particular H^+ ions; (2). The membrane contains an asymmetrically oriented, reversible ATPase which couples the flow of H^+ across the membrane to synthesis/hydrolysis of ATP. (3). The respiratory chain contains redox systems which are oriented for vectorial transfer across the membrane. As reducing equivalents are transferred through these systems from more negative to more positive redox potentials, protons are accepted on the inner side of the membrane (matrix) and released on the outer side. Thus, the redox reactions act to 'pump' the protons out of the mitochondrial matrix, leaving the inside alkaline with respect to the external medium and the membrane with a negative charge inside relative to the outside. The 'high energy' intermediate in oxidative phosphorylation is, by this mechanism, the electrochemical gradient of H^+ ions across the membrane ($\Delta\mu H^+$) expressed in volts:

$$\Delta\mu H^+ = -59.3\ \Delta pH + E \qquad (11)$$

where pH is the pH in the mitochondrial matrix minus that in the extramitochondrial space and E is the membrane potential in volts while 59.3 is constant consisting of the factor for converting logarithms to the base 10 to logarithms to the base e, the Faraday constant, the gas constant and the absolute temperature.

'Energy coupling' in oxidative phosphorylation is therefore combination of a redox driven, outwardly directed proton transport with a separate inwardly directed proton movement coupled to ATP synthesis. As long as the only mechanisms for proton translocation across the membrane are those of the redox pump and ATP synthase, an efficient transfer of the free energy occurs. The stoichiometry of the reactions (H^+/electron transferred and H^+/ATP) determines the stoichiometry of oxidative phosphorylation (expressed as ATP/electron, ATP/oxygen atom consumed, etc.). In this mechanism, uncoupling of oxidation from phosphorylation occurs when the membrane is permeable to protons ('leaky') or an alternate pathway exists for proton transfer across the membrane. Among the latter 'leak' pathways are those provided by the presence of lipid soluble weak acids, such as 2,4-dinitrophenol, which can cross the membrane in either their protonated (uncharged) form or their unprotonated (anionic) form. Such compounds permit net transfer of protons across the membrane, dissipating the difference in pH across the membrane (uncoupling oxidation and phosphorylation).

The first direct experimental data implicating the hydrogen ion gradient in ATP synthesis came from the field of photosynthesis. Jagendorf and colleagues (Hind and Jagendorf, 1965; Jagendorf and Uribe, 1966) observed that if chloroplasts isolated from spinach were incubated in the dark in an acidic medium (from pH 5 to 3.8) and diluted into an alkaline medium (pH 6.7–9.5) containing inorganic phosphate and ADP, there was synthesis of substantial amounts of ATP. The direction of the proton gradient (high inside, low outside) was opposite to that proposed for mitochondria, consistent with the observation that light induced uptake of protons by chloroplasts while respiration induced proton extrusion from mitochondria. The amount of ATP synthesized was dependent on the difference in pH between the two media and the buffer used in the medium. The dependence on these and other parameters were consistent with ATP synthesis being coupled to the flow of hydrogen ions from the internal compartment (acidic) to the external medium (alkaline). Incubation in acidic media containing increasing concentrations of weak acids such as succinate increased the amount of ATP synthesized. Thus, the weak acids appeared to enter the chloroplasts and buffer the internal hydrogen ion concentration, increasing the number of hydrogen ions transferred following dilution into the alkaline

medium, and thereby the amount of ATP synthesized. Uncou-
plers of photosynthetic phosphorylation also blocked ATP
synthesis in these 'pH jump' experiments, consistent with the
coupling occurring by the same mechanism as during normal
photosynthetic phosphorylation. The chemiosmotic model has
been the subject of many reviews (see for example Skulachev
(1991) and the readers may also find the Nobel Lecture by
Mitchell (1979) particularly interesting.

3. CHEMIOSMOTIC VERSUS CHEMICAL COUPLING

The initial formulations of chemiosmotic coupling provided
strong contrast to chemical coupling and the differences were
clear and unambiguous. In the chemiosmotic hypothesis the
electron transport chain functioned as a simple multistage
fuel cell in which a redox component is reduced and protonated
on the inside of the mitochondrial membrane and oxidized and
deprotonated on the outside of the membrane:

$$Q + 2e^- + 2H_i^+ \rightarrow QH_2 \tag{12}$$

$$QH_2 \rightarrow Q + 2H_e^+ + 2e^- \tag{13}$$

where the subscripts i and e are used to indicate the H^+ ions
are taken up from (equation 12) or released into (equation 13)
the intramitochondrial (matrix) or extramitochondrial space,
respectively. Both Q and QH_2 could, in this model, freely cross
the membrane but the reductant (equation 12) would be at a
lower (more negative) electrical potential than the oxidant
(equation 13). The difference in electrical potential would
thermodynamically 'drive' movement of H^+ ions from the inside to the outside of the mitochondria, leaving the inside medium alkaline with respect to the outside medium and the
membrane with an excess of negative charges on the inside
compared to the outside. ATP was then synthesized by H^+ ions

as they return from outside to inside through an ATP coupled H^+ transporter (ATP synthase). Thus, no chemical 'high energy compounds' were involved in energy conservation, even ATP synthesis was considered to occur through reactions involving H^+ ions.

As the chemiosmotic hypothesis was used to generate more specific, and therefore experimentally testable mechanisms, it underwent significant changes in response to the experimental results. Hydrogen ions were shown to be transported from the mitochondrial matrix to the extramitochondrial space leaving the matrix alkaline, and measurements of the membrane potential indicated it was large (−140−−180 mV) and of the predicted polarity. On the other hand, some properties of the energy coupling process were difficult to explain and a high energy chemical intermediate ($X \sim Y$) was added to the mechanism of ATP synthesis by Mitchell (1966). Similarly, to meet the thermodynamic requirements of ATP synthesis, the proposed H^+/e^- for each energy conservation site was increased from 1 to more than 1, and the H^+/ATP ratios were increased from 2 to 3 or 4. The latter two changes suggested the mechanism(s) of energy conservation in the respiratory chain were quite different from the original fuel cell model. Vectorial movements of electrons within the redox complexes became an important consideration and local charge separations were suggested as part of the first events in energy conservation. The central role of the transmembrane electrical potential was retained, although questions remained concerning the relative contributions of pH and E to the energy for ATP synthesis. Suggestions were made that the 'local' values for these parameters differed from the traditional bulk phase values ('local' was not clearly defined). Thus by the end of the 1970s, the chemical and chemiosmotic models for oxidative phosphorylation had converged significantly. A hybrid model based on a combination of electron transport driven charge separation and active H^+ pumps for primary energy conservation replaced the original fuel cell model. A central role for the

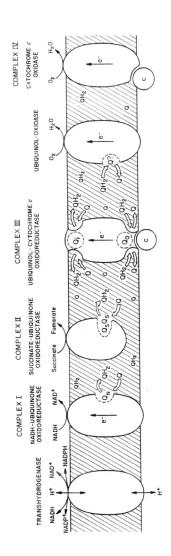

Fig. 1. A schematic representation of key charge translocating proteins in the membrane. The protein complexes extend across the membrane and flow of electrons occurs from one side to the other of the permeability barrier.

membrane potential and ATP formation coupled to an H^+ pump by conformation changes in the protein was generally accepted.

To the present time, great progress has been made in determining the general physical structure and peptide composition of the components of the respiratory chain and their orientation in the mitochondrial membrane. Current views on the approximate global structures of the key charge translocating proteins and their positions in the membrane has schematically summarized in a drawing by P.L. Dutton (Fig. 1). Although more is known about the internal structure than could be shown in this schematic, this is the primary area of needed research effort. Understanding oxidative phosphorylation will require much more detailed knowledge of the electron transfer and energy coupling reactions than is currently available. In addition, the mechanisms by which electron transfer occurs between the large transmembrane complexes remain to be established.

REFERENCES

Ball, E.G. (1938) Über die Oxydation und Reduktion der drei Cytochrom-Komponenten. Biochem. Z. 295, 262–264.

Battelli, F. and Stern, L. (1911) Die Oxydation der Bernsteinsäure durch Tiergewebe. Biochem. Z. 30, 172–194.

Battelli, F. and Stern, L. (1914) Einfluss der mechanischen Zerstörung der Zellstruktur auf die verschiedenen Oxydationsprozesse der Tiergewebe. Biochem. Z. 67, 443–471.

Beinert, H. and Sands, R.H. (1959) On the function of iron in DPNH cytochrome c reductase. Biochem. Biophys. Res. Comm. 1, 171–178.

Belitzer, V.A. and Tsybakova, E.T. (1939) The mechanism of phosphorylation associated with respiration. Biokhimiya 4, 516–534. English Translation in Biological Oxidations, H.M. Kalckar, ed. Prentice Hall, N.J. (1969), pp. 211–228.

Bensley, R.R. and Hoerr, N.L. (1934) Studies on cell structure by the freezing-drying method. V. The chemical basis of the organization of the cell. Anat. Rec. 60, 251–266.

Blasie, J.K., Erecinska, M., Samuels, S. and Leigh, J.S. Jr. (1978) The structure of a cytochrome oxidase-lipid model membrane. Biochim. Biophys. Acta. 501, 33–52.

Boyer, P.D. (1965) Carboxyl activation as a possible common reaction in substrate-level and oxidative phosphorylation. In *Oxidases and Related Redox Systems* (T.E. King, H.S. Mason and M. Morrison, eds.) Wiley, New York, Vol. 2, pp. 994–1017.

Boyer, P.D. (1974) Conformational coupling in biological energy transductions. In *Dynamics of Energy Transducing Membranes* (L. Ernster, R.W. Estabrook and E.C. Slater, eds.) Elsevier, Amsterdam, pp. 289–301.

Bumm, E., Appel, H. and Fehrenbach, K. (1934) Über die Beziehung zwischen Glykolyse und Atmung im tierischen Gewebe. Hoppe-Seyler's Z. 223, 207–213.

Cain, D.F. and Davies, R.E. (1962) Breakdown of adenosine triphosphate during a single contraction of working muscle. Biochem. Biophys. Res. Commun. 8, 361–366.

Chance, B. and Fugmann, U. (1961) ATP induced oxidation of exogenous cytochrome c in terminally inhibited phosphorylating particles. Biochem. Biophys. Res. Commun. 4, 317–327.

Chance, B. and Hollunger, G. (1960) Energy-linked reduction of mitochondrial pyridine nucleotide. Nature (Lond.) 185, 666–672.

Chance, B. and Pappenheimer, A.M. (1954) Kinetic and spectrophotometric studies of cytochrome b_5 in midgut homogenates of *Cecropia*. J. Biol. Chem. 209, 931–943.

Clark, W.M. (1960) *Oxidation-Reduction Potentials of Organic Systems*, Williams and Wilkins, Baltimore, MD.

Claude, A. (1938) A fraction from normal chick embryo similar to the tumor producing fraction of chicken tumor I. Proc. Soc. Exp. Biol. Med. 39, 398–403.

Claude, A. (1946) Fractionation of mammalian cells by differential centrifugation II. Experimental procedures and results. J. Exptl. Med. 84, 61–89.

Crane, F.L., Hatefi, Y., Lester, R.L. and Widmer, C. (1957) Isolation of a quinone from beef heart mitochondria. Biochim. Biophys. Acta 25, 220–221.

Crane, R.K., Miller, D. and Bihler, I. (1961) The restrictions on possible mechanisms of intestinal active transport of sugars. In *Membrane Transport and Metabolism* (A. Kleinzeller and A. Kotyk, eds.) pp. 434–449.

Coopenhaver, J.H. Jr. and Lardy, H.A. (1952) Oxidative phosphorylations: pathways and yield in mitochondrial preparations. J. Biol. Chem. 195, 225–238.

Davies, R.E. and Ogston, A.G. (1950) On the mechanism of secretion of ions by gastric mucosa and by other tissues. Biochem. J. 46, 324–333.

Dutton, P.L. (1978) Redox potentiometry: determination of midpoint potentials of oxidation-reduction components of biological electron-transfer systems. Methods Enzymol. 54, 411–435.

Eggleton, P. and Eggleton, G.P. (1927) XXV. The inorganic phosphate and a labile form of organic phosphate in the gastrocnemius of the frog. Biochem. J. 21, 190–195.

Ehrlich, P. (1885) *Das Sauerstoff-Bedürfniss des Organismus. Eine farbenanalytische Studie*. Berlin, Hirschwald.

Elvehjem, C.A. (1931) The role of iron and copper in the growth and metabolism of yeast. J. Biol. Chem. 90, 111–132.

Erecinska, M., Wilson, D.F. and Blasie, J.K. (1978) Studies on the orientations of the mitochondrial redox carriers. II. Orientation of the mitochondrial chromophores with respect to the plane of the membrane in hydrated, oriented mitochondrial multilayers. Biochim. Biophys. Acta 501, 63–71.

Fahmy, N.I., Hemming, F.W., Morton, R.A., Paterson, J.Y.F. and Pennock, F.L. (1958) Ubiquinone. Biochem. J. 70: 1P.

Fiske, C.H. and Subbarow, Y. (1929) Phosphorous compounds of muscle and liver. Science 70, 381–382.

Fernandez-Moran, H. (1962) Cell-membrane ultrastructure. Low temperature electron microscopy and X-ray studies of lipoprotein components in lamellar systems. Circulation 26, 1039–1065.

Fleisch, A. (1924) Some oxidation processes of normal and cancer tissue. Biochem. J. 18, 294–311.

Green, D.E. (1956) Structure and enzymatic pattern of the electron transfer system. In *Enzymes: Units of Biological Structure and Function* (O.H. Gaebler, ed.). Academic Press, N.Y., pp. 465–481.

Green, D.E. (1962) Structure and function in subcellular particles. Comp. Biochem. Physiol. 4, 81–122.

Hackenbrock, C.R., Gupta, S.C. and Chazotte, B. (1987) Mitochondrial electron transport: A random collision model. In *Advances in Membrane Biochemistry and Bioenergetics* (C.H. Kim, H. Tedeschi, J.J. Diwan and J.C. Salerno, eds.). Plenum Press, New York, pp. 61–74.

Haldane, J. and Smith, J.L. (1896) The oxygen tension in arterial blood, J. Physiol. 20, 497–520.

Haldane, J.B.S. (1927) Carbon monoxide poisoning in the absence of hemoglobin. Nature (Lond). 119, 352.

Hind, G. and Jagendorf, A.T. (1965) Light scattering changes associated with the production of a possible intermediate in photophosphorylation. J. Biol. Chem. 240, 3195–3201.

Hogeboom, G.H., Claude, A. and Hotchkiss, R.D. (1946) The distribution of cytochrome oxidase and succinoxidase in the mammalian liver cell. J. Biol. Chem. 165, 615–629.

Hogeboom, G.H., Schneider, W.C. and Pallade, G.E. (1948) Cytochemical studies of mammalian tissues. I. Isolation of intact mitochondria from rat liver; some biochemical properties of mitochondria and submicroscopic particulate matter. J. Biol. Chem. 172, 619–636.

Hopkins, F.G., Lutwak-Mann, C. and Morgan, E.J. (1939) Activity of succinic dehydrogenase. Nature (Lond.) 143, 556–557.

Hoppe-Seyler, F. (1879) Erregung des Sauerstoffs durch nascirenden Wasserstoff. Ber. dtsch. chem. Ges. 12, 1551–1555.

Jagendorf, A.T. and Uribe, E. (1966) ATP formation caused by acid-base transition of spinach chloroplasts. Proc. Natl. Acad. Sci. USA 55, 170–177.

Kagawa, Y. and Racker, E. (1966) Partial resolution of the enzymes catalysing oxidative phosphorylation: X. Correlation of morphology and function in submitochondrial particles. J. Biol. Chem. 241, 2475–2482.

Kalckar, H.H. (1969) Phosphorylation in kidney cortex. In *Biological Phosphorylations: Development of Concepts*. (H.H. Kalckar, ed.). Prentice-Hall, pp. 208–210.

Kearney, E.B. and Singer, T.P. (1955) On the prosthetic group of succinate dehydrogenase. Biochim. Biophys. Acta 17, 596–597.

Keilin, D. (1925) On cytochrome, a respiratory pigment, common to animals, yeast, and higher plants. Proc. Roy. Soc. B 98, 312–339.

Keilin, D. (1949) Effect of low temperature on the absorption spectra of haemoproteins; with observations on the absorption spectrum of oxygen. Nature (Lond.) 164, 254–259.

Keilin, D. (1966) *The History of Cell Respiration and Cytochrome*. Cambridge University Press.

Keilin, D. and Hartree, E.F. (1938) Cytochrome oxidase. Proc. Roy. Soc. B 125, 171–186.

Keilin, D. and Hartree, E.F. (1939) Cytochrome and cytochrome oxidase. Proc. Roy. Soc. B 127, 167–191.

Keilin, D. and Hartree, E.F. (1955) Relationship between certain components of the cytochrome system. Nature (Lond.). 176, 200–206.

Keilin, D. and King, T.E. (1958) Reconstitution of the succinic oxidase system from soluble succinic dehydrogenase and a particulate cytochrome system preparation. Nature (Lond.) 181, 1520–1522.

Kennedy, E.P. and Lehninger, A.L. (1948) Intracellular structures and the fatty acid oxidase of liver. J. Biol. Chem. 172, 847–848.

Kennedy, E.P. and Lehninger, A.L. (1949) Oxidation of fatty acids and tricarboxylic acid cycle intermediates by isolated rat liver mitochondria. J. Biol. Chem. 179, 957–972.

King, T.E., Kuboyama, M. and Takemori, S. (1965) On cardiac cytochrome oxidase: a cytochrome c-cytochrome oxidase complex. In *Oxidases and Related Redox Systems* (T.E. King, H.S. Mason and M. Morrison, eds.) Wiley, New York, Vol. 2, pp. 707–744.

King, T.E. (1963) Reconstitution of respiratory chain enzyme systems. XII. Some observations on the reconstitution of the succinate oxidase system from heart muscle. J. Biol. Chem. 238, 4037–4051.

Klingenberg, M., Slenczka, W. and Ritt, E. (1959) Comparative biochemistry of the pyridine nucleotide system in mitochondria of various organs. Biochem. Z. 332, 47–66.

Lardy, H.A. and Wellman, H. (1952) Oxidative phosphorylations: role of inorganic phosphate and acceptor systems in control of metabolic rates. J. Biol. Chem. 195, 215–224.

Laser, H. (1937a) Tissue metabolism under the influence of low oxygen tension. Biochem. J. 31, 1671–1676.

Laser, H. (1937b) Tissue metabolism under the influence of carbon monoxide. Biochem. J. 31, 1677–1682.

Lehninger, A.L. (1946) A quantitative study of the products of fatty acid oxidation in liver suspensions. J. Biol. Chem. 164, 291–306.

Lehninger, A.L. (1951) Phosphorylation coupled to oxidation of dihydrodiphosphopyridine nucleotide. J. Biol. Chem. 190, 345–459.

Levy, L. (1889) Über Farbstoffe in den Muskeln. Hoppe-Seyler's Z. 13, 309–325.

Lindsay, J.G., Owen, C.S. and Wilson, D.F. (1975) The invisible copper of cytochrome c oxidase: pH and ATP dependence of its midpoint potential and its role in the oxygen reaction. Arch. Biochem. Biophys. 169, 492–505.

Lindsay, J.G. and Wilson, D.F. (1974) Reaction of cytochrome c oxidase with CO: involvement of the invisible copper. FEBS Letters 48, 45–49.

Linossier, G. (1898) Contribution à l'étude des ferments oxydants. Sur la peroxydase du pus. C.R. Soc. Biol. (Paris) 50, 373–375.

Loew, O. (1901) Catalase, a new enzyme of general occurrence. Rep. U.S. Dept. Agric. no. 68, Washington.

Lohman, K. (1934) Über den Chemismus der Muskel Kontraktion. Naturwiss. 22, 409–413. English translation in *Biological Oxidations*, (H.M. Kalckar, ed.) Prentice-Hall, N.J. (1969), pp. 57–61.

Lundsgaard, E. (1930a) Untersuchungen über Muskelkontraktionen ohne Milchsäurebildung. Biochem. Z. 217, 162–177.

Lundsgaard, E. (1930b) Weitere Untersuchungen über Muskelkontraktionen ohne Milchsäurebildung. Biochem. Z. 227, 51–83.

Lundsgaard, E. (1931) Über die Energetik der anaeroben Muskelkontraktion. Biochem. Z. 233, 322–343.

Lundegårdh, H. (1945) Bleeding and sap movement. Arkiv. Bot. 32A, 12, 1–56.

MacMunn, C.A. (1884) On myohaematin, an intrinsic muscle pigment of vertebrates and invertebrates, on histohaematin, and on the spectrum of the supra-renal bodies. J. Physiol. (Lond.) 5, xxiv.

MacMunn, C.A. (1887) Further observations on myohaematins. J. Physiol. (Lond.) 8, 51–65.

Meyerhoff, O. (1925) '*Chemical Dynamics of Life Phenomena*' J.B. Lippincott CO. Philadelphia, PA.

Mitchell, P.D. (1961) Coupling of phosphorylation to electron and hydrogen transfer by a chemiosmotic type mechanism. Nature (Lond.) 191, 144–148.

Mitchell, P.D. (1966) Chemiosmotic coupling in oxidative and photosynthetic phosphorylation. Biol. Rev. 41, 445–502.

Mitchell, P.D. (1979) *David Keilin's Respiratory Chain Concept and its Chemiosmotic Consequences*. Reimpression de Les Prix Nobel en 1978, The Nobel Foundation, Norstedts Trycheri, Stockholm, Sweden.

Moore, C. and Pressman, B.C. (1964) Mechanism of action of valinomycin on mitochondria. Biochem. Biophys. Res. Commun. 15, 562–567.

Morton, R.A. (1958) Ubiquinone. Nature (Lond.) 182, 1764–1767.

Ohnishi, T. and Salerno, J.C. (1982) Iron-sulfur clusters in the mitochondrial electron-transport chain. In *Iron-Sulfur Proteins* (T.G. Spiro, ed.) vol. 4. New York, John Wiley.

Person, P., Wainio, W.W. and Eichel, B. (1953) The prosthetic groups of cytochrome oxidase and cytochrome *b*. J. Biol. Chem. 202, 369–381.

Pressman, B.C., Harris, E.J., Jagger, W. and Johnson, J. (1967) Antibiotic-mediated transport of alkali ions across lipid barriers. Proc. Natl. Acad. Sci. USA 58, 1949.

Raaflaub, J. (1955) Komplexbildner als Cofaktoren isolierter Zellgranula. Helv. Chem. Acta 38, 27–37. As translated into English and reprinted in *Biological Oxidations*, H.M. Kalckar, ed., pp. 262–272.

Rieske, J.S., McLennon, D.H. and Coleman, R. (1964) Isolation and properties of an iron-protein from the (reduced coenzyme Q)-cytochrome c reductase complex of the respiratory chain. Biochem. Biophys. Res. Communs. 15, 338–344.

Sands, R.H. and Beinert, H. (1959) On the function of copper in cytochrome oxidase. Biochem. Biophys. Res. Communs. 1, 175–178.

Schneider, W.C. (1946) Intracellular distribution of enzymes. I. The distribution of succinic dehydrogenase, cytochrome oxidase, adenosinetriphosphatase, and phosphorous compounds in normal liver. J. Biol. Chem. 165, 585–593.

Skulachev, V.P. (1991) Chemiosmotic systems in bioenergetics: H^+-cycles. Bioscience Reports 11, 387–441.

Slater, E.C. (1949) Cytochrome c_1 of Yakushiji and Okunuki. Nature (Lond.) 163, 532.

Slater, E.C. (1953) Mechanism of phosphorylation in the respiratory chain. Nature (Lond.) 172, 975–978.

Slater, E.C. and Cleland, W.K. (1953) The effect of calcium on the respiratory and phosphorylative activities of heart muscle sarcosomes. Biochem. J. 55, 566–580.

Stern, K.G. and Melnick, J.L. (1941) The photochemical action spectrum of the Pasteur enzyme in retina. J. Biol. Chem. 139, 301–323.

Sun, F.F., Prezbindowski, K.S., Crane, F.L. and Jacobs, E.E. (1968) Physical state of cytochrome oxidase: relationship between membrane formation and ionic strength. Biochim. Biophys. Acta 153, 804–818.

Thunberg, T. (1930) The hydrogen activating enzymes of the cells. Quart. Rev. Biol. 5, 318–347.

Traube, M. (1882) Über Aktivierung des Sauerstoffs. Ber. dtsch. chem. Ges. 15, 659–675.

Unwin, N. and Henderson, R. (1984) The structure of proteins in biological membranes. Sci. American 250, 78–95.

Vanderkooi, J., Maniara, G. and Erecinska, M. (1985) Mobility of fluorescent derivatives of cytochrome c in mitochondria. J. Cell Biol. 100, 435–441.

Van Gelder, B.F. and Beinert, H. (1969) Studies of the heme components of cytochrome c oxidase by EPR spectroscopy. Biochim. Biophys. Acta 189, 1–24.

Wainio, W.W., Cooperstein, S.J., Kollen, S. and Eichel, B. (1947) Cytochrome oxidase. Science 106, 471.

Warburg, O. (1912) Über Beziehungen zwischen Zellstruktur und biochemischen Reactionen. I. Pflügers Arch. 145, 277–282.

Warburg, O. (1915) Über die Empfindlichkeit der Sauerstoffatmung gegenüber indifferenten Narkotika. Pflüger's Arch. 158, 19–28.

Warburg, O. and Negelein, E. (1928) Über die photochemische Dissoziation bei intermittierender Belichtung und das absolute Absorptionsspektrum des Atmungsferments. Biochem. Z. 202, 202–228.

Warburg, O. and Negelein, E. (1929) Über das Absorptionsspektrum des Atmungsferments. Biochem. Z. 214, 64–100.

Wieland, H. (1912a) Über Hydrierung und Dehydrierung. Ber. dtsch. chem. Ges. 45, 484–499.

Wieland, H. (1912b) Studien über den Mechanismus der Oxydationsvorgänge. Ber. dtsch. chem. Ges. 45, 484–499.

Wilson, D.F. and Dutton, P.L. (1970) Energy-dependent changes in the oxidation-reduction potential of cytochrome b. Arch. Biochem. Biophys. 136, 583–584.

Wilson, D.F., Erecinska, M. and Dutton, P.L. (1974) Thermodynamic relationships in mitochondrial oxidative phosphorylation. Ann. Rev. Biophys. Bioeng. 3, 203–230.

Wilson, D.F., Owen, C.S. and Holian, A. (1977) Control of mitochondrial respiration: a quantitative evaluation of the roles of cytochrome c and oxygen. Arch. Biochem. Biophys. 182, 749–762.

Wilson, D.F., Stubbs, M., Veech, R.L., Erecinska, M. and Krebs, H.A. (1974) Equilibrium relations between the oxidation-reduction reactions and the adenosine triphosphate synthesis in suspensions of isolated liver cells. Biochem. J. 140: 57–64.

Yakushiji, E. and Okunuki, K. (1940) Über eine neue Cytochromkomponente und ihre Funktion. Proc. Imp. Acad. Japan 16, 299–302.

Yu, C.A., Nagaoka, S., Yu, L. and King, T.E. (1978) Evidence for the existence of a ubiquinone protein and its radical in the cytochrome b and c_1 region in the mitochondrial electron transport chain. Biochem. Biophys. Res. Communs. 82, 1070–1078.

A. Kleinzeller (Editor) A History of Biochemistry: Exploring the Cell Membrane.
(Comprehensive Biochemistry Vol. 39) © 1995 Elsevier Science B.V.

Chapter 7

The Epithelial Membrane

ROLF K.H. KINNE[1] and ARNOST KLEINZELLER[2]

[1]The Max-Planck Institute for Molecular Physiology,
44139 Dortmund, Germany;
[2]Department of Physiology, University of Pennsylvania, Philadelphia,
PA 19104-6085, U.S.A.

I. Introduction

In the simplest case, a single layer of epithelial cells interconnected by intercellular cement is atttached to a loose meshwork of connective fibers (called the 'basement membrane'). This is an 'epithelial membrane', a structure with distinctive physical properties[1], which is capable of producing a net selective unidirectional flux of an aqueous solution of solutes to flow between two compartments of the organism. Such a structure implies a polar nature of the epithelial cells, the portion attached to the basement membrane (blood side) differing functionally and often structurally from that at the apical cell face.

Exploration of transepithelial transport has greatly contributed to the development of the concept of solute pumps (see also Chapters 1 and 4): For the thermodynamic concept of a pump, metabolically-dependent changes in solute activity on both sides of the separating membrane have to be defined (Chapter 4). It was experimentally relatively easy to demonstrate the uphill absorptive and secretory processes by an

---

1 e.g. the capability of withstanding major osmotic gradients; thus, in the mammalian kidney, such gradients may correspond to a pressure gradient of more than 20 atm.

analysis of the changes occurring in the aqueous compartments separated by an epithelial membrane. The well-known
Ussing equation (see Chapter 2), when first derived for the
epithelial membranes such as the frog skin, defined the measured short-circuit current as a phenomenon occurring between
two well-stirred homogenous bathing solutions separated by a
'black box' cellular membrane; an analysis of the underlying
cellular events then followed. On the other hand, the study of
such phenomena in non-polar cells also required the definition
of solute changes in the cellular compartment, raising conten-
tious questions as to possible structural (cellular and subcellular compartments) as well as chemical compartmentation
(possible limitations as to availability of solvent, and binding
of solutes to cellular components); these problems rendered
the experimental approach and theoretical treatment much
more complex.

   This chapter aims to delineate the development of the concepts that transepithelial transport is composed of multiple
pathways, *i.e.* that both para- and transcellular routes exist
and that the transcellular pathway can be subdivided into two
transport steps occurring sequentially across two different cell
membrane borders arranged in series. It is hoped that this
article transmits some of the excitement with regard to mul-
tidisciplinary thinking and experimental improvements that
accompanied the development of one of the most fundamental
concepts in biology.

II. The establishment of the existence of epithelial cell layers

In 1838, the German comparative anatomist and physiologist
Jacob Henle published his studies entitled '*Über die Ausbreitung des Epitheliums im menschlichen Körper*' (on the distribution of epithelia in the human body) (Henle, 1838). In these
studies, based mainly on microscopic observations, he pro-

posed the existence of epithelia, *i.e.* cellular tissues that cover the free surfaces of the body and also line the ductular and tubular systems in glands and the body cavities (gallbladder, peritoneum, etc.). It is interesting to note that intestinal scrapings (Henle, 1837) and scrapings from other free surfaces of the body as well as excretory products were his main objects of investigation. The analysis of these specimens by light-microscopic techniques, which Henle systematically applied and refined, made epithelial cells for the first time visible and opened a new dimension in anatomy and histology. In this publication he introduced the basic terminology which is still used to the present time: '*Pflasterepithel*' (squamous epithelium), '*Zylinderepithel*' (columnar epithelium), and '*Flimmerepithel*' (ciliated columnar epithelium). Henle also offered some suggestions on the functional role of these epithelia (Henle, 1841) in attributing to them – at least in the glands – an involvement in the production and in the final composition of the secreted fluid either by 'pulling substances from the blood or retrieving substances from the secreted fluid back into the cells and even transforming them somehow'.

In the first illustrations of epithelia Henle focussed on the appearance of sheets of epithelial cells; the question of how these cells are interconnected is not discussed. Indications of some structures which looked like connections between epithelial cells were noted in 1870 (Rizzozero, *cf.* Fawcett, 1961) and in 1875 Arnold described the '*Kittsubstanz*' or 'cement' that filled the intercellular space in epithelia, thus adding a para- or intercellular pathway to the transcellular pathway postulated for epithelial function (see below).

III. The transcellular route in epithelial transport

1. DRIVING FORCES

Ludwig's 1861 studies (*cf.* Chapter 2) showed that the secretion of the salivary glands is a cellular process. The idea of absorptive and secretory processes was at first limited to glandular tissue. Thus, du Bois-Reymond (1849), when describing the transepithelial electrical potential in the frog skin, related this to the function of the glandular skin components. Only later was the more general concept put forward that epithelial cells are responsible for absorptive and secretory processes. In this respect, Hermann (1882) established this point for the frog skin, and Heidenhain (cf. 1888) for the unidirectional flow of water (and solutes) in the intestine. It ought to be mentioned here that with regard to intestinal absorption, Hoppe-Seyler (1878) appeared to be the first who had misgivings about the then prevalent view (cf. the scholarly paper of Reid (1900) on the older literature) that osmotic forces (i.e. diffusion) alone are responsible for transepithelial transport. Heidenhain (1888, 1894) and his students Leubuscher, Gumilewski and Röhmann then provided the significant experimental evidence for the participation of cellular activity: using loops of dog intestine in vivo, the absorptive process was followed by measuring the volume of introduced solutions. Evidence for the role of osmotic forces in the absorptive process had been established (Heidenhain, 1888) by demonstrating the passage of methylene blue and solutions from the intestinal lumen into the mucosa through both the cells and the intercellular cement[2]. The role of cellular participation ('physiologische Triebkraft', i.e. physiological driving force) was mandated by observations (Heidenhain, 1894) of absorption of fluid (and NaCl) from hypotonic solutions (i.e. against its concentration gradient), the inhibitory effect of the metabolic poison sodium fluoride, and the specificity of the process; solutions other than sodium chloride were poorly absorbed (Hermann, 1882). Heidenhain's (1894) concluding concept (p. 631), that intestinal absorption is brought about by the parallel

---

2 'Kittsubstanz', Arnold (1875)

processes of filtration (diffusion), and cellular mechanisms withstood the tests of a subsequent 100 years:

'The epithelium is composed of cells and intercellular spaces. Both are resorption pathways since it has been shown that during the resorption of solutions of dyes the dye can be found in both the cells and the intercellular cement. That (part of the intestinal solution) which passes by the latter pathway from the intestine into the mucosa is most probably brought about by diffusion. The active role will have to be ascribed to the epithelial cells.'

In spite of some experimental flaws of Heidenhain's data, Reid (1900, 1892) essentially confirmed the participation of cellular mechanisms in the absorptive process. Moreover, Reid (1892) also found that the direction of the transepithelial transport could be altered by some drugs, *i.e.* a prior absorptive tissue now displayed secretory properties.

Further direct evidence for the participation of cellular driving forces in epithelial transport was provided much later in studies on the relation between cellular metabolism (O_2 consumption or CO_2 production) and transepithelial ion transport. Thus, Zehran (1956) and Leaf et al. (1959) demonstrated in frog skin and toad bladder, respectively, a reasonably constant relationship between the amount of sodium ions transported and the increase of oxygen consumption associated with ion transport. A ratio of close to 3 sodium ions transported per ATP split was found and supported the idea of the involvement of the $(Na^+ + K^+)$-ATPase in this process (see also Chapter 4). In other epithelia, where both transcellular and paracellular movement of sodium occurs, like the kidney, this ratio was found to be higher which led to speculations about the nature of this apparently more efficient coupling. This mystery was resolved when active and passive sodium fluxes could be distinguished, based on the thermodynamical analysis of their driving forces. The latter basic concepts were developed in the laboratories of Ullrich (Frömter et al., 1973) and by Sauer (1973) who applied thermodynamic principles devel-

oped for transport across single membranes by Kedem and
Katchalsky (1963a,b,c) to epithelial transport.

2. CELLULAR ASYMMETRY

The first notion on cellular asymmetry of epithelial (and
equivalent) cells may well have come from studies of Sachs
(1865) on the uptake of nutrients by plant roots. Sachs based
his considerations on experiments of Hales in 1727 (*cf.* Chapter 4) and since repeated by numerous plant physiologists,
showing a copious flow of sap under considerable pressure
from plant stems cut just above their roots. The nature of this
'root force' ('*Wurzelkraft*') was now the object of inquiry. The
model of Sachs (Fig. 1), considered the flow of an aqueous solution of nutrients derived from the soil (from the quoted data
0.0056 parts K/1000 parts water; *i.e.* approx. 0.1 mM) into the
cells, containing some 0.1 M K. Thus a considerable 'endosmotic' flow of water had to occur from the external compartment across the cell membrane a, increasing the intracellular
osmotic pressure. For a system containing a central duct allowing the flow of sap into other parts of the system, an exit of
nutrient solution from the cells into this central duct across
cell membrane b will take place in accordance with existing
hydrostatic pressure gradients. Sachs now suggests that if the
'filtration resistance' (we now would say permeability) of the
inner membrane b is smaller than that of membrane a, there
will be a transcellular fluid flow. In this rather mechanistic
system, Sachs did not yet consider the actual possible mechanisms by which the concentrative gradient of salts between
cells and their environment is established, although he did
mention the possibility of several cells in series concentrating
the sap prior to the exit into a 'vessel'. Still the model envisaged a vectorial flow of a solution brought about by two cell
membranes in series, with differing functional (permeability)
parameters.

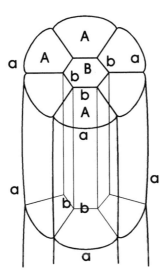

Fig. 1. Sachs' (1865) scheme of a root: A: absorbing cortical cells; B: central vessel for conducting sap; a and b: cell membranes (skins) facing the outer and central side, respectively.

A polar cell morphology was clearly also discerned by early anatomists who saw in intestinal and renal cells an apical structure composed of fine filaments (now called microvilli, organized in a brush border). In fact, Brücke (1861) considered this structure to be mobile structures which mechanically forced a flow of nutrients from the (intestinal) lumen directly into the cells, thus producing absorption; he took this arrangement to be evidence against the cell membrane concept as visualized by Schultz or Nägeli (*cf.* Chapter 2).

The pioneering studies on the cellular basis for the generation of a transmembranal potential in the frog skin by Ussing and his colleagues (Koefoed-Johnson and Ussing, 1958) represent the next very important step in the definition of vectorial transepithelial transport as a sequence of transport processes in the apical and basal cell membrane arranged in series. Their observations were facilitated by the fact that an appro-

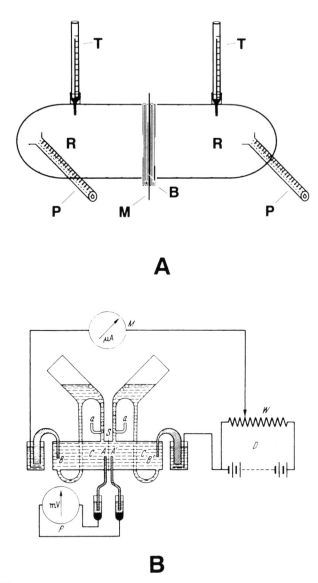

Fig. 2. The measurement of transepithelial transport: The chamber A: Reid (1892) for the study of fluid absorption or secretion: T, thermometers; R, reservoir; M, biological membrane, supported by porous material B; P, pipet for fluid volume measurements; and B: Ussing and Zehran (1951) for the study of the short-circuit current.

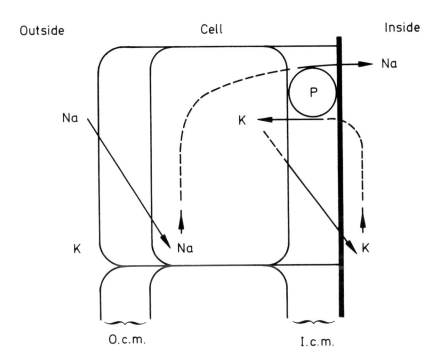

Fig. 3. Idealized epithelial cell, illustrating the two-membrane hypothesis for the origin of the frog skin potential. Outer membrane specifically but passively permeable to Na. Inner membrane specifically but passively permeable to K, also possessing a Na/K exchange pump. O.c.m. = outward-facing cell membrane; I.c.m. = inward-facing cell membrane. After Koefoed-Johnsen and Ussing (1958) with kind permission.

priate model epithelium was available – the frog skin which transported only sodium actively and by the ingenious and rigorous application of physico-chemical principles to natural membranes. The experimental setup used in these studies, the 'Ussing chamber' is shown in Fig. 2. It is actually a sophisticated major modification of the system used by Reid in his studies.

The sodium dependence of the transepithelial potential in the frog skin had already been observed by Galeotti (1904) but the combination of the concepts of flux ratio measurements and the short-circuit setup, both developed by Ussing (Ussing, 1949; Ussing et al., 1964), made the experimental observations possible which led to the two-membrane theory for transepithelial potentials shown in Fig. 3. This model has three elements: (1) The apical cell boundary is selectively but passively permeable to sodium. (2) The inward-facing cell membrane is virtually impermeable to sodium but is permeable to potassium and related cations like rubidium. (3) The inward-facing cell membrane is the seat of an active sodium transport mechanism (see Ussing and Leaf, 1978, for a review).

The two former conclusions had been mainly derived from ion substitution experiments, in which the apical membrane behaved like a sodium-selective electrode and the basal membrane very similar to a potassium-selective electrode. The latter was needed to explain the active transcellular sodium movement in view of the ion gradients that could be assumed from the electrical behaviour of the two membranes[3].

The proposal of Ussing was rapidly and widely accepted and led to numerous experiments that provided further evidence for the correctness of this model.

As reviewed by Ussing himself, not all of the experiments were, indeed, supportive (Ussing and Leaf, 1978). Thus, for a simple epithelium – with mostly transcellular transport and little paracellular movement – the vectorial transport of sodium across this epithelium could be explained. In the current literature all models for electrolyte transport across an epithelium contain the elements of asymmetry with regard to the permeability of the two plasma membranes, and for calcium and protons even the asymmetric arrangement of the active pump in the epithelial cell is maintained (Kinne, 1991).

---

3 The exact determination of intracellular ion concentrations and activities began with the development of ion sensitive microelectrodes (see Hinke, 1961). The difficulties encountered in such measurements are reviewed by Oschman (1978).

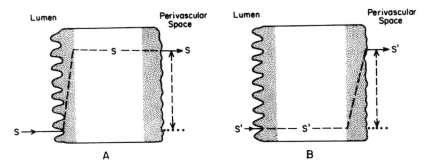

Fig. 4. Cellular localization of active (uphill) transport of solute S from the luminal to the peritubular space (absorption). A: push mechanism. Active step at the luminal membrane raises the intracellular concentration of S; substrate then exits passively at the basolateral membrane. B: pull mechanism. S' passes through the luminal cell membrane by a passive (equilibrating) mechanism; hence intracellular S' concentration is close to or lower than that in the luminal space. Active (uphill) mechanism at the basolateral membrane actually extrudes S' against its concentration gradient from the cell into the peritubular space. From Kleinzeller (1987) with kind permission.

A proposal of an asymmetry of the transport properties of the two plasma membranes arranged in series can already be found in the model proposed by Forster and Taggart (1950) for the coupling of metabolic processes to active cellular transport of solutes. They distinguished a push model and a pull model (Fig. 4) in which the active transport step is either located in the luminal membrane, thereby leading to an accumulation of the transported solute in the cell (push model) or in the contraluminal membrane – extruding the solute against its concentration gradient from the cell into the peritubular space (pull model). Thus, the Ussing model would correspond to the pull model. However, only a combination of asymmetry in the pumps *and* in the permeability properties of the two membranes could explain vectorial active translocation.

The two-membrane concept was significantly extended when Crane and coworkers (1961) advanced the idea of co-transport of sugar with sodium to explain the active transcel-

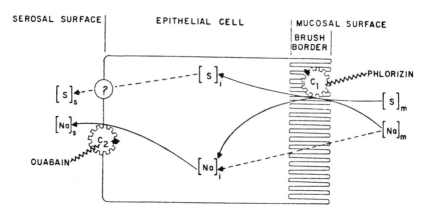

Fig. 5. Model of coupling between Na^+ and sugar transport in intestinal cells (Schultz and Zalusky, 1964). Subsequently, the mechanism of sugar exit at the basolateral cell membrane, denoted here by ?, was further defined as an equilibrating, carrier-mediated process (Kleinzeller and McAvoy, 1972; Kinne, 1976).

lular transport of organic solutes. Based on the observation of Csáky and Thale (1960) that the active uptake of the non-metabolized sugar 3-O-methyl-D-glucose in toad intestine depended strictly on the presence of sodium in the intestinal lumen and that ouabain inhibited sugar transport (Csáky et al., 1961; Crane et al., 1961), a model for transcellular sugar transport was conceived and refined (Fig. 5) (Schultz and Zalusky, 1964). This concept basically incorporated all three postulates of the Ussing model: the asymmetric distribution of an active pump and differences in the properties of the transport systems for the sugar in the luminal and contraluminal membrane, respectively. For the sugar this model is a push model which explains the accumulation of sugars inside the cell. The nature of the active 'sugar' pump represented a completely new concept (secondary active transport) in which the gradient of sodium across the luminal membrane acts as driving force for the accumulation of sugars – *i.e.* a pump that is only secondarily dependent on ATP hydrolysis which occurs when sodium, having entered the cell with the sugar, is actively extruded from the cell at the basal-lateral pole by the pull mechanism of the $(Na^+ + K^+)$-ATPase. In this model two interrelated, but separate thermodynamic forces have to be considered, *i.e.* the coupling of the uphill sugar transport to the flux of Na^+ in accordance with its concentration gradient, as well as the actual operation of the Na^+ pump which maintains the necessary Na^+ gradient. Thus, a coupling of the pull and push mechanisms is a better description of the model.

It is interesting to note that this model, probably for the sake of simplicity, does not consider the presence of sodium permeabilities at the luminal membrane competing for the sodium gradient, and also the potassium permeability shown by Ussing is not included. These elements were incorporated later when the 'crosstalk' between the luminal and contraluminal membranes by feedback mechanisms was considered. Such a crosstalk seemed necessary to coordinate the entry of osmolytes (*i.e.* sugar and sodium) across the apical membrane with their extrusion across the basal-lateral membrane; if not coordinated, severe changes in intracellular osmolality and sodium concentration could be envisaged. Changes in the electrical driving forces and alteration of channel activities induced by intracellular parameters were proposed (Diamond, 1982; Hudson and Schultz, 1984; Lau et al., 1984; Lewis and Wills, 1981; Schultz, 1977; 1981).

IV. The direct demonstration of different transport properties of the apical and basal-lateral plasma membrane

1. THE BIOCHEMICAL APPROACH

The models presented above were all consistent with the experimental findings accumulated thus far, but direct evidence for the differences in transport properties and information on the biochemical nature of the transport systems were lacking. Pharmacological studies using specific inhibitors suggested the asymmetric distribution of transport systems, the best example being ouabain. Autoradiographic studies promoted by Stirling and Kinter (Karnaky et al., 1976) demonstrated specific binding of this glycoside to the basal-lateral cell side in ion-transporting epithelia and were interpreted to indicate the presence of the (Na^++K^+)-ATPase only in basal-lateral plasma membranes (Eveloff et al., 1979; Kinne and Kinne-Saffran, 1978; Stirling 1972; Stirling and Landau, 1970). Definite proof required, however, the physical separation of the membranes covering the opposite cell poles and the separate evaluation of their transport properties. Such physical separation was first attempted at the microscale by Schmidt and Dubach (1971) in microdissected cross-sections of renal tubules combined with the enzymatic determination of the (Na^++K^+)-ATPase activity. These experiments showed, indeed, a predominant but not exclusive presence of this enzyme in the basal portion of the cells.

The biochemical approach to isolate the membranes after homogenization of epithelia turned out to be difficult with conventional methods (de Duve, 1965) because of the similar composition and thus density of the two plasma membranes. Therefore, the application of a new separation principle – differences in surface charge density – was necessary. This was done successfully by Heidrich et al. (1972) using free flow electrophoresis and later by Booth and Kenny (1974), who introduced the now widely used differential precipitation technique employing high concentrations of divalent cations. The development of material suited to generate very shallow density gradients (like Percoll) also aided in this physical separation (Mamelok et al., 1981). Using such techniques the localization of ATP-dependent ion pumps, receptor-stimulated adenylate cyclases, and phosphorylation reactions could be determined by classical biochemical procedures using membrane sheets or detergent dissolved membranes. The analysis of facilitated

diffusion transport systems in these membrane fractions required the creation of an experimental system where the
membrane separated two different compartments into which
or out of which solute movement could be followed. Such a
method existed for submitochondrial particles and was first
applied by Hopfer et al. (1973) to the study of sodium-dependent sugar transport in intestinal brush border membrane vesicles. In this seminal paper sodium-D-glucose cotransport was
demonstrated in the presence of defined ion gradients and in
the absence of ATP or other intracellular energy sources –
experimental conditions never achievable in the intact cell.
The use of isolated plasma membranes and vesicles rapidly
expanded and led to the characterization of a large variety of
transport systems in almost all epithelia that had been stud-

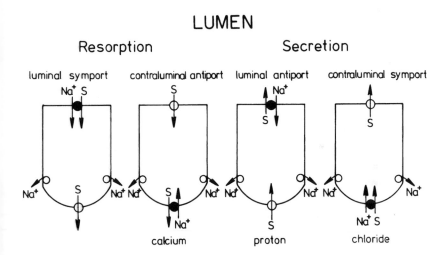

LUMEN

Resorption   Secretion

luminal symport   contraluminal antiport   luminal antiport   contraluminal symport

calcium   proton   chloride

INTERSTITIUM

*Fig. 6. Role of sodium cotransport systems in resorption and secretion. From
Kinne and Kinne-Saffran (1985) with kind permission.*

ied prior by classical biophysical techniques (Kinne and Kinne-Saffran, 1978; Kinne-Saffran and Kinne, 1989). In general it was found that the properties of the basal-lateral plasma membrane resembled more those observed in non-polarized cells, whereas the luminal membrane contains highly specialized transport systems (Evans, 1980; Kinne and Kinne-Saffran, 1978).

The work with isolated membrane vesicles had several other important implications: one was that reabsorption and secretion apparently involving similar transport systems could now be explained by the different location of relatively few secondary active sodium cotransport systems and sodium-independent facilitated diffusion systems (see Fig. 6) rather than postulating the involvement of a vast number of transporters to explain the different direction in transport (for a review see Kinne, 1991).

The vesicle studies also refined the analysis of transport energetics across the membranes. Secondary active sodium cotransport was demonstrated both as symport-sodium and the solute moving in the same direction, and as antiport-sodium and the solute moving in opposite directions (see also Chapter 3). In addition, tertiary active transport systems could be defined where accumulation of a solute inside the cell by a secondary active sodium cotransporter can, by flux coupling through an exchange system, lead to accumulation of a third solute inside the cell or its extrusion against an electrochemical potential difference (tertiary active transport of the latter solute). Such coupling of driving forces can occur at the same membrane as well as at the opposite cell poles (see Kinne, 1988, for review). Furthermore, new ATP-dependent and/or ATP-driven transport systems were discovered.

2. THE BIOPHYSICAL APPROACH

The molecular analysis of the different ionic permeabilities of the two membranes required another approach. In excitable cells Neher and Sakmann developed the patch-clamp technique by which the properties of single ion channels can be characterized (Hamill et al., 1981; Neher, 1982; Neher and Sakmann, 1976). The application of this technique to epithelial cells (Maruyama and Petersen, 1982) allowed to determine the location of ion channels in such cells, the description of their properties and the analysis of their regulation by intra- and extracellular parameters (for review see Koeppen, 1987; Palmer, 1990). Thus, tools are now available that can be used to describe in quite some detail the transport properties of the two plasma membranes arranged in series in the transporting epithelia.

All along during the focus on the transport across the intracellular compartment, which was considered to represent a fluid phase without any protein and organelles there was the additional hypothesis that transport of solutes across an epithelial cell can also involve vesicular transport. This transport mode, first observed in endothelial cells (Buck, 1958; Jennings et al., 1962; Moore and Ruska, 1957; Wissig, 1958), has recently been demonstrated to exist also in (other) epithelia. Unfortunately, our knowledge on the factors that determine specificity, directionality, and activity of such transport are still rather poorly understood.

V. The intercellular pathway

Once the indications of an intercellular transport pathway in epithelial membranes were established (and studies on intestinal fluxes were particularly instructive because of the relative experimental simplicity), the quantitative and qualitative aspects of this pathway were extensively studied. Höber (1901) demonstrated that the intercellular 'cement' had to be rather leaky in order to permit the entry of ammonium molybdate, which did not appear to enter the cells. Evidence that

net unidirectional fluxes were actually the resultant of unequal simultaneous fluxes in opposite directions (Goldschmidt
and Dayton, 1919; Visscher et al., 1944a,b) argued in favor of
a diffusion pathway, particularly when considering the respective rates as compared with fluxes across cell membranes.
These studies were greatly facilitated by the introduction of
radioisotopes as experimental tools (cf. Visscher et al.).

The advent of electron microscopy brought about a new
slant to the investigations of the concept of intercellular transport pathway. In an important study, Farquhar and Palade
(1963) described three structural elements of the intercellular
junction of epithelia, viz. (from the apical side downwards),
the Zonula occludens (tight junctions), Zonula adhaerens (intermediate junction), and Macula adhaerens (desmosome, apparently identical with the structure seen by Fawcett [1961]).
The whole length of the intercellular space represented a very
narrow channel. It was particularly the tight junction which
appeared to preclude a free passage of solutions between the
cells. Hence, hypotheses had to be advanced in order to explain
the movement of water together with actively transported solutes. The standing-gradient hypothesis (Fig. 7a) of Diamond
and Bossert (1967) was thought to be the answer, in effect
brushing aside physiological evidence available since the days
of Heidenhain in favor of a major intercellular passage of solutes. It took a great number of further studies (cf. Clarkson,
1967; Claude and Goodenough, 1973; Frizzell and Schultz,
1972; Ussing and Windhager, 1964) to make the tight junction
again leaky in many epithelia, including the intestine.

A particularly elegant demonstration of the existence of the
paracellular shunt pathway was provided by Frömter and
Diamond (1972) who, using the technique of voltage scanning,
could show that current flow through the Necturus gallbladder
was highest at the cell junctions. Also the careful electrophysiological and theoretical analyses of Boulpaep (1971), Frömter
(1972), Weinstein and Stephenson (1981a,b), and Guggino et
al. (1982) contributed significantly to the acceptance of the

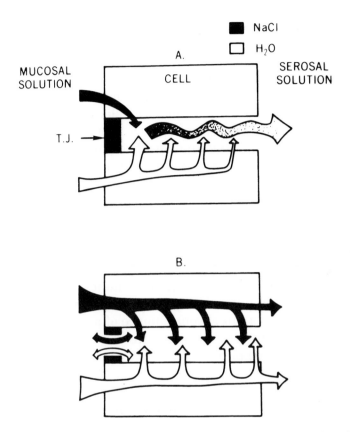

Fig. 7. Illustrations of (A) the original 'standing osmotic gradient' model for isotonic fluid absorption and (B) a modification of that model which takes into account the extreme 'leakiness' of the junctional complexes and the high hydraulic conductivities of the limiting membranes. From Schultz (1981) with kind permission.

paracellular route as (in some epithelia even major) pathway in transepithelial ion transport. A modified version of the standing osmotic gradient hypothesis which accounts for the extreme 'leakiness' of some junctional complexes is given in Fig. 7b.

The permeability of such structures obviously permits the

passage of aqueous solutions of inorganic salts, but prevents the entry of even relatively small proteins such as hemoglobin (*cf.* Miller, 1960, for the renal tubule). The pathway demonstrates some ionic selectivity, and the flux through it depends on the transmembrane potential. The magnitude of this shunt pathway is often a determinant of the electrical potential gradient across the epithelial membrane. For some time a direct relationship between the number of strands in the junctional morphology and the transepithelial resistance in Ohm/cm^2 was postulated (Claude and Goodenough, 1973). This assumption, however, was refuted later (Madara, 1989).

The narrow intercellular channel imparts to the shunt pathway another somewhat unexpected property. When first measured by several investigators, the bidirectional fluxes of D_2O showed an asymmetry which in terms of the Ussing equation (Chapter 4) was interpreted as indicative of an active transport system, until Ussing (1949) demonstrated that this phenomenon was a direct consequence of a continuous flow of water through the narrow structure. Thus, no active water transport had to be postulated.

VI. What's next?

With regard to the whole epithelium the generation of the intercellular connections as well as their control and the establishment and maintenance of the asymmetry of the cell membranes is an open question. This question can now be studied more easily because cell culture methods have been introduced, where polarized epithelial cells can be grown and studied (Handler, 1983; Mullin et al., 1986; Simons and Fuller, 1985; Simons and Wandinger-Ness, 1990). Also in vitro systems for examining cell membrane sorting will be very helpful.

Further developments can be best envisaged using again the subdivision of transepithelial flow into the various pathways. The paracellular pathway is still ill-defined with regard to its chemical components and its physical driving forces. The interaction of solutes with each other, with cell membrane components and with the compounds representing the paracellular 'cement', is poorly understood. A better definition is in particular necessary for leaky epithelia where the intercellular route determines the major transport characteristics of the whole epithelium.

With regard to the transcellular pathway the biochemical identification of transport proteins by cloning is under way (Coady et al., 1990; Hediger et al., 1987; 1989; Ikeda et al., 1989) and thus a better definition of the properties and the molecular structure underlying transporters and channels will be available (Mishina et al., 1984; Noda et al., 1982). 'Plasticity' of an epithelial cells, *i.e.* the insertion of these transporters into the correct membranes (Diamond, 1985; Simons and Wandinger-Ness, 1990; Stanton, 1984), the control of their activity and its biochemical basis is also an extremely interesting field in biology. Finally, we only start to appreciate the importance of the concepts of epithelial transport for the explanation of pathophysiological states of the cell which form the basis of disease processes (Molitoris et al., 1989).

Acknowledgements

The skillful and active collaboration of Mrs. D. Mägdefessel in preparing this manuscript is acknowledged most gratefully as is the support of all our colleagues and friends.

REFERENCES

Arnold, J. (1875) Die Kittsubstanz der Epithelien. Virchow's Arch. 64, 203–242.

300

Booth, A.G. and Kenny, A.J. (1974) A rapid method for the preparation of microvilli from rat kidney. Biochem. J. 142, 575–581.

Boulpaep, E. (1971) Electrophysiological properties of the proximal tubule: Importance of cellular and intercellular pathways. In *Electrophysiology of Epithelia* (Giebisch, G., ed.), pp. 91–112. Stuttgart, Schattauer.

Brücke, E. (1861) Die Elementorganismen. Sitzungsber. k. Akad. Wiss. Wien 16, 381–406.

Buck, T.C. (1958) The fine structure of endothelium of large arteries. J. Biophys. Biochem. Cytol. 4, 187–190.

Clarkson, T.W. (1967) The transport of salt and water across isolated rat ileum. Evidence for at least two distinct pathways. J. Gen. Physiol. 50, 595–727.

Claude, P. and Goodenough, D.A. (1973) Fracture faces of zonulae occludents from 'tight' and 'leaky' epithelia. J. Cell Biol. 58, 390–400.

Coady, M.J., Pajor, A.M., Toloza, E.M. and Wright, E.M. (1990) Expression of mammalian renal transporters in Xenopus laevis oocytes. Arch. Biochem. Biophys. 283, 130–134.

Crane, R.K. (1965) Na^+-dependent transport in the intestine and other animal tissues. Fed. Proc. 24, 1000–1006.

Crane, R.K., Miller, D. and Bihler, I. (1961) The restrictions on possible mechanisms of intestinal active transport of sugars. In *Membrane Transport and Metabolism*, (Kleinzeller, A. and Kotyk, A., eds.), pp. 439–449. New York, Academic Press.

Csáky, T.Z. and Thale, M. (1960) Effect of ionic environment on intestinal sugar transport. J. Physiol. (Lond.) 151, 59–65.

Csáky, T.Z., Hartzog, H.G. and Fernald, G.W. (1961) Effect of digitalis on active intestinal sugar transport. Am. J. Physiol. 200, 459–460.

De Duve, C. (1965) The separation and characterization of subcellular particles. Harvey Lect. 59, 49–87.

Diamond, J.M. (1982) Transcellular cross-talk between epithelial cell membrane. Nature (Lond.) 300, 683–685.

Diamond, J.M. (1985) A reversible epithelium. Nature (Lond.) 318, 311.

Diamond, J.M. and Bossert, W.H. (1967) Standing-gradient osmotic flow. A mechanism for coupling of water and solute transport in epithelia. J. Gen. Physiol. 50, 2061–2083.

Du Bois-Reymond, E. (1849) *Untersuchungen über thierische Electricität*, Vol. 2, pp. 1–884, Berlin 1860, G. Reimer.

Evans, W.H. (1980) A biochemical dissection of the functional polarity of the plasma membrane of the hepatocyte. Biochim. Biophys. Acta 604, 27–64.

Eveloff, J., Karnaky, Jr., K.J., Silva, P., Epstein, F.H. and Kinter, W.B. (1979) Elasmobranch rectal gland cell: autoradiographic localization of [³H]ouabain-sensitive Na,K-ATPase in rectal gland of dogfish, *Squalus acanthias*. J. Cell Biol. 83, 16–32.

Farquhar, M.G. and Palade, G.E. (1963) Junctional complexes in various epithelia. J. Cell Biol. 17, 375–412.

Fawcett, D.W. (1961) Intercellular bridges. Exp. Cell Res. Suppl. 8, 174–187.

Fisher, R.B. (1955) The absorption of water and some small solute molecules from the isolated small intestine of the rat. J. Physiol. (Lond.) 130, 855–864.

Forster, R.P. and Taggart, J.V. (1950) Use of isolated renal tubules for the examination of metabolic processes associated with active cellular transport. J. Cell. Comp. Physiol. 36, 251–270.

Frizzell, R.A. and Schultz, S.G. (1972) Ionic conductances of extracellular shunt pathway in rabbit ileum. J. Gen. Physiol. 59, 318–348.

Frömter, E. (1972) The route of passive ion movement through the epithelium of Necturus gallbladder. J. Membrane Biol. 8, 259–301.

Frömter, E. and Diamond, J.M. (1972) Route of passive ion permeation in epithelia. Nature New Biol. 235, 9–13.

Frömter, E., Rumrich, G. and Ullrich, K.J. (1973) Phenomenological description of Na$^+$, Cl$^-$ and HCO$_3^-$ absorption from proximal tubules of the rat kidney. Pflügers Arch. 343, 189–220.

Galeotti, G. (1904) Über die elektromotorischen Kräfte, welche an der Oberfläche thierischer Membranen bei der Berührung mit verschiedenen Elektrolyten zustande kommen. Zschr. phys. Chem. Stöchiom. Verwandtschaftslehre 49, 542–562.

Goldschmidt, S. and Dayton, A.B. (1919) Studies on the mechanism of absorption from the intestine. II. The colon. Am. J. Physiol. 48, 433–439.

Guggino, W.B., Windhager, E.E., Boulpaep, E.L. and Giebisch, G. (1982) Cellular and paracellular resistances of the Necturus proximal tubule. J. Membrane Biol. 67, 143–154.

Gumilewski, D. (1886) Über Resorption im Dünndarm. Pflügers Arch. 39, 556–592.

Hamill, O.P., Marty, A., Neher, E., Sakmann, B. and Sigworth, F.J. (1981) Improved patch-clamp techniques for high-resolution current recording from cells and cell-free membrane patches. Pflügers Arch. 391, 85–100.

Handler, J.S. (1983) Use of cultured epithelia to study transport and its regulation. J. Exp. Biol. 106, 55–69.

Hediger, M.A., Ikeda, T., Coady, M., Gundersen, C.B. and Wright, E.M. (1987) Expression of size-selected mRNA encoding the intestinal Na/glucose cotransporter in Xenopus laevis oocytes. Proc. Natl. Acad. Sci. USA 84, 2634–2637.

Hediger, M.A., Turk, E., Pajor, A.M. and Wright, E.M. (1989) Molecular genetics of the human Na$^+$/glucose cotransporter. Klin. Wochenschr. 67, 843–846.

Heidenhain, M. (1907) Plasma und Zelle. In Handbuch der Anatomie des Menschen, Vol. 8, p. 51. Jena, Gustav Fischer.

Heidenhain, R. (1883) Physiologie der Absonderung und Aufsaugung. I. Physiologie der Absonderungsvorgänge. In Handbuch der Physiologie, Vol. 5. (Hermann, L., ed.), p. 11. Leipzig, F.C.W. Vogel.

Heidenhain, R. (1888) Beiträge zur Histologie und Physiologie der Dünndarmschleimhaut. Pflügers Arch. 43 (Suppl.), 1–103.

Heidenhain, R. (1894) Neue Versuche über die Aufsaugung im Dünndarm. Pflügers Arch. 56, 579–631.

Heidrich, H.-G., Kinne, R., Kinne-Saffran, E. and Hannig, K. (1972) The polarity of the proximal tubule cell in rat kidney: Different surface charges for the brush-border microvilli and plasma membranes from the basal infoldings. J. Cell. Biol. 54, 232–245.

Henle, J. (1837) *Symbolae ad anatomiam villorum intestinalium, imprimis eorum epithelii et vasorum lacteorum.* Habilitationsschrift, Berlin.

Henle, J. (1838) Über die Ausbreitung des Epitheliums im menschlichen Körper. Müllers Arch. Anat. Physiol. wiss. Med., pp. 103–128.

Henle, J. (1841) Allgemeine Anatomie. In *Lehre von den Mischungs- und Formbestandteilen des menschlichen Körpers* (Voss, L., ed.), Leipzig.

Hermann, L. (1883) *Handbuch der Physiologie, Vol. V.* Leipzig, F.C.W. Vogel.

Hermann, L. (1882) Neue Untersuchungen über Hautströme. Pflügers Arch. 27, 280–290.

Hinke, J.A.M. (1961) The measurement of sodium and potassium activities in the squid axon by means of cation-selective glass microelectrodes. J. Physiol. (Lond.) 156, 314–335.

Hinke, J.A.M. (1970) Solvent water for electrolytes in the muscle fiber of the giant barnacle. J. Gen. Physiol. 56, 521–541.

Höber, R. (1901) Über Resorption im Darm. Dritte Mitteilung. Pflügers Arch. 86, 199–214.

Hopfer, U., Nelson, K., Perrotto, J. and Isselbacher, K.J. (1973) Glucose transport in isolated brush border membrane from rat small intestine. J. Biol. Chem. 248, 25–32.

Hoppe-Seyler, F. (1881) *Physiologische Chemie, Part II, Vol. 1, p. 349.* Berlin, A. Hirschwald.

Hudson, R.L. and Schultz, S.G. (1984) Sodium-coupled sugar transport: Effects on intracellular sodium activities and sodium-pump activity. Science 224, 1237–1239.

Ikeda, T.S., Hwang, E.-S., Coady, M.J., Hirayama, B.A., Hediger, M.A. and Wright, E.M. (1989) Characterization of Na^+/glucose cotransporter cloned from rabbit small intestine. J. Membrane Biol. 110, 87–95.

Ingraham, R.C. and Visscher, M.B. (1938) Further studies on intestinal absorption with performance of osmotic work. Am. J. Physiol. 121, 771–785.

Jennings, M.A., Marchesi, V.T. and Florey, H. (1962) The transport of particles across the walls of small blood vessels. Proc. R. Soc. B 156, 14–19.

Karnaky Jr., K.J., Kinter, L.B., Kinter, W.B. and Stirling, C.E. (1976) Teleost chloride cell. II. Autoradiographic localization of gill Na,K-ATPase in killifish *Fundulus heteroclitus* adapted to low and high salinity environments. J. Cell Biol. 70, 157–177.

Kedem, O. and Katchalsky, A. (1963a) Permeability of composite membranes. I. Electric current, volume flow of solute through membranes. Trans. Faraday Soc. 59, 1918–1930.

Kedem, O. and Katchalsky, A. (1963b) Permeability of composite membranes. II. Parallel elements. Trans. Faraday Soc. 59, 1931–1940.

Kedem, O. and Katchalsky, A. (1963c) Permeability of composite membranes. III. Series array of elements. Trans. Faraday Soc. 59, 1941–1953.

Kinne, R. (1976) Properties of the glucose transport system in the renal brush border membrane. Curr. Top. Membr. Transp. 8, 209–287.

Kinne, R.K.H. (1988) Sodium cotransport systems in epithelial secretion. Comp. Biochem. Physiol. 90A, 721–726.

Kinne, R.K.H. (1991) Selectivity and direction: plasma membranes in renal transport. Am. J. Physiol. 260, F153–F162.

Kinne, R. and Kinne-Saffran E. (1978) Differentiation of cell faces in epithelia. In *Molecular Specialization and Symmetry in Membrane Function* (Solomon, A.K., Karnovsky, M., eds.), pp. 272–315. Cambridge, Harvard University Press.

Kinne, R. and Kinne-Saffran, E. (1985) Renal metabolism: Coupling of luminal and antiluminal transport processes. In *The Kidney: Physiology and Pathophysiology* (Seldin, D.W., Giebisch, G., eds.), pp. 719–737. New York, Raven Press.

Kinne-Saffran, E. and Kinne, R.K.H. (1989) Membrane isolation: strategy, techniques, markers. Methods Enzymol. 172, 3–17.

Kleinzeller, A. (1967) From renal tissue slices to membrane vesicles. In *Renal Physiology: People and Ideas* (Gottschalk, C.W., Berliner, R.W. and Giebisch, G.H., eds.), pp. 131–163. Bethesda, American Physiological Society.

Kleinzeller, A. and McAvoy, E.M. (1972) Sugar transport across the peritubular face of renal cells of the flounder. J. Gen. Physiol. 62, 169–184.

Koefoed-Johnsen, V. and Ussing, H.H. (1958) The nature of the frog skin potential. Acta Physiol. Scand. 42, 298–308.

Koeppen, B.M. (1987) Electrophysiology of ion transport in renal tubule epithelia. Seminars Nephrol. 7, 37–47.

Lau, K.R., Hudson, R.L. and Schultz, S.G. (1984) Cell swelling increases a barium-inhibitable potassium conductance in the basolateral membrane of *Necturus* small intestine. Proc. Natl. Acad. Sci. USA 81, 3591–3594.

Leaf, A., Page, L.B. and Anderson, J. (1959) Respiration and active sodium transport of isolated toad bladder. J. Biol. Chem. 234, 1625–1629.

Lewis, S.A. and Wills, N.K. (1981) Interaction between apical and basolateral membranes during sodium transport across tight epithelia. Soc. Gen. Physiol. Series 36, 93–107.

Madara, J.L. (1989) Loosening tight junctions. Lessons from the intestine. J. Clin. Invest. 83, 1089–1094.

Mamelok, R.D., Macrae, D.R., Benet, L.Z. and Prusiner, S.B. (1981) Membrane population of bovine choroid plexus: Separation by density gradient centrifugation in modified colloidal silica. J. Neurochem. 37, 768–774.

Maruyama, Y. and Petersen, O.H. (1982) Single-channel current in isolated patches of plasma membrane from basal surface of pancreatic acini. Nature 299, 159–161.

Miller, F.L. (1960) Hemoglobin absorption by the cells of the proximal convoluted tubule of the mouse kidney. J. Biophys. Biochem. Cytol. 8, 689–718.

Mishina, M., Kurosaki, T., Tobimatsu, T., Morimoto, Y., Noda, M., Yamamoto, T., Tero, M., Lindstrom, J., Takahashi, T., Kuno, M. and Numa, S. (1984) Expression of functional acetylcholine receptor from cloned cDNAs. Nature (Lond.) 307, 604–608.

Molitoris, B.A., Chan, L.K., Shapiro, J.I., Conger, J.D. and Falk, S.A. (1989) Loss of epithelial polarity: A novel hypothesis for reduced proximal tubule Na^+ transport following ischemic injury. J. Membrane Biol. 107, 119–127.

Moore, D.H. and Ruska, H. (1957) The fine structure of capillaries and small arteries. J. Biophys. Biochem. Cytol. 3, 457–462.

Mullin, J.M., Fluk, L. and Kleinzeller, A. (1986) Basal-lateral transport and transcellular flux of methyl-α-D-glucoside across LLC-PK$_1$ renal epithelial cells. Biochim. Biophys. Acta 885, 233–239.

Neher, E. (1982) Unit conductance studies in biological membranes. In Techniques in Cellular Physiology Part II (Baker, P.F., ed.), pp. 1–16. Amsterdam, Elsevier.

Neher, E. and Sakmann, B. (1976) Single-channel currents recorded from membrane of denervated frog muscle fibers. Nature (Lond.) 260, 799–802.

Noda, M., Takahashi, H., Tanabe, T., Toyosato, M., Furutani, Y., Hirose, T., Asai, M., Inayama, S., Miyata, T. and Numa, S. (1982) Primary structure of α-subunit precursor of Torpedo californica acetylcholine receptor deduced from cDNA sequence. Nature (Lond.) 299, 793–797.

Oschman, J.L. (1978) Morphological correlates of transport. In Membrane Transport in Biology. Vol. III. Transport Across Multi-membrane Systems (Giebisch, G., Tosteson, D.C. and Ussing, H.H., eds.), pp. 55–93. Berlin/Heidelberg/New York, Springer.

Palmer, L.G. (1990) Epithelial Na channels: The nature of the conducting pore. Renal Physiol. Biochem. 13, 51–58.

Reid, E.W. (1892) Preliminary report on experiments upon intestinal absorption. Brit. J. Med. pp. 1133–1134.

Reid, E.W. (1900) On intestinal absorption of serum, peptone and glucose. Phil. Trans. R. Soc. London B 192, 211–297.

Sachs, J. (1865) Untersuchungen über die allgemeinen Lebensbedingungen der Pflanzen und die Funktionen ihrer Organe. In Handbuch der Experimental-Physiologie der Pflanzen, p. 178. Leipzig, W. Engelmann.

Sauer, F.A. (1973) Nonequilibrium thermodynamics of kidney tubule transport. In Handbook of Physiology, Section 8, Renal Physiology (Orloff, J. and Berliner, R.W., eds.), pp. 399–414, Washington, American Physiological Society.

Schmidt, U. and Dubach, U.C. (1971) Na-K-stimulated adenosine-triphosphatase: intracellular localization within the proximal tubule of the rat nephron. Pflügers Arch. 330, 265–270.

Schultz, S.G. (1977) Sodium-coupled solute transport by small intestine: a status report. Am. J. Physiol. 233, E249–E254.

Schultz, S.G. (1981) Homocellular regulatory mechanisms in sodium transporting epithelia: avoidance of extinction by 'flush-through'. Am. J. Physiol. 241, F579–F590.

Schultz, S.G. (1981) Salt and water absorption by mammalian small intestine. In *Physiology of the Gastrointestinal Tract* (Johnson, L.R., ed.), pp. 983–989. New York, Raven Press.

Schultz, S.G. and Zaluski, R. (1964) Ion transport in isolated rabbit ileum. II. The interaction between active sodion and active sugar transport. J. Gen. Physiol. 47, 1043–1059.

Simons, K. and Fuller, S.D. (1985) Cell surface polarity in epithelia. Annu. Rev. Cell Biol. 1, 295–340.

Simons, K. and Wandinger-Ness, A. (1990) Polarized sorting in epithelia. Cell 62, 207–210.

Stanton, B.A. (1984) Regulation of ion transport in epithelia: Role of membrane recruitment from cytoplasmic vesicles. Lab. Invest. 51, 255–257.

Stirling, C.E. (1972) Radioautography localization of sodium pump sites in rabbit intestine. J. Cell Biol. 53, 704–714.

Stirling, C.E. and Landay, B.R. (1970) High resolution radioautography of ouabain-^{3}H and phlorizin-^{3}H in in vitro kidney and intestine. Fed. Proc. 29, 595.

Ussing, H.H. (1949) The distinction by means of tracers between active transport and diffusion. The transfer of iodide across the isolated frog skin. Acta Physiol. Scand. 19, 43–56.

Ussing, H.H. and Windhager, E.E. (1964) Nature of shunt path and active sodium transport path through frog skin epithelium. Acta Physiol. Scand. 61, 484–504.

Ussing, H.H. and Leaf, A. (1978) Transport across multimembrane Systems. In *Membrane Transport in Biology. Vol. III. Transport Across Multi-Membrane Systems* (Giebisch, G., Tosteson, D.C. and Ussing, H.H., eds.), pp. 1–26. Berlin/Heidelberg/New York, Springer.

Ussing, H.H. and Zehran, K. (1951) Active transport of sodium as the source of electric current in the short-circuited isolated frog skin. Acta Physiol. Scand. 23, 110–136.

Visscher, M.B., Varco, R.H., Carr, C.W., Dean, R.B. and Erickson, D. (1944a) Sodium ion movement between the intestinal lumen and the blood. Am. J. Physiol. 141, 488–505.

Visscher, M.B., Fetcher, E.S., Carr, C.W., Gregor, H.P., Bushey, M.S. and Baker, D.E. (1944b) Isotopic tracer studies on the movement of water and ions between intestinal lumen and blood. Am. J. Physiol. 144, 550–575.

Weinstein, A.M. and Stephenson, J.L. (1981a) Models of coupled salt and water transport across leaky epithelia. J. Membrane Biol. 60, 1–20.

Weinstein, A.M. and Stephenson, J.L. (1981b) Coupled water transport in standing gradient models of the lateral intercellular space. Biophys. J. 35, 167–191.

Wissig, S.L. (1958) An electron microscope study of the permeability of capillaries in muscle. Anat. Rec. 130, 467–488.
Zehran, K. (1956) Oxygen consumption and active sodium transport in the isolated and short-circuited frog skin. Acta Physiol. Scand. 36, 300–318.

A. Kleinzeller (Editor) A History of Biochemistry: Exploring the Cell Membrane.
(Comprehensive Biochemistry Vol. 39) © 1995 Elsevier Science B.V.

Chapter 8

The Role of Membranes in Excitability

D.E. GOLDMAN

63, Loop Road, Falmouth, MA 02540, U.S.A.

This chapter deals with the structure and function of membranes insofar as they are essential participants in excitation processes. These processes are essential to living organisms, which spend most of their lives in self-maintenance, and reproduction. Their operations are carried out by activities which react to environmental changes, *i.e.*, respond to stimuli, and are superimposed on those activities which maintain steady states. When the amplitude and speed of the responses are large enough to be of immediate significance the system is said to be excitable.

I. Excitability

Present views of some of the underlying concepts of excitability are more or less as follows: any system (organization), excitable or not, is an arrangement of operating elements which carries out one or more functions. It may be characterized by its size, its effectiveness and its efficiency. Its elements may be simple or complex and need not be fixed in relation to each other. They may change from time to time and be controlled by external or internal factors or both. These elements are often called organs and the system an organism, especially if it shows signs of self-maintenance, biological or behavioral. The size is given by the number of irreducibly simple elements, regardless of how they may be combined or arranged. The effectiveness is determined by how well the functions are carried out and the efficiency is the ratio of the energy consumed to the energy supplied. If, in the presence of a suitable environment, it is also able to maintain itself as an independent entity and perform its inherent functions for a significant length of time it can be said to be alive. If it can also reproduce itself it can contribute to the maintenance and evolution of the species, if any, to which it belongs. Such organisms may have growth and decay phases, have finite life spans and may be susceptible to modification at any time.

The basic units of living organisms are cells and almost all of them are involved in excitability phenomena. The realization of what now seems obvious dates from the 1830's when Schleiden (1838) and Schwann (1839) developed the idea that was already being accepted for plant cells and that Purkinje (1837) suggested should be equally applicable to animal cells. This was followed by the observation by Remack (1852) that new tissue growth occurred only by division of preexisting cells which consist, as seems to have been noted by Schultze (1863) of a kind of slimy mixture called protoplasm (?cytoplasm). Cells may have one or more functions such as electrical or mechanical activity, the storage and protection of certain materials, the rearrangements of intracellular elements, as in mitosis, or the production, consumption, or both, of any of a very large number of substances. Their role in communication is of particular interest.

1. DEVELOPMENT OF THE CONCEPT

The awareness of active responses by living systems is, of course, much older than mankind. While primitive organisms may show no signs of awareness as human beings understand it, a mammal detecting the presence of a stalking predator can hardly be said to be unaware of it and may quickly develop an urge to flee or to counterattack. The earliest perceptions by human observers of excitation seem to have been primarily of subjective responses (feelings) and the occurrence of mechanical motion of part or all of the organism. It is interesting to note (see Thompson, 1974) that it was recorded by the ancient Greeks that contact with the electric fish *Torpedo* produced a painful shock and that this could be felt through a rod or spear. However, until the 17th century there appears to have been very little attention given to excitation phenomena; earlier views were exceedingly vague and were often based on belief in supernatural sources. The concept of irritability, a term later replaced by excitability[1], appears to have been introduced by Glisson (1677). von Haller (1750) distinguished irritability from sensitivity; incidentally he seems not to have believed that nerves were irritable. Reil (1796) pointed out that irritability is a specific property of living organisms. In a discussion of heart muscle, Fontana (1767) stated that

'The irritability of the fibre can be activated by a small cause and by a feeble impression but once activated it has a power proportional to its own forces, which can be much greater that the exciting cause'.

Also, according to Lucas (1910),

'the word excitation has by somewhat loose usage become applicable to all or any of the successive processes which constitute the connecting link between the application of a stimulus to a nerve or muscle and the appropriate final response'.

The definition employed in this chapter seems to be somewhat more useful. In accordance with the function of the system the appropriate stimulus may be simple or multiple. The response

---

1 The word excitation has usually referred to an increase in activity whereas the word depression has referred to a decrease. Nowadays the distinction, though real, is usually ignored and the context is relied on for clarity.

may have a threshold, be all-or-none, be single or repetitive, and may show adaptation. The system may act as an independent unit or as part of a chain or tree of sequential or parallel activities. In addition, any or all of these characteristics may be subject to modification by interaction with other systems. It may be structurally simple or fantastically complex. In fact, the reader must not be surprised at the frequent use of 'may', 'can', 'sometimes', etc. in this discussion. The thermodynamic and logical structural aspects of excitable systems in general have been discussed in considerable detail by Franck (1956).

Excitability is ubiquitous in the animal kingdom from the motility of protozoa to the chemical and metabolic responses produced by hormones in mammals, but seem to be rare in plants. However, examples of the latter which might come readily to mind are the action potential of the alga, *Nitella*, which has attracted considerable interest because of its electrical response (Osterhout, 1936; Blinks, 1936; Cole and Curtis, 1938; Kishimoto, 1964) and the mechanism of Venus Fly Trap, *Dionaea muscipula*, because of its special response to touch (Burdon-Sanderson, 1882, 1888; Jacobson, 1965). Further, excitability is not limited to living organisms. There is a wide variety of excitable inanimate systems now in use involving such devices as transistors, flush toilets, thermostats, industrial process control systems, internal combustion engines, electronic computers, lipid bilayers to which small amounts of special substances have been added, etc. Suppose, as a very simple example, that a voltage is applied to a circuit and produces an electric current. If the circuit contains a solenoid the magnetic field thus generated can be used to move an iron rod and so open a valve. This is a passive response, *i.e.*, a direct result of the energy supplied by the stimulus. However, if the opening of the valve permits the flow of water through a tube under the influence of a head of pressure, an active response will have occurred since the energy which drives the water flow will have come from a source other than that of the original stimulus. In this particular case the application of the voltage has served as a trigger, its primary function being to initiate an activity. The amplitude and speed of the response may or may not be related to the amplitude or speed of the stimulus. If in addition the pressure has been reduced by the flow and is to be restored following the cessation of the current, another active process must be present and if, still further, the pressure is maintained in the face of leakage losses there must also be a process constituting a trickle charger, in this case of the hydraulic variety, operating either continuously or intermittently. These energy sources need not all be separate nor need they be related in any way. A particularly interesting historical example is the iron wire in nitric acid. In the late 19th century it was noticed that certain metals could be made chemically passive by the presence of a stable surface layer formed by immersing the metal in an acid solution and that an appropriate stimulus could activate it temporarily (Ostwald, 1890). A study by Lillie (1918) of the iron wire in nitric acid showed that when the layer was destroyed at some point by mechanical or chemical means, the destruction spread rapidly along the wire in both directions. As the disturbance spread, the surface layer was rapidly reformed. The obvious analogy with nerve propagation led to a number of further studies by both physical chemists and biologists which were abandoned by the latter only after considerable advances in the understanding of nerve excitation had occurred.

2. THE ROLE OF THE MEMBRANE

Cell membranes act as material separators of two regions. They are extensively discussed in Chapter 2. They are distensible, are very thin relative to their extent and may have a very complex structure. They exert a powerful influence on the transfer of materials and energy from one region to the other and may contain substances which can interact with the environment and react to stimuli. They do not, themselves, ordinarily contain stores of energy sufficient to generate active responses but may control local material and energy exchange using such channels, enzymes and transport proteins as may be present. They are found everywhere in living systems and are of special importance in encompassing and supporting cells as well as intracellular elements such as mitochondria, nuclei or enzyme complexes, also certain cell composites which, themselves, are often treated as organs. They are subject to replacement and modification of their molecular elements for maintenance, repair, growth, reproduction, etc. Their role in excitation is primarily the control of the passage of substances in and out of the cell and their consequent ability to support stores of energy and materials, often in the form of concentration differences across the membrane which are provided by metabolic processes, or the making available to the cell interior substances which can release energy through their enzymatic activity. All this is accomplished in two ways. First, they consist basically of a lipid bilayer which blocks the passage of most non-lipid materials. The bilayer is interspersed with protein and lipoprotein structures which act as channels for the passage of a considerable number of substances including not only many small ions but also complex molecules. Second, they also contain molecular structures which act as receptors to appropriate chemical stimuli and whose participation is necessary for the penetration of selected substances through channels or for facilitation of exo- and endocytosis. These complexes are usually arranged in some orderly fashion, sometimes in clusters, e.g., for voltage gated ion channels in myelinated nerve or for enzymatic activities (Almers and Stirling, 1984). Surface layers, both external and internal, of specially constructed materials may provide protective agents or selective barriers and links in a chain of excitatory activity. An excitable membrane may then be visualized as one whose permeability characteristics are sensitive to appropriate applied forces which trigger active responses though not in the membrane itself. There are instances in which the energy of the stimulus can itself generate the output but these are passive systems and are apt to be found only in very small or unicellular organisms. Clearly the term 'excitable membrane' is something of a misnomer. It is the system which is excitable even though the biological burden borne by the membrane is a heavy one.

To rephrase: cells, like all objects, respond passively to the application of external forces but may also be excitable. Certain stimuli may be adequate to set off active responses for which separate, usually intracellular, energy supplies are provided. Cell excitability thus depends primarily on stimulus-induced changes in membrane characteristics, often permeability to selected substances, and on the presence of an energy source, often a concentration difference, for driving the cell's response. Alternatively the membrane's response might be the activation of some enzyme or enzymatic chain which could then liberate energy by catalyzing reactions involving stored substances. A backup source of energy is also required for maintainance. For example, the application of a suitable voltage change to a nerve or muscle fiber produces a gating current which is a passive response and which includes a rearrangement of electric charges within the membrane and thus results in a sequence of changes in its permeability to certain ions. This permits, as an active response, a current flow driven by the energy of ion concentration differences across the membrane as well as by that of any applied voltage which may be present. It may also result in the release of a transmitter molecule. A chain of operations then occurs including a sequence of permeability changes and an eventual return to the inactive state, possibly only after the cessation of the stimulus. Treatment with a variety of chemical agents can, of course, have a powerful influence on the response. However, neither the external stimulus nor the driving energy need be electrical. Electromagnetic radiation, a magnetic field, a mechanical distortion, a change in temperature, the activation of some chemical reaction or the application of certain compounds, all may produce important changes in the configurations of molecular elements, whether charged or not. Chemical substances, indigenous or applied, by their interaction with membrane or other cellular elements may modify certain permeabilities, strengthen or weaken some responses, generate new ones or abolish preexisting ones. Membranes involved in active responses include those of axons, dendrites, muscle fibers, secretory cells and a wide variety of others; the role of membranes in sensory systems is particularly important when they preside over the transduction of energy forms.

II. Excitation processes

Developments in science, both experimetal and conceptual, often pass from one area to another with considerable interaction and feedback so it is not surprising to find that biology has depended heavily on the physical sciences or that important advances in the physical sciences have often developed from observations on living systems. In addition, observations and experiments made on a single system are very soon apt to be extended to other systems and thus lead to important generalizations. While excitable systems can be extremely complicated, they often consist of relatively simple interacting subsystems although few have been studied to the extent of locating and identifying all their molecular elements and specifying their precise methods of operation. Developments in the understanding of excitable systems have occurred only fairly recently especially in problems involving the complex, nonlinear processes so often encountered by the biologist. The study of the ways in which such systems operate has become a very large field. Subjects of interest include excitatory agents, membrane elements on which they act, mechanisms of excitation, energy sources, the resulting intracellular processes and their influence on the behavior of the cell, interactions with other cells and the role of all these in the functioning of the organism. These functions include sensation, mechanical or electrical activity, the generation of heat and the manufacture, secretion, absorption or destruction of a wide variety of substances. Membranes are integral elements in all these activities and occur in a fantastic variety. In this discussion examples are given of some of the simpler systems. They have been selected primarily to illustrate the wide variety of types and mechanisms used by living organisms. Emphasis has been placed on the excitation processes leading to the responses rather than on the responses themselves since the latter are generated by the former and have a separate structure.

1. EXCITATION BY PHYSICAL STIMULI

1.1. Electrical excitation

The study of electrical excitation in nerve is an excellent example partly because the basic system has only one passive and one active element and partly because of the precision and ease with which electrical variables can be manipulated. Galvani (1791) noticed that when a frictional electric machine was operating, nearby freshly prepared frog's legs twitched as if alive and that if they were suspended with brass hooks from an iron balcony railing a contraction could be produced. These observations and a subsequent dispute with Volta (1800) as to their interpretation led to the creation by the latter of the voltaic pile and so to its modern successor, the galvanic cell. Together with Faraday's law (1833), this provided the basis for the development of electrochemistry. Müller (1838) had emphasized that functional activity in living organisms was a physicochemical phenomenon and Matteucci in the same year reported that there was an electric current flow between the cut surface of a nerve and its undamaged surface. Following a

TABLE I
Development of ideas of nerve excitation

Year	Biology	Physical Chemistry	Physics
1800	Galvani		
		Volta	
		Faraday	
	Matteucci		
	Müller		
	du Bois-Reymond		Ohm
1850	Hermann	Fick	Kelvin
	Bernstein	Kohlrausch	
		Helmholtz	
		Arrhenius	
		Nernst	
		Planck	
1900	Bernstein		Einstein
	Overton	Gouy	
	Hill		
	Donnan		
	Fricke		
	Cole		Mott
	Blink		
		Goldman	
	Hodgkin, Katz	Cole	
1950	Hodgkin, Huxley		
	Neher, Sakmann		
	Noda		
2000			

suggestion of Müller, du Bois-Reymond (1848), who worked
from the 1840s onwards, became deeply involved in studies on
nerve and muscle. Among his findings was that the application of a voltage of proper polarity, or its release, to a nerve

could produce a contraction of the muscle to which the nerve was attached and also that the muscle contracted when the nerve was appropriately subjected to a sudden stretch. Later, he offered a visualization of the nerve as having a layer of dipolar elements ('electrical molecules') under its surface with their poles lined up in opposing pairs. In the electrotonic state these became lined up with the dipoles all pointing in the same direction. A significant advance was made when Hermann in 1872 (see Hermann, 1899) proposed that

> 'polarization of the interface between the envelope and the core substance of the nerve fiber from which the electrotonus is derived plays a possible part in the progagation of excitation. Namely, an excited area of the core behaves negatively with respect to its unexcited neighborhood as a result of a current, which is equalized through the envelope material, but has an anelectrotonic (inhibitory) effect on the excited area and a catelectrotonic (excitatory) effect on the unexcited neighborhood'.

Parallel to these developments, there was steady experimental work along with progress in the physical chemistry and thermodynamics of electrolytic solutions. Ohm (1843) found a linear relation between electric current and its driving voltage. Fick (1855) discovered the quantitative relation between the rate of diffusion of a solute and the concentration gradient which drives it. Kohlrausch (1876), (see also Kohlrausch and Holborn, 1916). Arrhenius (1887), Nernst (1889), Gibbs (1906) and others were making considerable progress in the understanding of ionic solutions much of which was immediately applied to such electrochemical properties of membranes as were related to their permeability to ions, e.g., the Donnan (1911) effect on equilibrium potentials (see also Overbeek, 1956). Extensive studies on membrane permeability in the electrical sense became quite popular after developments in electrical techniques became available following the first World War.

In 1890, Fick's law of diffusion was expanded to become the Nernst-Planck equation, now recognized as governing the flow of ions through structureless media. It consists of three ele-

ments: the current-voltage relation of Ohm (1827), the diffusion equation of Fick (1855), and the Einstein (1906) relation
between the mobility and the diffusion coefficient of an ion.
Planck (1890a,b) obtained a solution for the steady state potential of a liquid junction but since the equation is non-linear
and the parameters of the system are not well known it was
necessary to make a number of assumptions among which is
that of microscopic electroneutrality. A modified solution introduced by Henderson (1907) has been widely used in electrochemistry. Unfortunately, while the liquid junction is essentially a resistance, a cell membrane is more nearly a leaky
condenser so that the Planck approach is not suitable. However the membrane is very thin and so Goldman (1943) was
able to extend Mott's (1939) treatment of the Cu-CuO rectifier
to the steady state potential and transport of ions. Also the
charge accumulation at the boundaries due to the potential
difference makes it very unlikely that microscopic electroneutrality can occur and the concept of the electrical double layer
(Helmholtz, 1879; Gouy, 1910) is therefore applicable. It was
thus possible to obtain a simple solution for the resting cell
membrane potential. Further development of electrical techniques and the advent of electronics made possible the direct
measurement of membrane resistance, (see Blinks, 1930) on
plant cells, and of membrane capacitance, first made by Fricke
(1923) on cell suspensions and thus to the realization that the
ion permeability changes produced by excitation could be accurately measured in electrical terms. All this, while not deal-
ing directly with the excitation process, provided a basis for
detailed analysis of the electrical properties of nerves and a
variety of other cells. These studies have been thoroughly discussed by Cole (1968).

   In the early 1900s, Bernstein (1902) proposed, on the basis
of Nernst's (1889) formulation of concentration cell potentials,
that the interior of nerve and muscle cells consisted of an electrolytic solution with a surface layer selectively permeable to
potassium and that the action potential was the result of a

temporary loss of that semipermeability brought about by an electrical stimulus. This appears to have been the first explicit expression of the idea of an 'excitable membrane'. Nernst's (1908) model treated the membrane as a layer of ionic solution with a pair of semipermeable boundaries. He was able to solve the diffusion equation for the system and thus offer explanations for some of the passive response characteristics of the nerve membrane as related to excitability. The treatment was extended by Hill (1910) and provided a considerable improvement as it could also explain phenomena generated by constant, alternating, logarithmic and progressive currents. Lapique (1909) made use of the observed intensity-time curves for electrical stimulation to develop his concept of chronaxie, defined as the duration of a constant current necessary for excitation when its strength is twice the rheobase. This concept, because of its apparent simplicity and broad applicability, became very popular within the medical profession and remained so for a long time. Later on, Blair (1932), Rashevsky (1933), Monnier (1934) and Hill (1936), on the basis of much data which had accumulated in the previous decades, developed the idea that the initiation of nerve excitation could be visualized as involving a two element linear process in which one excited and the other inhibited. Rushton (1937) and Young (1937) extended that formulation to a more general pair of linear differential equations. The result was a helpful reduction in the apparent complexity of the electrical responses although there remained a large number of observations which could not be accounted for particularly in the non-linear behavior of the system. None of these proposals were able to proceed beyond the observation that excitation occurred when the ion concentration changes reached some critical, not understood, value. An excellent thorough discussion of these problems has been given by Katz (1939).

Young (1936) brought the attention of electrophysiologists to the squid giant axon, which rapidly became the preparation with the use of which the excitable behavior was to be unraveled primarily because of its simple geometrical properties (up to as much as 1 mm in diameter, several cm in length with a more or less uniform cylindrical cross section). This permitted the use of internal electrodes with which it was possible to obtain direct measurements of both the membrane potential (Cole and Hodgkin, 1939) and, later, to determine the permeability sequences during the action potential using the space clamp of Marmont (1949) and the voltage clamp procedure of Cole (1949) who, with his collaborators, had earlier determined many of the electrical impedance characteristics of axonal and a number of other membranes, *i.e.*, resistance and capacitance along with negative resistance and inductance (see Cole, 1941) as related to excitation and oscillatory behavior. Also, Bernstein had long ago (1868) found that the potential across the membrane during activity fell not merely to zero but overshot it by a considerable amount. This finding was, however, largely ignored until Hodgkin and Huxley (1939) obtained it with the squid giant axon and Curtis and Cole (1942) confirmed it. In addition, Overton (1902) had pointed out that there must be an exchange between potassium and sodium during nerve excitation and the contraction of muscle[2]. Hodgkin and Katz (1949) showed that the action potential of nerve was primarily due to a sudden, large, temporary increase in membrane permeability to sodium. Hodgkin and Huxley then carried out a series of experiments on the squid giant axon using the voltage clamp procedure. This resulted (in 1952) in a demonstration that the values and time dependence of membrane ionic conductances, particularly of sodium and potassium, with the former chiefly responsible for excitation and the latter for recovery, were functions primar-

---

2 In 1986 (Kleinzeller, personal communication) Andrew Huxley remarked in conversation that 'If people had listened to what Overton had to say, the work of Alan (Hodgkin) and myself would have been obsolete.' This is a tribute to Overton but seems a bit modest in view of what they accomplished in both their thinking and their experimental work.

ily of the membrane potential and that their values could be represented by usable empirical equations. With this data and formulation they were able to reproduce the action potential of the squid giant axon by calculation to a highly satisfactory degree. The formulation was also able to account very well for the propagation of the action potential by extending the earlier ideas of Hermann (see Hermann, 1899) and Cremer (1900) with the axon behaving very much like the electrical cable of Kelvin but which also, as was later suggested by Cole (1968), had a built-in repeater. Also, Tasaki (1948) had demonstrated that two action potentials proceeding in opposite directions along an axon annihilated each other when they collided, thus showing that an axon is certainly not a simple, linear cable and that impulse propagation requires that the system be reset before another impulse can appear. While all this did not determine the underlying molecular mechanisms, it established the essential part played by the membrane conductances and thus provided a solid basis for further analytical studies of excitation and propagation. This led to the application of the voltage clamp to many other preparations such as myelinated nerve fibers, neuron cell bodies, muscle fibers, electroplaques, algal cells, etc. and to the use of other experimental conditions. The effects of varying the environmental concentrations, both internal and external, of sodium and potassium, and of calcium (see Frankenhaeuser and Hodgkin, 1957), etc., as well as the role of ion flow blockers and modifiers such as tetrodotoxin (Narahashi et al., 1964), and tetraethylammonium (Draper and Weideman, 1951) could now be interpreted in terms of detailed phenomenological characteristics of the membrane as could also the action of many narcotics and heavy metals for which an extensive literature already existed reaching back well into the 19th century (see Höber, 1945). However, the molecular basis of the action potential was not worked out until a number of years later following the development of the patch clamp (Neher and Sakmann, 1976) with which single ion channels could be isolated and of the

application of molecular biological techniques to the characterization of the structures of the voltage-gated membrane
channels which are responsible for the passage of sodium and
potassium (Noda et al., 1984). The precise process by which
the ions pass through the membrane is, however, still under
study at the present time (see Andersen and Kouppe, 1992).

A few words with respect to the backup maintenance mechanisms, *i.e.*, the so-called sodium and potassium pumps (*cf.*
Chapter 4). Their existence had been postulated by Steinbach
(1940) and Dean (1941) who pointed out that there had to be
some mechanism for replacing the sodium and potassium
transferred during muscular activity. The role of adenosine
triphosphate (ATP) as an energy source was worked out by
Hodgkin and Keynes (1955) followed by Skou (1957) who
showed that a sodium- and potassium-activated ATP hy-
drolyzing system fulfilled all the requirements for a transport
mechanism. It was later found (Caldwell et al., 1960) that certain other phosphagens, *e.g.*, arginine phosphate, could also be
active in the pumping process. Some of the responsible macro-
molecular complexes have now been isolated and characterized (Shull et al., 1985; Stahl, 1986).

1.2. Electrical cell-cell communication

Synapses are of two types: electrical and chemical. In this section only the former is treated. Chemical synapses, which are
much more common, will be discussed later. The electrical
synapse represents a somewhat more complex system than a
single cell in that it involves two voltage sensitive membranes
connected by specialized channels to form what is known as a
nexus or gap junction (Barr et al., 1965). A very simple, artificial, example is that of two squid giant axons in contact (an
ephapse) with electrical activity in one axon resulting in the
excitation of the other (Arvanitaki, 1942). The gap itself appears to contain structures which, acting as channels, permit

the exchange of some ions and small molecules between the cells (Loewenstein, 1967). As might then be expected, the ion transfer can be strongly affected by such conditions as the presence of hypertonic glucose or of an electrical resistor in parallel. In some cases current flow can occur in both directions, in others, not (Furshpan and Potter, 1959). The system thus has electrical transport characteristics rather similar to those of nerve and muscle fiber membranes.

Hopefully, the above examples should not only help to clarify the kinds of roles played by membranes in electrical excitability but provide a demonstration of some of the interactions of biology with physical science and technology. They may also help illustrate the recent rapid increase in the rate of formulation and interaction of new concepts and their experimental concomitants. The understanding of electrical excitability in terms of physical chemistry and the molecular structure of the membrane may thus be considered to be well advanced at the present time though hardly complete.

1.3. Mechanical excitation

Most living organisms are known to contain mechanosensitive elements whose outputs may be electrical, mechanical, electromagnetic or chemical.

1.3.1. Nerve

As in the case of electrical excitation the nerve appears to have been the first preparation to be actively studied since it is the simplest to deal with. Incidentally, the reader may remember that a blow to the ulnar nerve at the elbow produces not only pain at the site of the blow but also an odd sensation in the forearm and part of the hand. Tigerstedt (1880), following earlier observations of du Bois-Reymond, published an extensive study of the mechanical stimulation of nerve demonstrating that careful lateral blows to the nerve produces contractions of the attached muscle. This was later taken up by Blair (1936) using an air jet as a stimulus and then by Schmitz and Wiebe (1938) using an electromagnetic drive. Blair found a strength-duration curve very much like that due to electrical stimulation. Rosenblueth et al. (1953) observed that subthreshold electrical and mechanical pulses could be added in either order to produce an action potential. However, these findings were made on amphibian nerves with which only limited quantitative data could be obtained. Julian and Goldman (1962) were able to get more precision using the giant axon of the lobster, e.g., it no longer seemed necessary to destroy a fiber in order to generate an electrical response. Using a crystal-driven stylus as a stimulator they confirmed what had been previously found about mechanical stimulation. They also showed that the rate of increase of transverse mechanical distortion had a strong influence on the buildup of depolarization, that the increase in membrane conductance behaved in the same way with both electrical and mechanical stimulation and that the conductance increase occurred even when depolarization was prevented electrically. However when sodium was removed from the external medium compression produced almost no depolarization. More recently, the giant axon of the squid has been subjected to rapid stretches over a wide range of stretch rates (Wells and Goldman, unpublished). The results showed a remarkable similarity to what has been obtained both with electrical stimulation and mechanical compression although the depolarizing response was found not to be affected by the application of tetrodotoxin as it is with electrical stimulation, thereby suggesting that the mechanosensitive transfer element was not the same as the electrosensitive one. This turned out to be consistent with the above mentioned findings of Julian and Goldman and of Albuquerque et al. (1969) that tetrodotoxin blocks conducted activity in the isolated muscle spindle but not the receptor potential. The parameters of the axon, treated as a mechanical transmission line or cable, were then determined and calculations of the action potential were carried out using the Hodgkin-Huxley equations, modified by Adelman and Fitzhugh (1975) to include the effects of potassium accumulation, with the addition of a term which expressed a linear dependence of ion conductance on the stretch. In this way it was possible to show the effect of both stretch and stretch rate on the action potential and to duplicate quite well the observed action potentials with their development and recovery.

1.3.2. Mechanical sensory mechanisms

Some important mechanical and acoustical sensory systems, such as the auditory and vestibular organs of mammals and the lateral line organs of fish, appear to operate using mechanisms closely related to those of nerve fibers (de Vries et al. 1952; Gray 1959; Davis, 1961). The extensive work of von Bekesy (1960) on the role of the cochlea and the hair cells in hearing and recent studies by Hudspeth (1983) and his collaborators on the mechanism of operation of the hair cells have provided strong evidence that their bending is the primary stimulus analogous to the stretching or compression of an axon. Also, certain insect hairs have been shown to generate electrical responses to mechanical deformation (Wolbarsht, 1960). An extensive discussion of sensory reception as of 1965 can be found in the 1965 Cold Spring Harbor Symposia on Quantitative Biology. More recently the patch clamp has been used to demonstrate the presence of mechanosensitive ion channels in single fibers of tissue cultured, embryonic chick skeletal muscle (Guharay and Sachs, 1984). Mechanoreceptors have also been found in many other organisms from bacteria to mammals. Several types have been found including stretch inactivated as well as stretch activated. It has also been proposed that there may be some which are shear stress activated (Olesen et al., 1988) and that there are cases in which the cytoskeleton may be involved and possibly the extracellular matrix as well. Among these channels some are permeable to one or more ions such as sodium, potassium, calcium and chloride, some have directional properties and some are voltage sensitive as well. Up until now there appear to have been no molecular biological studies providing information as to their structure nor has the precise mechanism of their action yet been worked out. Much of this recent work has been summarized by Morris (1990) and Sachs (1991).

1.4. Visual excitation

Visual systems, of which there are a great variety, have been subjected to intensive study for a long time. They are exceedingly complex and involve a large number of steps, some of which have been analyzed, though only in part. An essential role played by membranes, in addition to their usual actions in the cellular chains of visual systems, appears to be providing support for photosensitive pigments. Boll (1876) seems to have provided the first demonstration that the retina contains a photosensitive pigment. An early study of retinal rods and visual purple made by Edridge-Green (1902) showed that bleaching of the retina occurred on exposure to light[3]. More recently, Sjöstrand (1949, 1953) has demonstrated the position of photosensitive pigments with respect to membranes in the rods and cones of a vertebrate eye. There is now considerable evidence that the photoreceptor cells of vertebrates at least support, by means of pump systems and by a separation of position of the sodium and potassium channels of the membrane, a steady flow (dark current) of sodium and potassium through it (see Korenbrot and Cohn, 1972). This is reduced by the indirect biochemical response to the incidence of light on rhodopsin to produce hyperpolarization through the following com-

---

3 Please note that this work was done with a government grant of 10 pounds Sterling!

plex sequence: photoisomerization of rhodopsin, activation of the G-protein, transducin, activation of a phosphodiesterase which then hydrolyzes cGMP which in turn closes sodium-specific channels and thus produces hyperpolarization of the cell (see Stryer, 1991).

1.5. Other types of physical receptor mechanisms

In electroreceptors, baroreceptors, thermoreceptors, etc., membranes also appear to act as key controllers of excitation processes. It will be evident that since receptor system stimuli act, although in many different ways, to control electric current flow through membrane channels which in turn produce electrical responses in axons, these latter may be visualized as a special kind of electroreceptor. How closely baroreceptors may be related to the commonly encountered mechanoreceptors remains to be determined (see Hajduczok et al., 1990). The exact roles played by all these membranes do not appear to have been well worked out at the present time, nor are the mechanisms of pain receptors at all understood.

2. THE ROLE OF CHEMICAL SUBSTANCES IN EXCITATORY PROCESSES

Living organisms are composed entirely of chemical substances, many of which play essential roles in excitatory processes[4]. These substances serve as stimuli, as membrane receptors, as channels or as guardians (activators or blockers) of channels, as carriers, as sources of energy for responses, as intermediates (second messengers), as enzymes, as modifiers of various reactions, as end products of an excitatory process,

---

4 The term chemical does not necessarily refer to synthetic materials!

etc. Their actions tend to be highly specific and are generally
slower than purely physical processes such as electrical, mechanical or electromagnetic. Their specificity provides an important biological advantage in that similar processes can, if
necessary, be kept separate and their form adjusted to the
requirements of the function to be carried out. A very simple
example of chemical stimulation (an electrochemical stimulus
in fact) is the application of potassium ions to the environment
of nerve fibers, thus producing depolarization. This, however,
is trivial compared to the absorption of, say, foodstuffs applied
to the mucosa of parts of the gastrointestinal tract or the widespread action of certain hormones, to say nothing of the intri-
cacies of reproductive processes. The detailed study of the participation of membranes in these phenomena is relatively
recent since it has largely depended on highly developed physical and organic chemical techniques and the use of the latest
procedures of molecular biology such as recombinant DNA
manipulation. The cloning of DNA seems to have been first
used by Cohen et al. (1973) in *E. coli* and by Morrow et al.
(1974) in eukaryocytes. DNA sequencing has been developed
by Maxam and Gilbert (1977) using base specific chemical
cleavage and by Sanger et al. (1977) with chain terminating
inhibitors. Considerable progress has been made in identifying and characterizing important molecules, notably membrane transport G proteins and ion channels as well as the
activities to which they belong but, as has already been mentioned, almost all excitable systems are not only exceedingly
intricate but also interact with each other in ways which are
still far from being understood[5].

---

5 Current activity in this field is now so rapid that parts of this chapter may be
obsolete by the time this volume goes to press.

2.1. Chemoreceptors

These are widely distributed among living organisms from the most primitive to the most complex. They serve as attractors, as repellants and as contributors to the information pool on which the organism acts (*cf.* Chapter 5). The role played by membranes seems quite varied but little is known in most cases about their specific structure or precise function.

   The mechanisms of taste and smell have been studied for a long time with great interest but until very recently the studies have necessarily been confined to anatomy and phenomenology. Access to the molecular mechanisms of reception and transduction has now become possible and the sequence of participants from receptor molecules (*e.g.*, glycoproteins gp^{95} and gp^{56}) to ion channels appears to have been worked out. Attention has also been directed to the apparent homology of the sequences of smell, taste, vision and synaptic transmitter receptors (see Lancet et al., 1988). Mammalian carotid and aortic bodies, which are concerned with the control of respiration, act as detectors of blood PO_2, PCO_2 and pH. De Castro (1926) seeems to have been the first to call attention to the possibility that this might be the case and Heymans et al. (1930) then worked out the physiological behavior of the carotid body.

2.2. Chemical synaptic transmission

The earliest reference to chemical synapses, the term 'synapse' having been proposed by Sherrington (1897), seems to be that of du Bois-Reymond (1877) who suggested that secretion by the nerve of some chemical compound could be the cause of neuromuscular excitation. This was later shown to be the case by the studies of Dale and Feldberg (1936), Loewi (1921, 1922, 1924a,b,c), Feldberg (1945) and others (*cf.* Chapter 5). Many of these transmitters have now been identified.

The basic work of Eccles (1964), Katz (1969) and their respective collaborators has gone a long way toward unraveling the structure and behavior of the chemical synapse. This mechanism for the transfer of information from nerve cell to nerve cell provides an example worth describing. In its simplest form two cell membranes are involved; one being primarily electrosensitive and the other chemosensitive. Although there are many variations, it operates more or less as follows: (a) The stimulus, an action potential, travels along a nerve fiber to the terminal where it depolarizes the membrane. Note that this is a case where the direct stimulus arises within the cell;

(b) The depolarization opens voltage gated calcium channels in the membrane thus permitting a concentration driven calcium influx;

(c) The increase of calcium in the terminal triggers exocytosis of transmitter molecules which have accumulated in the presynaptic vesicles. A number of neurotransmitter molecules, among them acetylcholine, γ-aminobutyric acid, norepinephrine and serotonin, have now been identified;

(d) The transmitter diffuses across the intercellular space and binds to a receptor molecule on the membrane of the postsynaptic terminal;

(e) The transmitter-receptor complex acts to modify ion conductances in the postsynaptic membrane by controlling ion channels either directly or via intermediates (second messengers);

(f) The result is a depolarization, or hyperpolarization, of the postsynaptic membrane thereby producing, or suppressing by interaction with another synaptic connection, electrical activity in the postsynaptic nerve fiber;

(g) The transmitter is then released from the receptor and reabsorbed by the presynaptic terminal, inactivated or destroyed;

(h) The accumulated calcium is pumped back out of the presynaptic terminal;

(*i*) More vesicles loaded with transmitter, supplied by the cell body, appear in the presynaptic terminal thus restoring the system to its original steady state.

It can readily be seen that these steps are not yet completely understood in molecular terms although the general role of the membranes is clear and a number of the molecular elements have been identified and characterized (see Hille, 1992; Schwartz and Kandell, 1991). The mechanism of the neuromuscular junction is essentially similar although it has, of course, the added complication of its relation to the excitation-contraction coupling process. Excitation of the muscle fiber itself, at least in vertebrate skeletal muscle, has been shown to be due to the release of calcium from the endoplasmic reticulum as a result of depolarization of the transverse tubules (Porter and Palade, 1957; Huxley and Taylor, 1958). The system has been well reviewed by del Castillo and Katz (1965), (see also Aidley, 1971). The structure of the acetylcholine receptor has recently been characterized (Noda et al., 1983; Mishina et al., 1985).

Next in complexity are the spinal reflexes such as the sharp withdrawal of a hand from a hot plate or a charged electric wire. This 'simple' system contains a receptor, at least three (usually more) chemical synapses and one or more effector organs. On the other hand, the number of elements involved in cortical operations can be mind-boggling.

Finally, it should be evident that the great number and variety of excitable systems used by living organisms requires a great number and variety of membrane mediated transfer processes as well as energy storage and supply mechanisms. It should also be evident that a single type of stimulus can generate one or more kinds of response, *e.g.*, epinephrine acts on several systems, and, that, conversely, several kinds of stimuli can in many cases generate a particular response, *e.g.*, any of a large number of stimuli can activate pain receptors.

Perhaps it will some day be possible to construct a detailed flow diagram for the operations of a living organism in molecular terms including both excitation and steady state processes bu the completion of the diagram will obviously take a long, long time and its presentation will require a great deal of paper.

Acknowledgments

The author is greatly indebted to Drs. S.I. Feinstein, J.E. Goldman, J.A.B. Gray, B. Hille, A.L. Hodgkin, and A. Kleinzeller for valuable comments and for help in locating important references.

REFERENCES[6]

Adelman, W.J. Jr. and Fitzhugh, R. (1975) Solutions of the Hodgkin-Huxley equations modified for potassium accumulation in a periaxonal space. Fed. Proc. 34, 1322–1329.
Aidley, D.J. (1971) *The Physiology of Excitable Cells.* Cambridge Univ. Press.
Albuquerque, E., Chung, S.H. and Ottoson, D. (1969) Impulse generation in the isolated muscle spindle under the action of tetrodotoxin. Acta Physiol. Scand. 75, 301–312.
Almers, W. and Stirling, C. (1984) Distribution of transport proteins over animal cell membranes. J. Memb. Biol. 77, 169–186.
Arrhenius, S. (1887) Über die Dissociation in wassergelösten Stoffe. Z. Phys. Chem. 1, 631–648.
Arvanitaki, A. (1942) Interactions électriques entre deux cellules nerveuses contigues. Arch. Int. Physiol. 52, 381–407.
Barr, L., Dewey, M.M., Berger, W. (1965) Propagation of action potentials and the structure of the nexus in cardiac muscle. J. Gen. Physiol. 48, 797–823.
Bekesy, G. von (1960) *Experiments in Hearing.* New York, McGraw-Hill.
Bernstein, J. (1868) Über den zeitlichen Verlauf der negativen Schwankung

6 The extreme difficulty of obtaining access to some of the above references has sometimes made it necessary to rely on second hand sources. I offer an apology and a warning (see, e.g., Physics Today, May 1952, p. 15).

des Nervenstroms. Pflüger's Arch. Physiol. des Menschen und der Thiere 1, 173–203.

Bernstein, J. (1902) Untersuchungen zur Thermodynamik der bioelektrischen Ströme. Pflüger's Arch. ges. Physiol. 92, 522–562.

Blair, H.A. (1932) On the intensity-time relation for stimulation by electric currents. J. Cell. Comp. Physiol. 15, 147–157.

Blair, H.A. (1936) The time-intensity curve and latent addition in the mechanical stimulation of nerve. Am. J. Physiol. 114, 586–593.

Blinks, L.R. (1930) D.C. resistance of Nitella. J. Gen. Physiol. 13, 495–508.

Blinks, L.R. (1936) The effects of current flow in large plant cells. Cold Spring Harbor Symp. Quant. Biol. 4, 34–42.

Boll, F. (1876) Zur Anatomie und Physiologie der Retina. Mbr. Acad. Wiss. Berlin. 41, 383–404.

Brazier, M.A.B. (1959) The Historical Development of Neurophysiology. In Handbook of Physiology, Sect. 1, Neurophysiology, Chapt. 1. Washington, Am. Physiol. Soc.

Brown-Séquard, C.E. (1852) Experimental researches applied to physiology and pathology. Med. Exam. Philadelphia. 8, 481–504.

Burdon-Sanderson, J. On the electromotive properties of the leaf of Dionaea in the excited and unexcited states. Phil. Trans. Roy. Soc. (Lond.) B. I (1882), 173, 1–55. II (1888), 179, 417–449.

Caldwell, P.C., Hodgkin, A.L., Keynes, R.D. and Shaw, T.I. (1960) The effects of injecting 'energy-rich' phosphate compounds on the active transport of ions in the giant axons of Loligo. J. Physiol. (Lond.) 152, 7329–752.

Cohen, S.N., Chang, A.C.Y., Boyer, H.W. and Helling, R.B. (1973) Construction of biologically functional plasmids in vitro. Proc. Nat. Acad. Sci. USA 70, 3240–3244.

Cole, K.S. (1968) Membranes, Ions, and Impulses. Berkeley, Univ. of Cal. Press.

Cole, K.S. (1949) Dynamic electrical characteristics of the squid axon membrane. Arch. Sci. Physiol. 3, 253–258.

Cole K.S. and Curtis H.J. (1938) Electric impedance of Nitella during activity. J. Gen. Physiol. 22, 37–64.

Cremer, M. (1900) Über Wellen und Pseudowellen. Zeit. Biol. 40, 393–418.

Curtis, H.J. and Cole, K.S. (1942) Membrane resting and action potentials from the squid giant axon. J. Cell. Comp. Physiol. 19, 5–144.

Dale, H.H. and Feldberg, W. (1936) Release of acetylcholine at voluntary motor nerve endings. J. Physiol. (Lond.) 86, 353–380.

Davis, H. (1961) Some principles of sensory receptor action. Physiol. Rev. 41, 391–416.

Dean, R.B. (1941) Theories of electrolyte equilibrium in muscle. Biol. Symp. 3, 331–348.

De Castro, F. (1926) Sur la structure et l'innervation de la glande inter-carotidienne (glomus caroticum) de l'homme et des mammifères, et sur un nouveau système d'innervation autonome du nerf glossopharingien. Trav. Lab. Rech. Biol. (Univ. Madr.) 24, 365–432.

Del Castillo, J. and Katz, B. (1961) Biophysical aspects of neuromuscular transmission. Prog. in Biophys. 6, 121–170.

De Vries, H., Jielof, R. and Spoor, A. (1952) The microphonic activity of the lateral line. J. Physiol. (Lond.) 116, 137–157.

Donnan, F.G. (1911) Theorie der Membrangleichgewichte und Membranpotentiale bei Vorhandensein von nicht dialysierende Electrolyten. Z. Elektrochem. 17, 572–581.

Draper, M.H. and Weidmann, S. (1951) Cardiac resting and action potentials recorded with an intracellular electrode. J. Physiol. (Lond.) 115, 74–94.

Du Bois-Reymond, E. (1848) Untersuchungen über thierische Electricität. Vol I, Berling, Springer.

Du Bois-Reymond, E. (1877) Gesammelte Abhandlung zur algemeinen Muskel-und-Nerven Physik. vol. II, Leipzig, Veit.

Eccles, J.C. (1964) The Physiology of Synapses. Berlin, Springer.

Edridge-Green, F.W. (1902) Some observations on the visual purple of the retina. Trans. Ophth. Soc. 22, 300–302.

Einstein, A. (1906) Zur Theorie der Brownschen Bewegung. Ann. Phys. 19, 289–306.

Faraday, M. (1833) IV. Experimental Researches in Electricity - Third Series. Phil. Trans. 123, 23–53.

Feldberg, W. (1945) Present views on the mode of action of acetylcholine in the central nervous system. Physiol. Rev. 25, 596–642.

Fick, A. (1855) Über Diffusion. Ann. Phys. 94, 59–87.

Fontana, F.G.F. (1767) De legibus irritabilitatis. Quoted from H.E. Hoff (1942). The history of the refractory period. Yale J. Biol. Med. 14, 635–672.

Franck, U.F. (1956) Models for biological excitation processes. Prog. Biophys. 6, 173–206.

Frankenhaeuser, B. and Hodgkin, A.L. (1957) The action of calcium on the electrical properties of squid axons. J. Physiol. (Lond.) 137, 218–244.

Fricke, H. (1923) The electric capacity of call suspensions. Physical Rev. 21, 708–709.

Furshpan, E.J. and Potter, D.D. (1959) Transmission of the giant motor synapses of the crayfish. J. Physiol. (Lond.) 145, 289–325.

Galvani, A.L. (1791) De viribus electricitatis in motu musculari. Bonon. Sci. Art. Inst. Com. Bologna 7, 363–418.

Gibbs, J.W. (1906) On the equilibrium of heterogeneous substances. In: The collected works of J. Willard Gibbs. Vol I., New Haven, Yale Univ. Press.

Glisson, F. (1677) Tractatus de ventriculo et intestinis. London, Henry Brome.

Goldman, D.E. (1943) Potential, impedance and rectification in membranes. J. Gen. Physiol. 27, 36–60.

Gouy, M.G. (1910) Sur la constitution de la charge électrique à la surface d'un électrolyte. J. Phys. Chim. (4), 9, 457–468.

Gray, J.A.B. (1959) Mechanical into electrical energy in certain mechanoreceptors. Prog. Biophys. 9, 285–324.

REFERENCES

Grundfest, H. (1965) Electrophysiology and pharmacology of different components of bioelectric transducers. Cold Spring Harbor Symp. Quant. Biol. 30, 1–14.
Guharay, F. and Sachs, F. (1984) Stretch activated singe ion channel currents in tissue-cultured embryoic chick skeletal muscle. J. Physiol. (Lond.) 352, 685–701.
Hajduczok, G., Ferlic, R.J., Chapleau, M.W. and Abboud, F.M. (1990) Mechanism of transduction of arterial baroreceptors. FASEB J. 4, A2,85.
Haller, A. von (1753) De partibus corporis humani sensibilibus et irritabilibus. Comm. soc. reg. sci. Gottingae. 2, 114–158.
Helmholtz, H. von (1879) I. Studien der electrischen Grenzschichten. Wied. Ann. Phys. Chem. 7, 337–382.
Henderson, P. (1907) Zur Thermodynamik der Flüssigkeitsketten. Z. Phys. Chem. 75, 118–127.
Hermann, L. (1899) Zur Theorie der Erregungsleitung und der elektrischen Erregung. Arch. ges. Physiol. 75, 574–590.
Heymans, C., Boukaert, J.J. and Dautrebande, L. (1930), Sinus carotidien et reflexes respiratoires. II. Influence respiratoires reflexes de l'acidose de l'alcalose, de l'anhydride carbonique, de l'ion hydrogène et de l'anoxémie. Sinus carotidien et changes respiratoires dans les poumons et au dela des poumons. Arch. Int. Pharmacodyn. 39, 400–450.
Hill, A.V. (1910) A new mathematical treatment of changes of ionic concentration in muscle and nerve under the action of electric currents, with a theory as to their mode of excitation. J. Physiol. (Lond.) 40, 190–224.
Hill, A.V. (1936) Excitation and accomodation in nerve. Proc. Roy. Soc. (London) Sect. B, 119, 305–355.
Hill, B. (1992) G protein-coupled mechanisms and nervous signalling. Neuron 9, 187–195.
Hodgkin, A.L. and Huxley, A.F. (1939) Action potentials recorded from inside a nerve fibre. Nature (Lond.) 144, 710–711.
Hodgkin, A.L. and Huxley, A.F. (1952) A quantitative description of membrane current and its application to conduction and excitation in nerve. J. Physiol. (Lond.) 117, 500–544.
Hodgkin, A.L. and Katz, B. (1949) The effect of sodium ions on the electrical activity of the giant axon of the squid. J. Physiol. (Lond.) 108, 37–77.
Hodgkin, A.L. and Keynes, R.D. (1955) Active transport of cations in giant axons from *Sepia* and *Loligo*. J. Physiol. (Lond.) 128, 61–88.
Höber, R. (1945) *Physical Chemistry of Cells and Tissues*. Philadelphia, Blakiston.
Hudspeth, A.J. (1983) Mechanoelectrical transduction by hair cells in the acousticolateralis sensory system. Ann. Rev. Neurosci. 6, 187–215.
Jacobson, S.L. (1965) Receptor response in Venus's Fly-Trap. J. Gen. Physiol. 49, 117–129.
Julian, F.J. and Goldman, D.E. (1962) The effects of mechanical stimulation on some electrical properties of axons. J. Gen. Physiol. 46, 297–313.

336   REFERENCES

Katz, B. (1939) *Electrical excitation in nerve*. London, Oxford University Press.
Katz, B. (1969) *The release of neurotransmitter substances*. Liverpool University Press.
Kelvin, Lord (Wm. Thompson) (1855) On the theory of the electrical telegraph. Proc. Roy. Soc. (London) 7, 382–399.
Kishimoto, U. (1964) Current voltage relations in *Nitella*. Japan J. Physiol. 14, 515–527.
Kohlrausch, F. (1876) Über das Leitungsvermögen der in Wasser gelösten Elektrolyte im Zusammenhang mit der Wanderung ihrer Bestandtheile. nach. K. ges. de Wiss. (Göttingen), 213–244.
Kohlrausch, F. and Holborn, L. *Das Leitvermögen der Elektrolyte*. Teubner, Leipzig, 1916.
Korenbrot, J.I. and Cone, R.A. (1972) Dark ionic flux and the effects of light in isolated rod outer segments. J. Gen. Physiol. 60, 20–45.
Lancet, D., Lazard, D., Heldman, J., Khen, M. and Nef, P. (1988) Molecular transduction in smell and taste. Cold Spring Harbour Symp. Quant. Biol. 53, 343–348.
Lapique, L. (1907) Recherches quantitatives sur l'excitation électrique des nerfs, traitée comme une polarization. 2', J. Physiol. path. gén. 9, 620–635.
Lapique, L. (1935) La chronaxie en biologie générale. Biol. Rev. 10, 483–514.
Leake, C.D. (Compiler and Editor) (1956) *Some Founders of Physiology*. Washington, XXth International Physiological Congress.
Lillie, R.S. (1918) Transmission of activation in passive metals as a model of the protoplasmic or nervous transmission. Science 48, 51.
Loewenstein, W.R. (1967) On the genesis of cellular communication. Dev. Biol. 15, 503–520.
Loewi, O. (1921, 1922, 1924a,b,c) Über humorale Übertragbarkeit der Herznervenwirkung. Pflügers Arch. ges. Physiol. des Menschen und Thiere. I, 189, 239–242; II, 193, 201–213; III, 203, 408–412; IV, 204, 361–367; V, 204, 628–640.
Lucas, K. (1910) An analysis of changes and differences in the excitatory process of nerves and muscles based on the physical theory of excitation. J. Physiol. (Lond.) 40, 225–249.
Marmont, G. (1949) Studies on the axon membrane; I. A new method. J. Cell and Comp. Physiol. 34, 351–382.
Matteucci, C. (1838) Sur le courant électrique de grenouille. Ann. chim. phys. 68, 93–106.
Maxam, A.M. and Gilbert, W. (1977) A new method for sequencing DNA. Proc. Nat. Acad. Sci. USA 74, 560–564.
Mishina, M., Tobimatsu, T., Imoto, K., Tanaka, K., Fujita, Y., Fukadu, K., Kurasaki, M., Takahashi, H., Morimoto, Y., Hirose, T., Inayama, S., Takahashi, T., Kuno, M. and Numa, S. (1985) Location of functional regions of acetylcholine receptor alpha-subunit by site-directed mutagenesis. Nature 313, 364–369.

Monnier, A.M. (1934) *L'excitation Electrique des Tissues*. Paris, Hermann.
Morris, C.E. (1990) Mechanosensitive ion channels. J. Memb. Biol. 113, 93–107.
Morrow, J.F., Cohen, S.N., Chang, A.C.Y., Boyer, H.W., Goodman, H.M. and Helling, R.B. (1974) Replication and transcription of eukaryotic DNA in *Escherichia coli*. Proc. Nat. Acad. Sci. USA 71, 5436–5437.
Mott, N.F. (1939) The theory of crystal rectifiers. Proc. Roy. Soc. (Lond.) A 171, 27–28.
Müller, J. (1838) *Handbuch der Physiologie des Menschen*. Coblentz, Holscher. Vol. I, Eng. translation Wm. Baly. 1838.
Narahashi, T., Moore, J.W. and Scott, W.R. (1964) Tetrodotoxin blockage of sodium conductance increase in lobster giants axons. J. Gen. Physiol. 47, 965–974.
Neher, E. and Sakmann, B. (1976) Single channel currents recorded from membranes of denervated frog muscle fibres. Nature (Lond.) 260, 799–802.
Nernst, W. (1889) Die elektromotorische Wirksamkeit der Jonen. Zeit. phys. Chem. 4, 129–181.
Nernst, W. (1908) Zur Theorie des elektrischen Reizes. Pflüger's Arch. 122, 275–314.
Noda, M., Furutani, Y., Takahashi, H., Toyosato, M., Tanabe, T., Shimizu, S., Kikyotani, S., Kayano, T., Hirose, T., Inayama, S. and Numa, S. (1983). Cloning and sequence analysis of calf cDNA and human genomic DNA encoding alpha subunit precursor of muscle acetylcholine receptor. Nature 305, 818–823.
Noda, M., Shimizu, S., Tanabe, T., Takai, T., Kayamo, T., Ikeda, T., Takahashi, H., Nakayama, H., Kanoaka, Y., Minamino, N., Kangawa, K., Matsuo, H., Raftery, M.A., Hirose, T., Inayama, S., Hayashida, H., Miyata, T. and Numa, S. (1984) Primary structure of *Electrophorus electricus* sodium channel deduced from cDNA sequence. Nature (Lond.) 312, 121–127.
Ohm, G.S. (1827) Die galvanische Kette mathematisch gearbeitet. Berlin.
Olesen, S.-P., Clapham, D. and Davies, P.F. (1988) Haemodynamic shear stress activates a K^+ current in vascular endothelial cells. Nature (Lond.) 331, 168–270.
Osterhout, W.J.V. (1936) Electrical phenomena in large cells. Physiol. Rev. 16, 216–237.
Ostwald, W. (1890) Elektrische Eigenschaften halbdurchlässiger Scheidewände. Zeit. phys. Chem. 6, 71–82.
Overbeek, J.T.G. (1956) The Donnan Equilibrium. Prog. Biophys. 6, 58–84.
Overton, E. (1902) Beiträge zur allgemeine Muskel- und Nervenphysiologie. Pflüger's Arch ges Physiol. 92, 346–386.
Planck, M. (1890a) Über die Erregung von Elektricität and Wärme in Elektrolyten. Ann. Phys. Chem. 39, 161–186.
Planck, M. (1890b) Über die Potentialdifferenz zwischen zwei verdünnten Lösungen binärer Elektrolyte. Ann. Phys. Chem. 40, 561–576.

Purkinje, J.E. (1937) Über die gangliöse Natur bestimmter Kerntheile. Ber. Versam. deutsche Naturforschung Ärzte. Prague.

Rashevsky, N. (1933) Outline of a physico-mathematical theory of excitation. Protoplasma 20, 42–56.

Reil, J. (1796) Von der Lebenskraft. Arch. Physiol. 1, 8–162.

Remack, R. (1852) Über extracelluläre Entstehung thierischer Zellen. Arch. Anat. Physiol. Wiss. Med. 45–57.

Rosenblueth, A., Alvarez-Buylla, R. and Garcia Ramos, J. (1953) The responses of axons to mechanical stimuli. Acta Physiol. Latinoamer. 3, 204–215.

Ross, E.M. (1989) Signal sorting and amplification through G protein-coupled receptors. Neuron 3, 141–152.

Rushton, W.A.H. (1937) A graphical solution of a differential equation with application to Hill's treatment of nerve excitation. Proc. Roy. Soc. (Lond.) B 123, 382–395.

Sachs, F. (1991) Stress sensitive ion channels: an update. In: *Sensory Transduction* Soc. Gen. Physiol. 45th Ann. Symp., New York, Rockefeller University Press.

Sanger, F., Nicklen, S. and Coulson, A.R. (1977) DNA sequencing with chain-termination inhibitors. Proc. Nat. Acad. Sci. USA 74, 5463–5467.

Schleiden, M. (1838) Beiträge zur Phytogenesis. Arch. Anat. Physiol. Wiss. Med. 137–176.

Schmitz, W. and Wiebe, W. (1938) Zur Frage der mechanischen Nervenreizung. Pflüger's Arch. 240, 289–299.

Schultze, M. (1863) *Das Protoplasma der Rhizopoden und der Pflanzenzellen.* Leipzig, Engelmann.

Schwann, T. (1863) *Mikroskopische Untersuchungen ueber die Übereinstimmung in der Struktur und dem Wachsthum der Thiere und Pflanzen.* Berlin, Sander.

Schwartz, J.H. and Kandel, E.R. Synaptic transmission mediated by second messengers. In: *Principles of Neural Science,* 3rd. ed., Amsterdam, Elsevier.

Sherrington, C.S. (1987) Textbook of Physiology, 7th ed., M. Foster, Ed., London, Macmillan.

Shull, G.E., Schwartz, A. and Lingrel, J.B. (1985) Amino acid sequence of the catalytic subunit of the $(Na^+ + K^+)$ ATPase deduced from a complementary DNA. Nature (Lond.) 316, 691–695.

Singer, S.J. and Nicolson, G.L. (1972) The fluid mosaic model of the structure of cell membranes. Science 175, 720–731.

Sjöstrand, F.S. (1949) An electron microscope study of the retinal rods of the guinea pig eye. J. Cell. and Comp. Physiol. 33, 383–404.

Sjöstrand, F.S. (1953) The ultrastructure of the outer segments of rods and cones of the eye as revealed by the electron microscope. J. Cell. and Comp. Physiol. 42, 15–44.

Skou, J.C. (1957) The influence of some cations on the adenosine triphosphatase from peripheral nerves. Biochem. Biophys. Acta. 23, 394.

Steinbach, H.B. (1940) Electrolyte balance of animal cells. In Cold Spring Harbor Symp. Quant. Biol. 8, 242–252.

Stryer, L. (1991) Visual excitation and recovery. J. Biol. Chem. 266, 10711–10714.

Stahl, W.L. (1986) Na/K ATPase of nervous tissue; a review. Neurochem. Int. 8, 449–476.

Tasaki, I. (1949) Collision of two nerve impulses in the nerve fiber. Biochim. Biophys. Acta 3, 494–497.

Thompson, D'A.W. (1947) A Glossary of Greek Fishes. Oxford University Press.

Tigerstedt, R. (1880) Studien über mechanische Nervenreizung. Helsingfors, Finn. Litt. Ges.

Volta, A. (1800) On electricity excited by the mere contact of conducting substances of different kinds. Phil. Trans. Roy. Soc. (Lond.) 90, 408.

Wolbarscht, M.L. (1960) Electrical characteristics of insect mechanoreceptors. J. Gen. Physiol. 44, 105–122.

Young, G. (1937) Note on excitation theories. Psychometrica 2, 103–106.

Young, J.Z. (1936) Structure of nerve fivers and synapses in some invertebrates. Cold Spring Harbor Symp. Quant. Biol. 4, 1–6.

Name Index

Adelman Jr, W.J., 325, *332*
Adler, J., *185*, 212, 213, *218*, *223*
Aidley, D.J., 331, *332*
Albuquerque, 324, *335*
Alexander, R.S., 110, *129*
Almers, W., 70, *78*, 312, *332*
Andersen, J.P., 166, 168, *179*, 322, *332*
Anderson, M.P., 72, 77, 166, *175*
Anraku, Y., 212, *218*
Anthes, J.C., 193, *219*
Appleman, J.R., 104, *123*
Ariëns, E.J., 190, *218*
Armstrong, C.M., 68, 70, 77
Arnold, J., 281, 282, *299*
Aronson, P.S., 109, *123*, 160, *175*
Arrhenius, S., 316, 317, *332*
Arunlakshana, O., 190, *218*
Arvanitaki, A., 322, *332*

Baker, G.F., 118, 122, *132*
Ball, E.G., 259, *272*
Baly-Horuk, 120, *123*
Bangham, A.D., 55, *78*
Bar-Sinai, A., 205, *225*
Bärlund, H., 38, 52, 59, *79*
Barnett, J.E.G., 103, *123*
Barr, L., 45, 62, *80*, 322, *332*
Battelli, F., 245, *272*
Bayliss, W.M., 17, *23*
Baylor, D.A., 72, 77
Bean, R.C., 66, 77

Bear, A., 43, *87*
Beck, J.C., 109, *123*
Behn, V., 55, 77
Beinert, H., 241, 244, *272*, 277
Belitzer, V.A., 252, 253, *272*
Bell, G.I., 118, *126*
Bell, R.M., 48, 77
Bennet, V., 194, *218*
Bennett, V., 46, 77
Bensley, R.R., 262, *272*
Bernard, C., 200, *219*
Bernstein, J., 4, 17, *23*, 36, 61, 77, 144, *175*, 316, 318, 320, *332*, *333*
Berridge, M.J., 48, 77, 193, 209, *219*
Beutner, R., 52, 77, *83*
Billah, M.M., 193, *219*
Bishop, W.R., 48, 77
Blair, H.A., 319, 324, *333*
Blakeley, K.R., 94, 95, *123*
Blasie, J.K., 48, 77, 249, *273*
Blaustein, M.P., 111, *123*, 157, *181*
Blinks, L.R., 37, 77, 310, 316, 318, *333*
Blobel, G., 49, *78*
Blostein, R., 166, *182*
Bodansky, M., 19, *23*
Boeynaems, J.M., 194, 196, *219*
Bogomolny, R.A., 166, *184*
Boll, F., 326, *333*
Bonting, S.L., 163, *175*
Booth, A.G., 292, *300*
Boron, W., 115, *123*

Boron, W.F., 110, *130*
Bossert, W.H., 296, *300*
Bothwell, M., 199, *219*
Boulpaep, E., 296, *300*
Boulpaep, E.J., 115, *123*
Bourne, H.R., 210, *219*
Boyer, P.D., 94, 95, *123*, 265, *273*
Boyle, P.J., 144, *175*
Brandley, B.K., 46, *78*
Branton, D., 44, *78*
Breckenridge, L.J., 70, *78*
Bretscher, M.S., 45, 62, *78*
Briggs, G.E., 155, *175*
Brisson, A., 201, 202, *219*
Brodie, T.G., 136, *175*
Brooks, M.M., 139, 140, 143, *175*
Brooks, S.C., 39, 61, *78*, 139, 140, 143, *175*
Brown, R., 10, *23*
Brücke, E., 33, 74, *78*, 285, *300*
Bubanovic, F., 61, *81*
Buck, T.C., 295, *300*
Bull, H.B., 19, *23*
Bumm, E., 253, *273*
Bunge, G., 19, *23*
Burdon-Sanderson, J., 310, *333*
Bütschli, O., 36, *78*
Buzhinski, 67, *83*

Cabantchik, Z.I., 112, 113, *123*
Cain, D.F., 252, *273*
Cala, P.M., 111, 115, *124*
Caldwell, P.C., 322, *333*
Cantarow, A., 19, *23*
Carafoli, E., 166, *181*
Caron, M.G., 207, *225*
Carpenter, G., 203, *219*
Carruthers, A., 101, 119, *124*
Cartaud, 200, *219*
Caryk, T., 115, *128*
Cass, A., 112, *124*
Cassell, D.J., 203, *228*

Catterall, W.A., 69, 70, *78*
Chambers, R., 36, *78*
Chance, B., 243, 260, 261, *273*
Changeux, J.-P., 201, 202, *219*, *220*, *229*
Chapman, D., 63, *78*
Chappel, J.B., 67, *78*, 111, *124*
Chen, C.-C., 108, *124*
Chen, C.-J., 165, *175*
Chevalier, J., 65, *78*
Chinkers, M., 204, *220*
Christensen, H.N., 20, *23*, 105, 108, *124*, 137, 148, 150, 158, *175*
Civan, M.M., 41, 60, *78*
Clark, A.J., 1, 5, *23*, 39, *78*, 97, *124*, 189–191, 195, 216, 217, *220*
Clark, G.A., 137, 147, *176*
Clark, W.M., 259, *273*
Clarkson, E.M., 155, *176*
Clarkson, T.W., 296, *300*
Claude, A., 21, *23*, 255, 262, *273*
Claude, P., 296, 298, *300*
Cleland, W.K., 263, *277*
Coady, M.J., 299, *300*
Cohen, B.E., 55, *78*
Cohen, G.N., 22, *23*, 96, 104, *124*, 146, 148, *176*
Cohen, I.B., 11, *23*
Cohen, P., 193, 203, *220*
Cohen, S., 203, *219*
Cohen, S.N., 328, *333*
Cohn, E.T., 142, *176*
Cohn, M., 22, *23*
Cohn, R.A., 326, *336*
Cohn, W.E., 142, *176*
Cohn, Z.A., 49, *78*
Cohnheim, O., 135, *176*
Cole, K.C., 42, *78*
Cole, K.S., 37, 43, 69, *78*, *79*, 310, 316, 318, 320, 321, *333*
Cole, R.H., 43, *79*
Collander, R., 38, 52, 59, *79*, 137, 139, *176*

Colowick, S.P., 22, *23*
Conn, P.M., 198, *220*
Conolly, T.J., 64, *79*
Conti-Troconti, B.M., 202, *220*
Conway, E.J., 65, *79*, 110, *124*, 144, 166, *175*, *180*
Cook, J.S., 48, *79*
Cook, R.P., 39, *79*, 191, *220*
Coopenhaver Jr, J.H., 257, *273*
Cori, C.F., 22, *23*, 93, 95, *124*, 146, *176*
Cori, G.F., *23*
Corsen, S.A., 94, *127*
Cramer, W.A., 71, *79*
Crane, R.K., 6, 22, *23*, 100, 108, 122, *124*, 147, 157, 158, 168, *176*, *180*, 243, 266, *273*, 290, *300*
Cremer, M., 321, *333*
Crofts, 67, *78*
Crompton, M., 111, *125*
Cronan, J.E., 48, *79*
Crum-Brown, A., 200, *220*
Csáky, T.Z., 100, *125*, 147, 157, *176*, 290, *300*
Cuatrecasas, P., 192, 194, 196, *220*, *221*, *224*
Cunningham, 75, *79*
Curran, P.F., 108, *130*
Curtis, H.J., 310, 320, *333*
Cushman-Wardzala, 119, *125*

Dale, H.H., 200, *221*, 329, *333*
Dalmark, M., 112, *124*, *125*
Danielli, J.F., 5, 9, 10, 20, *23*, *24*, 35, 39, 40, 42, 44, 45, 49, 50, 52, 59, 62, 63, 73, *79*, *88*, 91, 93, 94, 97, 99, 102, 103, 123, *125*, *131*, 145, 151, *176*
Danowski, T.S., 142, *176*
Davies, R.E., 167, *176*, 252, 266, *273*
Davies, R.I., 95, *128*
Davis, H., 325, *333*

Davson, H., 9, 15, *23*, *24*, 27, 35, 39, 40, 42, 44, 49, 50, 52, 59, 61, 63, 65, *79*, 91, 93, 94, 96, 97, *125*, 143, *176*
Dayton, A.B., 296, *301*
De Castro, F., 329, *333*
de Duve, C., 63, *79*, 292, *300*
de Gier, J., 63, 64, *79*
De Haen, C., 194, 196, *221*
de Vries, H., 27, 32, 57, *79*, 325, *333*
de Weer, P., 164, *177*
Dean, R.B., 139, 143, 152, *177*, 322, *333*
DeGowin, E.L., 142, *177*
Deitmer, J.W., 115, *125*
del Castillo, J., 200, *221*, 331, *334*
Devaux, P.F., 48, *79*
Dewey, M.M., 45, 62, *80*
Diamond, J.M., 291, 296, 299, *300*, *301*
DiBona, D.R., 65, *81*
Dick, D.A.T., 41, *80*
Dimroth, P., 167, *177*
Ding, G., 65, *80*
Dixon, J.F., 165, *177*
Dixon, M., 3, *24*
Dobberstein, B., 49, *78*
Dohlmann, H.G., 207, *221*
Donnan, F.G., 5, 14, 16, 18, *24*, 36, *80*, 137, *177*, 316, 317, *334*
Dowman, C.B.B., 142, *177*
Draper, M.H., 321, *334*
du Bois-Reymond, E., 2, 4, *24*, 33, 282, *300*, 316, 323, 329, *334*
Dubach, U.C., 292, *304*
Duke, W.W., 200, *224*
Dumont, J.E., 194, 196, *219*
Durbin, R.P., 165, *179*
Dutrochet, R.J.H., 28, *80*
Dutton, P.L., 243, 260, 272, *273*, *278*
Dziemian, A.J., 43, *86*

Eccles, J.C., 330, *334*
Eddy, A.A., 108, 109, *125*

Edelman, G.M., 46, *80*
Edidin, M., 48, *80*, 193, *221*, *222*
Edridge-Green, F.W., 326, *334*
Ege, R., 94, *125*
Eggleton, G.P., 251, *273*
Eggleton, P., 251, *273*
Ehrenstein, G., 67, *80*
Ehrlich, P., 1, 4, 15, *24*, 39, *80*, 188,
  191, 192, 216, 217, *221*, 234, 235,
  274
Einstein, A., 316, 318, *334*
Eisenman, G., 53, 67, *80*
El-Moatassim, C., 72, *80*
Ellory, J.C., 163, *177*
Elvehjem, C.A., 241, *274*
Engelhardt, M., 173, *177*
Ennis, P.D., 45, 68, *88*
Enns, L.H., 110, *130*
Erecinska, M., 160, *177*, 249, *274*
Evans, E.A., 46, *80*
Evans, W.H., 294, *300*
Eveloff, J., 292, *300*

Fahmy, N.I., 243, *274*
Fambrough, D.M., 167, *177*
Faraday, M., 315, 316, *334*
Farquhar, M.G., 296, *300*
Fatt, P., 200, 201, *221*
Fawcett, D.W., 281, 296, *301*
Feldberg, W., 200, *221*, 329, *333*, *334*
Fenn, W.O., 144, *177*
Fernandez-Moran, H., 248, *274*
Fick, A., 4, *24*, 31, 33, 52, 74, *80*,
  316–318, *334*
Field, R.A., *23*
Finean, J.B., 44, *80*
Finkelstein, A., 55, 56, 58, *80*, *85*, *87*
Fischer, E.H., 205, *221*
Fisher, R.B., 101, *125*
Fiske, C.H., 251, *274*
Fitzhugh, R., 325, *332*
Fleckenstein, A., 159, *177*
Fleisch, A., 235, *274*

Flexner, S., 191, *222*
Fontana, F.G.F., 309, *334*
Forbes, D.J., 63, *85*
Forbush, B., 164, *179*
Forster, R.P., 289, *301*
Forte, J.G., 165, *177*
Foster, M., 17, *24*
Fox, C.F., 104, *125*
Franck, J., 145, *177*
Franck, U.F., 310, *334*
Frankenhaeuser, B., 321, *334*
Fraser, C.M., 216, *228*
Fraser, T.R., 200, *220*
Frey-Wysseling, A., 42, *80*
Freychet, P., 192, *222*
Fricke, H., 37, *80*, 316, 318, *334*
Frizzell, R.A., 296, *301*
Frömter, E., 283, 296, *301*
Fruton, J.S., 17, 19, *24*
Frye, L.D., 48, *80*, 193, *222*
Fugmann, U., 261, *273*
Fujita, A., 53, *84*
Fuller, S.D., 298, *305*
Furlong, C.E., 212, *222*
Furshpan, E.J., 323, *334*

Gaddum, J.H., 189, 200, *221*, *222*
Gale, E.F., 20, *24*, 105, *125*, 147, 148,
  177
Galen, 231, *274*
Galeotti, G., 287, *301*
Galvani, A.L., 315, 316, *334*
Ganser, A.L., 165, *177*
Garbers, A.G., 204, *222*
Garcia-Romeu, F., 110, 112, *125*
Gardos, G., 156, *177*
Gautam, N., 209, 210, *222*
Geck, P., 101, 114, 115, *125*, 160, *177*
Gelman, E.P., 48, *79*
Gibbs, J.W., 14, *24*, 36, *81*, 317, *334*
Giebisch, G., 10, *24*
Gilbert, W., 328, *336*

Gilman, A.G., 205, 207–209, 222, 223, 229
Giordana, B., 171, 177
Glisson, F., 309, 334
Glynn, I.M., 111, 125, 153, 164, 178
Goldacre, R.J., 145, 178
Goldfarb, D., 63, 81
Goldfine, I.D., 203, 222
Goldman, D.E., 54, 81, 316, 318, 324, 334, 335
Goldschmidt, S., 296, 301
Goldshleger, R., 167, 178
Goodenough, D.A., 296, 298, 300
Gorter, E., 38, 42, 50, 81
Goto, K., 110, 126
Gould, G.W., 118, 126
Gouy, M.G., 316, 318, 334
Graham, T., 32, 81
Gray, J.A.B., 325, 334
Green, D.E., 115, 126, 244, 247, 274
Greenberg, S., 72, 81
Greengard, P., 193, 222
Gregor, P., 203, 222
Gregory, H., 198, 222
Grendel, F., 38, 39, 42, 43, 50, 81
Grenningloh, G., 203, 222
Grosclose, R., 109, 127
Gryns, 16, 24
Guggenheim, E.A., 36, 80
Guggino, W.B., 296, 301
Guharay, F., 72, 81, 325, 335
Gumilewski, D., 282, 301
Gunn, R.B., 112, 126

Hackenbrock, C.R., 248, 274
Hahn, L.A., 142, 178
Hajduczok, G., 327, 335
Haldane, J., 236, 274
Hales, S., 133, 140, 178, 284, 301
Halestrap, A.P., 113, 126
Haller, von, 309, 339
Hamburger, H.J., 16, 17, 24, 32, 36, 57, 61, 81, 134, 141, 142, 178

Hamill, G.F., 69, 81
Hamill, O.P., 295, 301
Hamilton, S.L., 211, 223
Handler, J.S., 298, 301
Handovsky, H., 15, 24
Hardy, M.A., 65, 81
Harris, E.J., 110, 112, 113, 126
Harris, J.E., 142, 153, 178
Hartree, E.F., 240, 243, 245, 255, 275
Harvey, E.N., 38, 42, 79, 81
Harvey, W.R., 165, 178
Hasselbach, W., 165, 178
Hausdorff, W.P., 214, 223
Häusler, H., 95, 121, 126
Haydon, D.A., 56, 66, 81
Hazelbauer, G.L., 212, 213, 223
Hazum, E., 198, 223
Hediger, M.A., 118, 126, 170, 178, 299, 301
Hedin, S.G., 37, 58, 81
Heidenhain, M., 296, 301
Heidenhain, R.P.H., 3, 4, 24, 134, 136, 137, 178, 282, 301, 302
Heidrich, H.-G., 292, 302
Heinemann, S.H., 69, 81
Heinz, E., 105, 108, 115, 125, 126, 157, 160, 170, 171, 178, 180
Helmholtz, H. v., 32, 81
Henderson, P., 318, 335
Henderson, P.J.F., 118, 126
Henderson, R., 45, 68, 81, 174, 178, 249, 277
Henle, J., 280, 281, 302
Hepler, J.R., 209, 223
Hermann, L., 33, 81, 282, 302, 316, 317, 321, 335
Hertwig, O., 36, 82
Hevesy, G., 57, 82
Hewson, W., 29, 30, 82
Heymans, C., 329, 335
Hill, A.V., 19, 24, 144, 177, 179, 189, 223, 316, 319, 335
Hille, B., 68–70, 82, 177, 331

Hinckle, 117, *127*
Hind, G., 268, *274*
Hinke, J.A.M., 288, *302*
Hladky, S.B., 99, *126*
Hoagland, D.R., 139, *179*
Höber, R., 4, 9, 16, 19, *24*, 27, 33, 35,
  37, 40, 51, 62, *82*, 92, 93, 95, 121,
  126, 136, 137, *179*, 295, *302*, 321,
  335
Hochmuth, R.M., 46, *80*
Hodgkin, A.L., 5, 22, *24*, 54, 55,
  66, 69, 72, 77, *82*, 97, 100, *126*,
  150, 153, 156, *179*, 316, 320–322,
  333–335
Hoerr, N.L., 262, *272*
Höfer, R., 19, *24*
Hoffman, C., 59, *87*
Hoffman, J.F., 163, 164, *179*
Hoffmann, N., 114, *126*
Hofmeister, F., 14, 16, 18, *24*
Hogeboom, G.H., 255, 262, *274*
Hokin, L.E., 8, *24*, 48, *86*, 155, 165,
  177, *179*, 192, 209, *223*
Hokin, M.B., 192, 209, *223*
Hokin, M.R., 155, *179*
Holborn, L., 317, *336*
Holle, 69, *82*
Hollenberg, M.D., 192, 196, 213, 216,
  217, *221*, *223*
Hollmann, M., 203, *223*
Hollunger, G., 260, *273*
Holzer, H., 20, *24*
Hopfer, U., 109, *127*, 159, 169, *181*,
  293, *302*
Hopkins, C.R., 198, *223*
Hopkins, F.G., 14, 18, 21, *24*, 246, *274*
Hoppe-Seyler, F., 233, 238–240, *274*,
  275, 282, *302*
Horio, M., 165, *179*
Horrecker, 96, *127*
Howell, W.H., 200, *224*
Hubell, W.C., 48, *82*
Hudson, R.L., 291, *302*

Hudspeth, A.J., 325, *335*
Huf, E., 138, *179*
Hülle, 33, *82*
Huxley, A.F., 5, *24*, 66, 69, *82*, 316,
  320, 331, *335*

Ikeda, T.S., 299, *302*

Jacobs, M., 141, *179*
Jacobs, M.H., 5, *24*, 27, 35, 41, 59, *82*,
  91, 93, 94, 112, *127*
Jacobs, S., 194, 196, *224*
Jacobson, S.L., 310, *335*
Jagendorf, A.T., 268, *274*, *275*
James, D.E., 120, *127*
Jauch, P., 109, *127*, 170, *179*
Jennings, M.A., 295, *302*
Jolly, P.C., 154, *182*
Jones, C.R., 46, *86*
Jordan, P.C., 70, *82*
Jørgensen, P.L., 166, 168, *179*
Julian, F.J., 324, *335*

Kaback, H.R., 108, *127*, 158, 171, 174,
  179
Kabat, 75, *82*
Kagawa, Y., 249, *275*
Kahn, C.R., 46, *82*, 198, *224*
Kalckar, H., 213, *224*
Kalckar, H.H., 252, *275*
Kalckar, H.M., 171, *179*
Kandel, E.R., 10, *24*
Kandell, E.R., 331, *338*
Karlin, A., 202, 203, *224*
Karlish, S.J.D., 111, *125*
Karlish, S.J.S., 164, *178*
Karnaky Jr, K.J., 292, *302*
Karnielli, E., 120, *127*
Kasahara, 117, *127*
Kasbekar, D.H., 165, *179*
Katchalsky, A., 54, *82*, 152, *179*, 284,
  303

Katz, B., 54, *82*, 200, 201, *221*, 316, 319, 320, 330, 331, *334–336*
Kearney, E.B., 244, *275*
Kedem, O., 54, *82*, 151, *179*, 284, *303*
Keilin, D., 232, 234, 237–241, 243, 245, 250, 255, *275*
Keinan, D., 198, *223*
Kelvin, Lord, 316, *336*
Kennedy, E.P., 104, *125*, 256, 262, *275*
Kenny, A.J., 292, *300*
Keyes, A.B., 114, *127*
Keyes, P., 191, *224*
Keynes, R.D., 55, 66, *82*, 150, 153, 156, 167, *176*, *179*, 322, *335*
Kim, J.W., 121, 122, *127*
Kimmich, G.A., 159, *179*
King, T.E., 250, 265, *275*
Kinne, R., 109, 114, *127*, 169, 170, *179*, *180*, *183*
Kinne, R.K.H., 288, 290, 292, 294, *303*
Kinne-Saffran, E., 292, 294, *303*
Kinsella, J.L., 121, *130*
Kinter, 292, *303*
Kipnis, D.M., 96, *127*
Kirkwood, J.G., 54, *82*
Kishimoto, U., 310, *336*
Kleinzeller, A., 10, *24*, 41, 46–48, *82*, *83*, 96, 103, *127*, 155, *180*, 289, 290, *303*, 320, *336*
Klenk, F., 46, *83*
Klingenberg, M., 260, *275*
Knauf, P.A., 113, *128*
Kobilka, B., 214, *224*
Koch, C.A., 204, *224*
Koefoed-Johnsen, V., 57, *83*
Koefoed-Johnson, V., 285, 287, *303*
Koeppe, H., 16, 17, 36, 61, *83*, 141, *180*
Koeppen, B.M., 295, *303*
Kohlrausch, F., 316, 317, *336*
Kölliker, M.A., 200, *224*
Kono, T., 119, *131*

Korenbrot, J.I., 326, *336*
Kornberg, H.L., 20, *83*, 96, *128*, 171, *180*
Kornberg, R.D., 48, *83*
Koshland Jr, D.E., 213, *229*
Kostyuk, P.G., 69, *83*
Kotyk, A., 10, *24*, 41, *82*, 96, *127*, 155, *180*
Kouppe, 322, *332*
Kozawa, S., 94, 95, *128*
Krebs, E.G., 193, *224*
Krebs, H.A., 18, 20, *24*
Kregenow, F.M., 115, *128*
Krnjevic, K., 38, *83*
Krogh, A., 7, *24*, 51, 138, 145, 152, *180*
Krulwich, T.A., 111, *128*
Kuhn, T., 11–13, *24*
Kundig, W., 75, *83*

Lancet, D., 329, *336*
Landau, B.R., 292, *305*
Lange-Carter, C.A., 217, *224*
Langley, J.N., 39, *83*, 187, 188, 192, 200, 216, 217, *224*, *225*
Langmuir, I., 3, 5, *24*, 29, 38, 50, 55, *83*
Lanyi, J.K., 166, *183*
Lapique, L., 319, *336*
Lardy, H.A., 257, *273*, *275*
Laser, H., 253, *276*
Laszt, 156, *180*
Lau, K.R., 291, *303*
Läuger, P., 67, 70, *83*, 109, *127*, 165, 170, *179*, *180*
Lawrence, A.S.C., 56, *83*
Leaf, A., 283, 288, *303*, *305*
LeFevre, M.E., 100, 102, *128*
LeFevre, P.G., 92, 95, 99, 100, 102, 104, *128*
Lefkowitz, R.J., 207, *225*
Lehninger, A.L., 256, 262, 263, *275*, *276*

Lenard, J., 47, *87*
Leslie, R.S., 63, *78*
Leubuscher, 282, *303*
Lev, A., 162, *180*
Lev, A.A., 67, *83*
Levine, R., 21, *25*, 96, *128*
Levitzki, A., 205, *225*
Levy, L., 238–240, *276*
Lewis, S.A., 291, *303*
Liebig, J., 15, 31, *83*, 141, *180*
Lienhard, G.E., 104, *123*
Lillie, R.S., 11, 16, *25*, 57, *83*, 311, *336*
Lindsay, J.G., 242, *276*
Lindstrom, J., 201, *225*
Ling, G.N., 40, 41, 58, *83*, 97, *128*, 161, 162, *180*
Linossier, G., 233, *276*
Lis, H., 191, *228*
Loeb, J., 17, *25*, 31, 34, 52, *83*
Loew, O., 233, *276*
Loewenstein, W.R., 323, *336*
Loewi, O., 95, 121, *126*, 200, *225*, 329, *336*
Logothetis, D.E., 211, *225*
Lohmann, K., 252, *276*
Lowenstein, W.R., 71, *84*
Lucas, K., 309, *336*
Lucké, B., 57, *84*
Ludwig, C., 17, *25*, 133, *180*, 281, *303*
Lundegårdh, H., 91, *128*, 137, 138, 140, 152, 155, 166, *180*, 266, *276*
Lundsgaard, E., 95, 96, *128*, 147, *180*, 251, 258, *276*
Luzzati, V., 44, *84*

Mackinnon, R., 70, *84*
MacMunn, C.A., 234, 238, 239, *276*
Madara, J.L., 298, *303*
Maiden, M.C.J., 118, *126*, 172, *180*
Maizels, M., 155, *176*
Makinose, M., 165, *178*
Makowski, L., 71, *84*

Maloney, P.C., 113, *128*
Mamelok, R.D., 292, *303*
Maricq, A.V., 203, *225*
Marmont, G., 69, *84*, 320, *336*
Marsland, D., 191, *225*
Maruyama, Y., 295, *303*
Marx, J.L., 201, *225*
Mathews, A.P., 19, *25*
Mathews, L.S., 205, *225*
Matteucci, C., 33, 315, 316, *336*
Maxam, A.M., 328, *336*
Mayer, J.E., 145, *177*
McAvoy, E.M., 290, *303*
McClendon, J.F., 19, *25*
McConnell, H.M., 48, *82*, *83*
McCutcheon, M., 57, *84*
McLaughlin, S., 47, *84*
Melnick, J.L., 253, *277*
Meunier, J.-C., 201, *226*
Meyer, G.M., 20, *25*, 137, *184*
Meyer, H., 34, *84*
Meyer, K.H., 53, *84*
Meyerhof, O., 20, *25*
Meyerhoff, O., 258, *276*
Michaelis, L., 52, 53, *84*
Michaud, N., 63, *81*
Michell, R.H., 193, 209, *226*
Miller, C., 70, *84*
Miller, D., 168, *180*
Miller, F.L., 297, *304*
Mills, J.W., 46–48, *83*
Mishina, M., 299, *304*, 331, *336*
Mitchell, P.D., 5, 20, *25*, 73, 75, *84*, 97, 98, 103, 106–108, 110, 114, 122, *128*, *129*, 145–149, 154, 157, 163, 167, 168, 171, *177*, *180*, *181*, 266, 269, 270, *276*
Mitchison, J.A., 43, 46, 50, *84*
Miyajima, A., 211, *226*
Molitoris, B.A., 299, *304*
Monnier, A.M., 319, *337*
Monod, J., 5, 22, *23*, *25*, 96, 104, *124*, 146, 148, 171, *176*, *181*

Moore, C., 67, *84*, 263, *276*
Moore, D.H., 295, *304*
Morávek, 9, *25*
Morris, C.E., 326, *337*
Morrow, J.F., 328, *337*
Mott, N.F., 318, *337*
Moyle, J., 110, 114, *129*
Mudge, G.H., 145, *181*
Mueckler, M., 117, *129*
Mueller, P., 55, 56, 66, 67, *84*
Müller, J., 30, *84*, 315, 316, *337*
Mullin, J.M., 298, *304*
Murer, H., 111, *129*, 159, 169, *181*
Myiamoto, N., 94, *128*

Nachmanson, D., 201, *226*
Naftalin, R.J., 102, 103, *129*
Nagano, J., 93, *129*
Nägeli, C., 1, 4, 11, 13, 17, *25*, 27, 28,
    34, 36, 45, 51, 57, *84*, *85*, 133, *181*,
    285
Nakazawa, 147, *181*
Narahashi, T., 321, *337*
Nasse, H., 111, *129*, 141, *181*
Nastuk, W.L., 200, 201, *226*
Nathanson, A., 37, *85*
Nedergaard 1963, S., 165, *178*
Negelein, E., 236, *278*
Neher, E., 69, *85*, 295, *304*, 316, 321,
    337
Nellans, H.N., 160, *181*
Nelson, M.T., 111, *123*, 157, *181*
Nernst, W., 4, 14, *25*, 32–34, 40, 52,
    58, 74, *85*, 316–319, *337*
Netter, H., 144, *181*
Neupert, W., 49, *89*
Newport, J.W., 63, *85*
Newton, I., 33, *85*, 174, *181*
Nicholson, G.L., 193, *228*
Nickerson, M., 190, *226*
Nicolson, G.L., 45, 62, *85*, *88*
Nishizuka, Y., 72, *85*

Noda, M., 299, *304*, 316, 322, 331,
    337
Noguchi, H., 191, *222*
Nolet, 28, *85*
Noma, A., 72, *85*
Numa, S., 70, *85*, 201, 203, *226*

Ocklind, C., 46, *85*
O'Dea, R.F., 213, *226*
Ogston, A.G., 266, *273*
Ohm, G.S., 316–318, *337*
Ohnishi, T., 244, *277*
Ohsawa, M., 171, *181*
Okunuki, K., 242, 246, *278*
Olesen, S.-P., 325, *337*
O'Maley, E., 110, *124*
Opie, E.L., 145, *181*
Orbach, E., 55, *85*
Oschman, J.L., 288, *304*
Osterhout, W.J.V., 63, *85*, 91, *129*,
    137, 139, 140, 145, *181*, 310, *337*
Ostwald, W., 14, 15, *25*, 32, 33, 40,
    74, *85*, 311, *337*
Ottoson, D., 72, *85*
Overbeek, J.T.G., 317, *337*
Overton, E., 4, 9, 11–13, 15–18, 20,
    25, 27, 33–35, 41, 57, 61, 77, *85*,
    86, 92, *129*, 134, 135, 144, *181*, 316,
    320, *337*
Oxender, D.I., 212, *227*

Paganelli, C.V., 58, *86*
Palade, G.E., 49, *86*, 296, *300*, 331,
    337
Palmer, L.G., 295, *304*
Pappenheimer, A.M., 243, *273*
Pappenheimer, J.R., 58, *86*
Pardee, A.B., 171, *181*, 212, *226*
Park, C.R., 22, *25*, 96, 101, *129*
Parnas, J.K., 21, *25*
Parpart, A.K., 43, *86*
Parsegian, V.A., 70, *86*
Parsons, D.S., 101, *125*

Passow, H., 95, 112, 113, *129*
Pastan, I.H., 199, *226*
Pasternak, A.C., 49, *86*
Pasteur, L., 253, 277
Patlak, C.S., 102, 103, *129*
Paton, W.D.M., 190, 192, *226*
Pedersen, P.L., 166, *181*
Pérez-Sala, O., 213, *226*
Person, P., 241, 277
Peters, R.A., 18, *25*, 195, 203, *226*, 227
Petersen, O.H., 295, *303*
Pethica, B.A., 50, *86*
Pfeffer, W.F., 135, *182*
Pfeffer, W.F.P., 3, 4, 17, 27–29, 31, 32, 34, 36, 51, *86*
Piperno, J.R., 212, *227*
Pitts, R.F., 110, *129*
Planck, M., 14, 32, 53, *86*, 316, 318, *337*
Polvani, C., 166, *182*
Ponder, E., 141, *182*
Pook, R.I., 165, 167, *182*
Porter, 43, *86*, 331, *337*
Posner, B.I., 199, *227*
Post, R.L., 154, 163, *182*
Potter, D.D., 323, *334*
Pressman, B.C., 67, *84*, *86*, 110, 112, 113, *126*, 263, *276*, 277
Preston, G.M., 64, *86*
Price, L., 13, *25*
Pringsheim, N., 27, 57, *86*
Pritchett, D.B., 203, *227*
Purkinje, J.E., 308, *338*

Quaranta, V., 46, *86*
Quastel, J.H., 147, *182*
Quincke, G., 33, *86*
Quiocho, F.A., 119, *129*, 171, *182*

Raaflaub, J., 263, 277
Rabon, E.C., 166, *182*
Racker, E., 3, *25*, 249, 275

Raftery, M.A., 201, 202, *220*, 227
Randle, P.J., 96, *129*
Randles, J., 159, *179*
Rang, H.P., 192, *226*
Ransom, F., 191, 227
Rashevsky, N., 319, *338*
Rea, P.A., 165, 167, *182*
Redman, C.M., 48, *86*
Regen, D.M., 101, *130*
Reid, E.W., 4, *25*, 135, 136, *182*, 282, 283, 286, *304*
Reil, J., 309, *338*
Reiner, J.M., 65, *79*, 94, 96, 97, *125*
Remack, R., 308, *338*
Reuben, M.A., 166, *182*
Richards, A.N., 147, *184*
Richet, C., 17, *25*
Rickenberg, H.V., 146, 148, *182*
Ricklis, E., 147, *182*
Rideal, E.K., 55, *87*
Rieske, J.S., 244, 277
Riggs, T.R., 108, *124*, 148, 170, *182*
Ringer, S., 16, *25*
Rini, J.M., 121, 122, *130*
Riordan, J.R., 71, *87*
Ritchie, A.M., 14, *25*
Ritchie, J.M., 153, 156, *182*
Rizzozero, 281, *304*
Robertson, J.D., 44, *87*
Robertson, R.N., 155, *182*
Robinson, J.R., 145, *182*
Rodbell, M., 205, *227*
Röhmann, 282, *304*
Roos, A., 110, *130*
Roseman, S., 97, *130*, 161, *182*
Rosen, O.M., 203, *227*
Rosenberg, P.A., 58, *87*
Rosenberg, T., 92, 95, 97, 99, 101, 104, 107, *130*, *132*, 148–151, *182*, *184*
Rosenblueth, A., 324, *338*
Ross, E.M., 205, *227*
Roth, R.A., 203, *228*

Rothman, J.E., 47, *87*
Rothstein, A., 110, 112, 113, *123, 130*
Rottem, S., 47, *87*
Rudin, D.O., 67, *84*
Ruhland, W., 59, *87*
Rushton, W.A.H., 319, *338*
Ruska, H., 295, *304*
Russell, J.M., 115, *130*

Sachs, F., 36, 72, *81, 87*, 325, 326, *335, 338*
Sachs, G., 133, 138, 165, *182, 183*
Sachs, J., 3, *25*, 284, 285, *304*
Sacktor, B., 109, 121, *123, 130*
Sakmann, L., 69, *85*, 295, *304*, 316, 321, *337*
Salerno, J.C., 244, *277*
Sands, R.H., 241, 244, *272, 277*
Sanger, F., 328, *338*
Satoh, T., 210, *228*
Sauer, F.A., 283, *304*
Savarese, T.M., 216, *228*
Scatchard, G., 192, *228*
Schäfer, E.A., 17, *25*
Schagina, L.V., 150, *183*
Schatzmann, H.J., 153, 165, *183*
Schild, H.O., 190, *218*
Schild, L., 116, *130*
Schleiden, M., 308, *338*
Schlessinger, J., 199, 204, *228, 229*
Schlögl, R., 54, *87*
Schlue, W.-R., 115, *125*
Schmid, G., 54, *87*
Schmidt, C., 15, 141, *183*
Schmidt, U., 292, *304*
Schmidt, W.J., 43, *87*
Schmitt, F.O., 43, *87*
Schmitz, W., 324, *338*
Schneider, W.C., 262, *277*
Schobert, B., 166, *183*
Schofield, P.R., 203, *228*
Schönheimer, R., 144, *183*
Schulman, J.H., 55, *87*

Schultz, C.H., 29, 30, *87*
Schultz, S.G., 108, *130*, 147, 158, *183*, 285, 290, 291, 296, 297, *301, 302, 304, 305*
Schultze, M., 308, *338*
Schwann, T., 30, *87*, 308, *338*
Schwartz, J.H., 331, *338*
Seddon, J.M., 44, *87*
Segel, I.H., 107, *130*
Semenza, G., 159, 169, *183*
Serrano, R., 165, 168, *183*
Shannon, J.A., 95, 100, 102, *130*, 138, *183*
Sharon, N., 191, *228*
Shemyakin, 67, *87*
Shepartz, B., 19, *23*
Sherrington, C.S., 329, *338*
Shooter, E.M., 199, *229*
Shporer, M., 41, 60, *78*
Shull, G.E., 322, *338*
Sieghart, W., 203, *228*
Sievers, J.F., 53, *84*
Simmonds, S., 19, *24*
Simon, M.I., 209, *228*
Simon, S.M., 49, *87*
Simons, K., 298, 299, *305*
Singer, S.J., 45, 49, 62, *88*, 119, *130*, 193, *228*
Singer, T.P., 244, *275*
Sjöstrand, F.S., 43, *88*, 326, *338*
Skou, J.C., 5, 22, *26*, 155, 163, *183*, 322, *338*
Skulachev, V.P., 167, 168, *183*, 269, *277*
Slater, E.C., 243, 263, 265, *277*
Slein, M.W., *23*
Smith, G.H., 96, *129*
Smith, H.W., 27, *88*
Smith, J.L., 236, *274*
Smrcka, A.V., 209, *228*
Sokol, P.P., 111, *130*
Soleimani, M., 115, *130*
Sollner, K., 53, *88*

Solomon, A.K., 20, *26*, 58, *86*, 153, *183*
Spiegel, A.M., 205, *228*
Springer, M.S., 193, 213, *228*
Stahl, W.L., *322, 338*
Stanton, B.A., 299, *305*
Starling, E.H., 17, *26*
Stavermann, A.J., 41, *88*
Steck, T.L., 47, 62, *88*, *89*, 113, *130*
Stein, W.D., 35, 45, 50, *88*, 99, 102, 103, 109, 122, *130, 131*
Steinbach, H.B., 139, 143, 144, 152, *183, 184*, 322, *339*
Steinman, R.M., 49, *78, 88*
Stephenson, J.L., 296, *305*
Stephenson, M., 20, *26*
Stephenson, R.P., 190, *228*
Stern, K.G., 253, *277*
Stern, L., 245, *272*
Stevens, C.F., 202, *228*
Stewart, D.R., 112, *127*
Stirling, C., 312, *332*
Stirling, C.E., 292, *305*
Stoeckenius, W., 44, *88*, 166, *184*
Stokes, J.B., 115, *131*
Straub, F.B., 43, *88*, 156, *184*
Straub, R.W., 153, 156, *182*
Strosberg, A.D., 211, *228*
Stroud, R.M., 203, *229*
Stryer, L., 210, *229*, 327, *339*
Stühmer, W., 69, *88*
Subbarow, Y., 251, *274*
Sun, F.F., 249, *277*
Sutherland, E.W., 46, 76, *88*, 193, 205, *229*
Suzuki, K., 119, *131*
Svaetichin, G., 72, *85*
Sweadner, K.J., 167, *184*
Szent-Györgyi, A., 143, *184*

Taggart, J.V., 289, *301*
Takeuchi, N., 201, *229*
Tang, W.-J., 208, *229*

Tank, D.W., 202, *229*
Tanner, M.J.A., 104, *131*
Tappeiner, H., 134, *184*
Tarpley, H.L., 101, *130*
Tasaki, I., 321, *339*
Taussig, R., 208, *229*
Taylor, P., 202, *229*, 331, *335*
Teorell, T., 5, *26*, 41, 53–55, 58, 60, *88*, 110, *131*, 137, 138, *184*
Thale, M., 147, *176*, 290, *300*
Thomas, R.C., 115, *131*
Thompson, D'A.W., 309, *339*
Thovert, 35, *88*
Thunberg, T., 235, *277*
Tigerstedt, E., 17, *26*
Tigerstedt, R., 323, *339*
Tokuda, H., 166, *184*
Tosteson, D.C., *24*, 67, *88*, 112, *131*
Traube, M., 3, 28, 31, 32, 34, *88*, 233, *277*
Troshin, A.S., 40, *88*, 161, *184*
Tschopp, J., 71, *88*
Tsybakova, E.T., 252, 253, *272*
Tucker, E.M., 163, *177*
Turner, J., 159, *184*
Turner, R.J., 107, *131*

Uhlenbruck, G., 46, *83*
Ullrich, A., 199, 204, *228, 229*, 283, *305*
Ullrich, K.J., 113, *131*
Unemoto, T., 166, *184*
Unwin, N., 249, *277*
Unwin, P.N.T., 45, 68, *88*, 201, 202, *219*
Uribe, E., 268, *275*
Ussing, H.H., 5, *24*, 51, 55, 57, *83, 88*, *89*, 92, 97, 98, 110, *131*, 149, 152, *184*, 285–288, 296, 298, *303, 305*

v. Thiman, K., 18, *88*
Vale, W.W., 205, *225*
van Deenen, L.L.M., 43, *89*

van Gelder, B.F., 241, *277*
Van Slyke, D.D., 137, *184*
van Slyke, D.D., 37, *88*
van 't Hoff, J.H., 4, 14, 28, 32, *88*, *89*
Vanderkooi, J., 248, *277*
Verkleij, A.J., 45, *89*
Verkman, A.S., 64, *89*
Verworn, M., 14, 17, *26*
Verzar, 156, *184*
Vidaver, G.A., 102, 103, *131*
Vile, R.G., 75, 76, *89*
Visscher, M.B., 296, *305*
Voelker, D.R., 45, 48, *89*
Vogt, H., 136, *175*
Volta, A., 308, 315, 316, *339*
von Bekesy, G., 325, *339*
von Helmholtz, H., 316, 318, *335*
von Mohl, H., 1, *25*, 27, 29, *89*

Wade, J.B., 65, *89*
Wainio, W.W., 241, 246, *277*, *278*
Wandinger-Ness, A., 298, 299, *305*
Wang, H., 122, *131*
Warburg, O., 19, *26*, 236, 245, *278*
Wearn, J.T., 147, *184*
Weaver, C.T., 205, *229*
Weber, M., 201, *229*
Weideman, S., 321, *334*
Weinhouse, S., 18, *26*
Weinstein, A.M., 296, *305*
Weis, R.M., 213, *229*
Weiss, R.A., 75, 76, *89*
Wellman, H., 257, *275*
West, I.C., 108, *131*, 158, 171, *184*
Westenbrink, H.G.K., 93, *131*
Whitehead, A.N., 13, *26*
Whittam, R., 163, *184*
Widdas, W.F., 6, *26*, 92, 95, 99–102, 118, 122, *131*, *132*, 171, *184*
Widnell, 121, *132*
Wiebe, W., 324, *338*

Wieland, H., 235, *278*
Wienhues, U., 49, *89*
Wieth, J., 112, *125*
Wikström, M.K.F., 165, 166, *184*
Wilbrandt, W., 40, 53, *89*, 92, 94, 95, 97, 99, 101, 104, *130*, *132*, 137, 150, 151, 156, *184*
Willingham, M.C., 199, *226*
Wills, N.K., 291, *303*
Wilson, D.F., 242, 243, 260–262, *276*, *278*
Wilson, D.M., 172, *185*
Wilson, J.E., 147, *176*
Wilson, T.H., 172, *185*
Windhager, E.E., 296, *305*
Winkler-Wilson, 97, 104, *132*
Winzler, R.J., 46, *89*
Wirtz, K.W.A., 48, *89*
Wissig, S.L., 295, *306*
Wolbarsht, M.L., 325, *339*
Wood, R.E., 94, 100, *132*
Worthington, C.R., 48, 77

Yaffe, M.P., 49, *88*, 119, *130*
Yakushiji, E., 242, 246, *278*
Yankner, B.A., 199, *229*
Yoda, A., 64, *89*
Yoda, S., 64, *89*
Young, G., 319, *339*
Young, J.D.-E., 71, *89*
Young, J.Z., 319, *339*
Yu, C.A., 247, *278*
Yu, J., 46, 47, *89*

Zadunaisky, J.A., 114, *132*, 160, *185*
Zalusky, R., 108, 147, *183*, 290, *305*
Zehran, K., 283, 286, *305*, *306*
Zerahn, K., 152, *184*
Zilversmit, D.B., 48, *89*
Zorzano, A., 118, 120, *132*
Zwaal, R.F.A., 38, 43, *90*

Subject Index

absorption, 1, 134, 135, 160, 282, 286, 293
acetylcholine, 39
  membrane phosphoinositides, 192
  receptors, 191, 200, 331
action potential, 66, 320
active transport, 6–8, 73, 98, 133, 289
  amino acids, 137, 146
  dyes, 136, 137
  electrogenicity, 153, 159
  energy requirement, 138, 149
  fluids, 135, 138
  ions, 137
  sugars, 146
adenylate cyclase, 46, 193, 205, 214, 292
  receptors, 207
aggrephores, 65
amino acid transport, electrogenicity, 158
anion exchangers, 111
anion respiration theory, 140
antibiotics as ionophores, 67
antigens, 1, 39, 46
antiport, 12, 106, 107
association-induction hypothesis, 161
asymmetry
  carrier, 101
  cellular, 284, 289
  electrical potential, 53
  membrane, 45, 47, 74
  membrane lipid, 48, 62

atrial natriuretic factor, receptor, 204

bacteriorhodopsin, 173
band 3 membrane proteins, 47, 113
baroreceptors, 327
bioelectric phenomena, 2, 32, 33, 36, 315

calcium channel, 69, 72
  G-protein regulation, 211
  voltage gating, 330
calcium pump, 165, 330
cardiac exchanges, 167
cardiac glycosides, 135, 153, 163
carrier, 63, 73, 91, 145
  amino acid, 105
  heterogeneity, 104
  mobility, 100, 102, 173
  proteins, 104, 113
  sugar transport, 104
carrier translocation, 119
carrier–solute complex, 98
carriers, 7, 12
  asymmetry, 122
  enzyme properties, 96, 97, 104
  flux coupling, 107
  heterogeneity, 118
  proteins, 117, 121
  sugars, 100
cation exchange, 61, 144, 320
cation exchangers, 110
cation–anion symport, 114

[355]

cell adhesion proteins, 46
cell–cell communication, 7, 322
channels, 7, 45, 64, 65, 68, 69
  electrolytes, 9
  ionic selectivity, 67
  protein kinase, 72
  proteins, 66–71
  stretch activation, 72
  sugar transport, 169
charged membrane, 53
chemiosmotic hypothesis, 158, 167, 266
chemotaxis, 171
  receptors, 213
chloride channel, 69, 71, 76
chloride pump, 114, 165
chronaxie, 319
colloid chemistry, membranes, 14, 15, 41
cotransport, 12, 99
counterflow, 99, 101, 103, 151
countertransport, 12
coupling of fluxes and metabolism, 154
cytochrome oxidase, 165, 166, 241
cytoskeleton, 46, 48, 62, 65, 74, 325

desmosome, 296
diffusion, 6, 12, 17, 41, 51, 92, 98
  intercellular, 283
  ions, 54
  lateral, in membrane, 48

electrical excitation, 315
electrogenicity of transport, 109, 111, 115, 154, 158, 169, 171
Electrophorus, 201
electroreceptors, 327
endocytosis, 49
endosmosis, 28, 284
energy transduction, 7
epidermal growth factor
  receptor, 196, 198
epinephrine receptor, 196

epithelial growth factor, 204
epithelial membrane, 279
equilibrium potentials, 317
evolution in science, 11
exchange diffusion, 97, 98, 110
excitability, 1, 8
  membranes, 66, 307
excitability-inducing material, 66
exocytosis, 49
exosmosis, 28

facilitated diffusion, 6, 12, 73, 97, 101, 292, 293
fluid-mosaic membrane, 45, 62, 193
flux coupling, 99, 107
  tertiary, 115
flux ratio equation, 54
fusion, 49

G-protein, 203, 327
G-protein receptors, 197, 205
GABA receptor, 203
gap junctions, 71, 322
gating
  channels, 66, 69, 70, 72
Gibbs–Donnan equilibrium, 19, 53
glucose transporter, 117
glutamine acid receptor, 203
glycine receptor, 203
glycoproteins, 12, 46, 62, 113, 191
group translocation, 6, 75, 97, 146, 154
GTP-ase receptors, 208

H^+-gradient hypothesis, 158
H^+-sugar cotransport, electrogenicity, 171
Hodgkin–Huxley equation, 320, 325
Hoffmeister's series, 14, 16, 60
hormone receptors, 46, 193, 194

inhibition analysis, 105
insulin
  receptor, 120, 192, 194, 196, 203
  sugar transport, 95, 96, 119

integrins, 46
intercellular fluxes, 136, 295
ion channels, 295, 321
  G-protein regulation, 211
  ligand gating, 200
  mechanosensitivity, 325
  receptor, 200, 329
  voltage gating, 312
ion-selective channels, 66
ionic antagonists, 16
ionic channels, ligand gating, 193
ionic distribution
  cells and tissues, 31, 138, 141, 163
ionic pumps, 139, 152, 292
ionic selectivity, 298
  channels, 67, 69
ionophores, 67, 263
iron wire nerve model, 311
irreversible thermodynamics, 54, 151

ligand–receptor complex, 189, 190
light receptors, 72
lipid bilayer membrane, 10, 42, 44,
    50, 55, 62, 73, 312
  mitochondria, 63
  nucleus, 63
lipid membrane, 13, 34, 39
  narcosis, 20, 34
lipid vesicles, 55
lipoid-sieve theory, 59
lipoproteins, membranes, 44
liposomes, 41, 55, 63, 162, 202

mechanical excitation, 323
membrane
  asymmetry, 47
  bioelectric phenomena, 2
  carriers, 7
  channels, 7
  elasticity, 31, 42
  excitability, 307
  fusion, 50
  glycoproteins, 12

microdissection, 36
$(Na^+ + K^+)$-ATP-ase, 22
osmotic phenomena, 27, 31
osmotic properties, 3, 6
permeability, 7
pores, 13
proteins, 8
receptors, 1, 7, 9, 39, 187
semipermeable, 15, 32
sieve theory, 38, 40
solute binding proteins, 171
turnover, 48
valves, 28, 135
membrane electrical potentials, 32,
    33
membrane fusion, 49
membrane lipids
  cholesterol, 39, 43
  cholesterol esters, 34
  phospholipids, 34, 39, 43
membrane potential, 52, 53, 109
  action, 66, 320
  cells, 36
  Donnan, 36
  epithelial, 287
  mitochondria, 270
membrane proteins, 42, 45, 246
  asymmetry, 47
  electrical asymmetry, 47
  methylation, 193
  mobility, 193, 248
  organization, 45, 46
membrane pumps, 133
membrane semipermeability, 256
membrane turnover, 74
membrane vesicles, 108, 168, 293
membranes potential
  cells, 318, 320
microdissection of membrane, 36
mitochondria, 255
mitochondrial membrane, 49, 245,
    263
mosaic membrane, 37, 44

multidrug resistance glycoprotein,
    71, 165
muscarinic receptor, 192, 200

Na⁺-gradient hypothesis, 108, 147,
    157, 266
Na⁺–amino acid cotransport, 169, 170
Na⁺–glucose cotransport
    electrogenicity, 169
(Na⁺–K⁺)-ATP-ase, 22, 64, 155, 161,
    162, 283, 291, 292
Na⁺–sugar cotransport, 290
narcosis, 20, 34
narcotics, 321
nicotinic cholinergic receptor, 193,
    200

osmosis, 3, 19, 40, 57, 136
    lipid solubility, 33
    mitochondria, 263
    red blood cells, 29
    water, 28
oxidative phosphorylation, 250, 252,
    264

paracellular shunt, 296
patch-clamp, 12, 69, 295, 321, 325
permeability, 6, 7, 31, 35, 41, 51
    electrolytes, 60
    ions, 144
    narcosis, 38
    physical, 35, 51
    physiological, 35, 51, 93, 136
    substrate competition, 95
    sugars, 94, 100
    water, 56
permease, 108, 171
phosphatidic acid hypothesis, 8
phosphoinositide turnover, 209
phosphoinositides, second messenger,
    193
phospholipases, 48, 193, 209
phospholipid, 47

phosphotransferase, 75, 161
photosynthetic phosphorylation, 269
pore proteins, 71, 103
pores, 13, 40, 50, 52–54, 57, 63, 75,
    92, 102
potassium channel, 69, 72
potassium pump, 165
precipitation membranes, 32, 53
protein kinase, 48, 214
proton-ATP-ase, 158
proton force, 167
proton pump, 165, 267, 270
proton–solute cotransport, 108

receptor methylation, 212
receptors, 1, 7–9, 39, 46, 63, 72, 74,
    188
    chemotaxis, 213
    enzyme interaction, 196, 203
    internalization, 120, 197, 199
    ion channels interaction, 196
    microclustering, 197
    mobility, 193, 195, 196, 202
    recycling, 214
    regulation, 213
    synaptic transmitter, 329
redox carrier, 247
redox pump, 140, 155, 266, 267
revolution in science, 11
rhodopsin, 209, 326, 327

scientific revolution, 11, 13
second messengers, 193, 206, 209
secondary active transport, 157, 159,
    169, 294
secretion, 1, 134, 135, 160, 281, 282,
    286, 293
selective membranes, 53
selective permeability, 34, 36
semipermeability, 7, 15, 31, 32
serine kinase receptors, 204
serotonine receptor, 203
short-circuit current, 152, 286

sieve theory, 38, 40, 52
signal transduction, 7, 9, 48, 76, 195,
     196
sodium, 69
sodium channels, 327
sodium pump, 108, 111, 135, 143,
     145, 152, 153, 322
sodium–solute cotransport, 108, 157,
     289, 290, 293
sorption theory, 161
squid giant axon, 319
standing-gradient hypothesis, 296,
     297
stretch-activated channels, 325
sugar transport, electrogenicity, 158
symport, 12, 106, 108
synapses, 10, 322
     chemical, 329

thermoreceptors, 327
threonine kinase receptor, 204

tight junctions, 296
Torpedo, 201, 309
transducin, 210
transmembrane enzymes, receptors,
     200, 204
tyrosine kinase, 211
     receptors, 203

ultrafiltration theory, 59
unit membrane, 44
unstirred layers, 58
Ussing equation, 55, 149, 280

visual excitation, 326
voltage clamp, 320, 321
voltage gating, 70

water channel, 64
water pump, 145